The Mycota

Edited by
K. Esser

Springer

Berlin
Heidelberg
New York
Barcelona
Hong Kong
London
Milan
Paris
Singapore
Tokyo

The Mycota

The Mycota

A Comprehensive Treatise
on Fungi as Experimental Systems
for Basic and Applied Research

Edited by K. Esser

IX *Fungal Associations*

Volume Editor:
B. Hock

With 69 Figures and 16 Tables

Springer

Series Editor

Professor Dr. Dr. h.c. mult. KARL ESSER
Allgemeine Botanik
Ruhr-Universität
44780 Bochum, Germany
Tel.: +49(234)32-22211
Fax: +49(234)32-14211
e-mail: Karl.Esser@ruhr-uni-bochum.de

Volume Editor

Professor Dr. Bertold Hock
Department of Botany
TU München at Weihenstephan
85350 Freising, Germany
Tel.: +49(8161)713396
Fax: +49(8161)714403
e-mail: hock@weihenstephan.de

ISBN 3-540-62872-X Springer-Verlag Berlin Heidelberg New York

Library of Congress Cataloging-in-Publication Data.

The Mycota. Includes bibliographical references and index. Contents: 1. Growth, differentiation, and sexuality/editors, J.G.H. Wessels and F. Meinhardt – 2. Genetics and biotechnology. 1. Mycology. 2. Fungi. 3. Mycology – Research. 4. Research. I. Esser, Karl, 1924– . II. Lemke, Paul A., 1937– . QK603.M87 1994 589.2 ISBN 3-540-57781-5 (v. 1: Berlin: alk. paper) ISBN 0-387-57781-5 (v. 1: New York: alk. paper) ISBN 3-540-58003-4 (v. 2: Berlin) ISBN 0-387-58003-4 (v. 2: New York)

Springer-Verlag Berlin Heidelberg New York
a member of BertelsmannSpringer Science+Business Media GmbH

© Springer-Verlag Berlin Heidelberg 2001
Printed in Germany

Production Editor: PRO EDIT GmbH, Heidelberg, Germany
Cover design: Springer-Verlag, E. Kirchner

Typesetting by Best-set Typesetter Ltd., Hong Kong

Printed on acid-free paper   SPIN 10537164   31/3136/Di   5 4 3 2 1 0

Series Preface

Mycology, the study of fungi, originated as a subdiscipline of botany and was a descriptive discipline, largely neglected as an experimental science until the early years of this century. A seminal paper by Blakeslee in 1904 provided evidence for self-incompatibility, termed "heterothallism", and stimulated interest in studies related to the control of sexual reproduction in fungi by mating-type specificities. Soon to follow was the demonstration that sexually reproducing fungi exhibit Mendelian inheritance and that it was possible to conduct formal genetic analysis with fungi. The names Burgeff, Kniep and Lindegren are all associated with this early period of fungal genetics research.

These studies and the discovery of penicillin by Fleming, who shared a Nobel Prize in 1945, provided further impetus for experimental research with fungi. Thus began a period of interest in mutation induction and analysis of mutants for biochemical traits. Such fundamental research, conducted largely with *Neurospora crassa*, led to the one gene: one enzyme hypothesis and to a second Nobel Prize for fungal research awarded to Beadle and Tatum in 1958. Fundamental research in biochemical genetics was extended to other fungi, especially to *Saccharomyces cerevisiae*, and by the mid-1960s fungal systems were much favored for studies in eukaryotic molecular biology and were soon able to compete with bacterial systems in the molecular arena.

The experimental achievements in research on the genetics and molecular biology of fungi have benefited more generally studies in the related fields of fungal biochemistry, plant pathology, medical mycology, and systematics. Today, there is much interest in the genetic manipulation of fungi for applied research. This current interest in biotechnical genetics has been augmented by the development of DNA-mediated transformation systems in fungi and by an understanding of gene expression and regulation at the molecular level. Applied research initiatives involving fungi extend broadly to areas of interest not only to industry but to agricultural and environmental sciences as well.

It is this burgeoning interest in fungi as experimental systems for applied as well as basic research that has prompted publication of this series of books under the title *The Mycota*. This title knowingly relegates fungi into a separate realm, distinct from that of either plants, animals, or protozoa. For consistency throughout this Series of Volumes the names adopted for major groups of fungi (representative genera in parentheses) are as follows:

Pseudomycota

Division: Oomycota (*Achlya, Phytophthora, Pythium*)
Division: Hyphochytriomycota

Eumycota

Division: Chytridiomycota (*Allomyces*)
Division: Zygomycota (*Mucor, Phycomyces, Blakeslea*)
Division: Dikaryomycota

Subdivision: Ascomycotina
    Class: Saccharomycetes (*Saccharomyces, Schizosaccharomyces*)
    Class: Ascomycetes (*Neurospora, Podospora, Aspergillus*)
Subdivision: Basidiomycotina
    Class: Heterobasidiomycetes (*Ustilago, Tremella*)
    Class: Homobasidiomycetes (*Schizophyllum, Coprinus*)

We have made the decision to exclude from *The Mycota* the slime molds which, although they have traditional and strong ties to mycology, truly represent nonfungal forms insofar as they ingest nutrients by phagocytosis, lack a cell wall during the assimilative phase, and clearly show affinities with certain protozoan taxa.

The Series throughout will address three basic questions: what are the fungi, what do they do, and what is their relevance to human affairs? Such a focused and comprehensive treatment of the fungi is long overdue in the opinion of the editors.

A volume devoted to systematics would ordinarily have been the first to appear in this Series. However, the scope of such a volume, coupled with the need to give serious and sustained consideration to any reclassification of major fungal groups, has delayed early publication. We wish, however, to provide a preamble on the nature of fungi, to acquaint readers who are unfamiliar with fungi with certain characteristics that are representative of these organisms and which make them attractive subjects for experimentation.

The fungi represent a heterogeneous assemblage of eukaryotic microorganisms. Fungal metabolism is characteristically heterotrophic or assimilative for organic carbon and some nonelemental source of nitrogen. Fungal cells characteristically imbibe or absorb, rather than ingest, nutrients and they have rigid cell walls. The vast majority of fungi are haploid organisms reproducing either sexually or asexually through spores. The spore forms and details on their method of production have been used to delineate most fungal taxa. Although there is a multitude of spore forms, fungal spores are basically only of two types: (i) asexual spores are formed following mitosis (mitospores) and culminate vegetative growth, and (ii) sexual spores are formed following meiosis (meiospores) and are borne in or upon specialized generative structures, the latter frequently clustered in a fruit body. The vegetative forms of fungi are either unicellular, yeasts are an example, or hyphal; the latter may be branched to form an extensive mycelium.

Regardless of these details, it is the accessibility of spores, especially the direct recovery of meiospores coupled with extended vegetative haploidy, that have made fungi especially attractive as objects for experimental research.

The ability of fungi, especially the saprobic fungi, to absorb and grow on rather simple and defined substrates and to convert these substances, not only into essential metabolites but into important secondary metabolites, is also noteworthy. The metabolic capacities of fungi have attracted much interest in natural products chemistry and in the production of antibiotics and other bioactive compounds. Fungi, especially yeasts, are important in fermentation processes. Other fungi are important in the production of enzymes, citric acid and other organic compounds as well as in the fermentation of foods.

Fungi have invaded every conceivable ecological niche. Saprobic forms abound, especially in the decay of organic debris. Pathogenic forms exist with both plant and animal hosts. Fungi even grow on other fungi. They are found in aquatic as well as soil environments, and their spores may pollute the air. Some are edible; others are poisonous. Many are variously associated with plants as copartners in the formation of lichens and mycorrhizae, as symbiotic endophytes or as overt pathogens. Association with animal systems varies; examples include the predaceous fungi that trap nematodes, the microfungi that grow in the anaerobic environment of the rumen, the many insectassociated fungi and the medically important pathogens afflicting humans. Yes, fungi are ubiquitous and important.

There are many fungi, conservative estimates are in the order of 100000 species, and there are many ways to study them, from descriptive accounts of organisms found in nature to laboratory experimentation at the cellular and molecular level. All such studies expand our knowledge of fungi and of fungal processes and improve our ability to utilize and to control fungi for the benefit of humankind.

We have invited leading research specialists in the field of mycology to contribute to this Series. We are especially indebted and grateful for the initiative and leadership shown by the Volume Editors in selecting topics and assembling the experts. We have all been a bit ambitious in producing these Volumes on a timely basis and therein lies the possibility of mistakes and oversights in this first edition. We encourage the readership to draw our attention to any error, omission or inconsistency in this Series in order that improvements can be made in any subsequent edition.

Finally, we wish to acknowledge the willingness of Springer-Verlag to host this project, which is envisioned to require more than 5 years of effort and the publication of at least nine Volumes.

Bochum, Germany   KARL ESSER
Auburn, AL, USA   PAUL A. LEMKE
April 1994   *Series Editors*

Addendum to the Series Preface

In early 1989, encouraged by Dieter Czeschlik, Springer-Verlag, Paul A. Lemke and I began to plan *The Mycota*. The first volume was released in 1994, five other volumes followed in the subsequent years. Also on behalf of Paul A. Lemke, I would like to take this opportunity to thank Dieter Czeschlik, his colleague Andrea Schlitzberger, and Springer-Verlag for their help in realizing the enterprise and for their excellent cooperation for many years.

Unfortunately, after a long and serious illness, *Paul A. Lemke* died in November 1995. Without his expertise, his talent for organization and his capability to grasp the essentials, we would not have been able to work out a concept for the volumes of the series and to acquire the current team of competent volume editors. He also knew how to cope with unexpected problems which occurred after the completion of the manuscripts. His particular concern was directed at Volume VII; in this volume, a posthumous publication of his is included.

Paul A. Lemke was an outstanding scientist interested in many fields. He was extremely wise, dedicated to his profession and a preeminent teacher and researcher. Together with the volume editors, authors, and Springer-Verlag, I mourn the loss of a very good and reliable friend and colleague.

Bochum, Germany   KARL ESSER
April 2000

Volume Preface

The vital role of fungal associations for all ecosystems was only recognized in the second half of the 20th century. The growing realization that the relevance of fungal associations goes beyond classical mycology has greatly accelerated research in this field. The availability of new tools, provided especially by molecular biology, has triggered new approaches to the study of fungi as hosts as well as symbionts.

Due to the enormous amount of new work in the field of fungal associations, it has been impossible to include all topics of interest in this Volume. Rather, it has been decided to concentrate in more depth on subjects such as mycorrhizae, lichens, as well as some new developments coming up more recently, e.g., *Geosiphon* and *Piriformospora* associations. Therefore a compromise had to be made, resulting in the omission of other important aspects such as fungal endophytes or symbioses with insects. Nevertheless, it is hoped that this Volume will contribute to a better understanding of fungal associations.

It has been a pleasure to edit this book, primarily due to the stimulating discussions with the series editor Prof. Karl Esser. I am indebted to Springer-Verlag for all the help and active cooperation during the preparation of this Volume. I am grateful to Stefanie Rauchalles who has eliminated many errors in the manuscripts and helped to attain consistency in the presentation of the chapters.

I hope that this book will help to answer questions concerning the complex fungal associations and provide guidance for future research.

Freising-Weihenstephan, April 2000   Bertold Hock

Contents

Fungal Bacterial Interactions

List of Contributors

BAREA, J.M., Dep. de Microbiología del Suelo y Sistemas Simbióticos, Estación Experimental del Zaidín, CSIC, Prof. Albareda 1, 18008 Granada, Spain

BENNETT, J.W., Department of Cell and Molecular Biology, Tulane University, New Orleans, Louisiana 70118, USA

BHARTI, K., School of Life Sciences, Jawaharlal Nehru University, New Delhi, 110067, India

BLECHERT, O., Max-Planck-Institut für Terrestrische Mikrobiologie, Philipps-Universität, Karl-von-Frisch-Straße, 35032 Marburg, Germany

BONFANTE, P., Dip. di Biol. Vegetale dell'Univ. di Torino, Centro di Studio sulla Micologia del Terreno del CNR, Viale Mattioli 25, 10125 Torino, Italy

DALPÉ, Y., Eastern Cereal and Oilseed Research Centre (ECORC), Wm. Saunders Building, Central Experimental Farm, 960 Carling Avenue, Ottawa, Ontario K1A 0C6, Canada

DEKA, D., School of Life Sciences, Jawaharlal Nehru University, New Delhi, 110067, India

DULIEU, H., Laboratoire de Phytoparasitologie INRA/CNRS, Centre de Microbiologie du Sol et de l'Environnement, Institut National de Recherche Agronomique, BV 1540, 21034 Dijon cédex, France

DUMAS-GAUDOT, E., Laboratoire de Phytoparasitologie INRA/CNRS, Centre de Microbiologie du Sol et de l'Environnment, Institut National de Recherche Agronomique, BV 1540, 21034 Dijon cédex, France

FEIBELMAN, T., Microbia, Inc., One Kendall Square, Bldg. 1400W, Suite 1418, Cambridge, MA 02139, USA

FRANKEN, P., Max-Planck-Institut für Terrestrische Mikrobiologie, Philipps-Universität, Karl-von-Frisch-Straße, 35032 Marburg, Germany

GIANINAZZI-PEARSON, V., Laboratoire de Phytoparasitologie INRA/CNRS, Centre de Microbiologie du Sol et de l'Environnement, Institut National de Recherche Agronomique, BV 1540, 21034 Dijon cédex, France

GRANDMOUGIN-FERJANI, A., Laboratoire Mycologie/Phytopathologie/Environnement (L.M.P.E.), Université du Littoral Côte d'Opale, B.P. 699-62228 Calais Cedex, France

HAHN, A., Technische Universität München in Freising-Weihenstephan, Botanisches Institut, Alte Akademie 12, 85350 Freising, Germany

HAMPP, R., Universität Tübingen, Botanisches Institut, Physiologische Ökologie der Pflanzen, Auf der Morgenstelle 1, 72076 Tübingen, Germany

HOCK, B., Technische Universität München in Freising-Weihenstephan, Botanisches Institut, Alte Akademie 12, 85350 Freising, Germany

HONEGGER, R., Institut für Pflanzenbiologie, Universität Zürich, Zollikerstr. 107, 8008 Zürich, Switzerland

HUREK, T., Max-Planck-Institut für Terrestrische Mikrobiologie, Philipps-Universität, Karl-von-Frisch-Straße, 35032 Marburg, Germany

JEFFRIES, P., Research School of Biosciences, University of Kent, Canterbury, Kent CT2 7NJ, UK

KLUGE, M., Institut für Botanik, Technische Universität Darmstadt, 64287 Darmstadt, Germany

KOST, G., Systematische Botanik und Mycologie, Fachbereich Biologie, Karl-von-Frisch-Straße, 35032 Marburg, Germany

KRANNER, I., Institut für Pflanzenphysiologie, Karl-Franzens Universität in Graz, Schubertstraße 51, 8010 Graz, Austria

KUMARI, M., School of Life Sciences, Jawaharlal Nehru University, New Delhi, 110067, India

LUMBSCH, H.T., Universiät Essen, Fachbereich 9/Botanik, 45117 Essen, Germany

MAIER, W., Institut für Pflanzenbiochemie, Weinberg 3, 06120 Halle (Saale), Germany

NEHLS, U., Universität Tübingen, Botanisches Institut, Physiologische Ökologie der Pflanzen, Auf der Morgenstelle 1, 72076 Tübingen, Germany

OBERWINKLER, F., Fakultät für Biologie, Lehrstuhl Spezielle Botanik und Mykologie, Universität Tübingen, Auf der Morgenstelle 1, 72076 Tübingen, Germany

OTT, S., Heinrich Heine Universität, Botanisches Institut, Universitätsstraße 1, 40225 Düsseldorf, Germany

RANA, D., School of Life Sciences, Jawaharlal Nehru University, New Delhi, 110067, India

REQUENA, N., Max-Planck-Institut für Terrestrische Mikrobiologie, Philipps-Universität, Karl-von-Frisch-Straße, 35032 Marburg, Germany

REXER, K.-H., Systematische Botanik und Mycologie, Fachbereich Biologie, Karl-von-Frisch-Straße, 35032 Marburg, Germany

Roy, A., School of Life Sciences, Jawaharlal Nehru University, New Delhi, 110067, India

Sahay, N.S., School of Life Sciences, Jawaharlal Nehru University, New Delhi, 110067, India

Sancholle, M., Laboratoire Mycologie/ Phytopathologie/Environnement (L.M.P.E.), Université du Littoral Côte d'Opale, B.P. 699-62228 Calais Cedex, France

Schüßler, A., Institut für Botanik, Technische Universität Darmstadt, 64287 Darmstadt, Germany

Sharma, J., School of Life Sciences, Jawaharlal Nehru University, New Delhi, 110067, India

Singh, A., School of Life Science, Jawaharlal Nehru University, New Delhi, 110067, India

Strack, D., Institut für Pflanzenbiochemie, Weinberg 3, 06120 Halle (Saale), Germany

Sudha, School of Life Sciences, Jawaharlal Nehru University, New Delhi, 110067, India

Thakran, S., School of Life Sciences, Jawaharlal Nehru University, New Delhi, 110067, India

van Tuinen, D., Laboratoire de Phytoparasitologie INRA/CNRS, Centre de Microbiologie du Sol et de l'Environnement, Institut National de Recherche Agronomique, BV 1540, 21034 Dijon cédex, France

Varma, A., School of Life Sciences, Jawaharlal Nehru University, New Delhi, 110067, India

Walter, M., Institut für Pflanzenbiochemie, Weinberg 3, 06120 Halle (Saale), Germany

Wiese, J., Universität Tübingen, Botanisches Institut, Physiologische Ökologie der Pflanzen, Auf der Morgenstelle 1, 72076 Tübingen, Germany

Wright, S., USDA-ARS, Soil Microbial Systems Laboratory, 10300 Baltimore Ave., Beltsville, Maryland 20705-2350, USA

Mycorrhizae

1 Exploring the Genome of Glomalean Fungi

V. Gianinazzi-Pearson, D. van Tuinen, E. Dumas-Gaudot, and H. Dulieu

CONTENTS

I. Introduction

The order Glomales has been created in the Zygomycota to encompass fungi forming arbuscular mycorrhiza (AM) in a mutualistic symbiosis with plant roots (Morton 1993). Glomalean fungi are complex but extremely successful organisms. Since their appearance about the time of the Devonian era, they have persisted through periods of important environmental change and spread to newly evolving plant species to become abundant symbionts in terrestrial ecosystems across the globe (Smith and Read 1997). They are, nevertheless, obligate symbionts and are assumed to reproduce asexually, which raises the question of how they have dealt with mutational problems and adaptation to new hosts or habitats during evolu-

tion. Reproductive structures are large, vegetative, unicellular spores which harbour several hundreds or thousands of nuclei (Fig. 1A), making Glomales an unusual group of fungi (Giovannetti and Gianinazzi-Pearson 1994). However, as in other organisms, the absence of a sexual stage in their life cycle does not necessarily mean that genetic flux between individuals is absent. Vegetative anastamosis occurs between hyphae originating from different germinating spores of a same species and in symbiotic mycelium growing within root tissues (Hepper and Mosse 1975; Casana and Bonfante-Fasolo 1988; Tommerup 1988), providing the possibility for somatic exchange of nuclei and therefore for a heterokaryotic state. Until recently, little was known about glomalean genomics and it is only with the advent of powerful molecular techniques that it has been possible to venture research into their genetic makeup. Likewise, because of the incalcitrance of these fungi to pure culture and consequent difficulties in obtaining sufficiently large quantities of material, the analyses of gene products has remained an extremely challenging but relatively unexplored area. Basic knowledge about genome structure, complexity and function in the Glomales is essential to an understanding of the reproductive biology of this important group of fungi and their apparent stability during coevolution in symbiosis with many different plant taxa.

II. Genome Size and Organisation

A. DNA Contents of Nuclei

The amount of DNA in a haploid genome is called the genome size or C value (Swift 1950). The ploidy levels in glomalean fungi are not well defined, but spore nuclei have been presumed to be haploid (Rosendahl and Sen 1992). Values obtained for DNA contents of nuclei of the Glo-

Laboratoire de Phytoparasitologie INRA/CNRS, Centre de Microbiologie du Sol et de l'Environnement, Institut National de Recherche Agronomique, Street address: 17, rue Sully 21000 Dijon France, BV 1540, 21034 Dijon cédex, France

The Mycota IX
Fungal Associations
Hock (Ed.)
© Springer-Verlag Berlin Heidelberg 2001

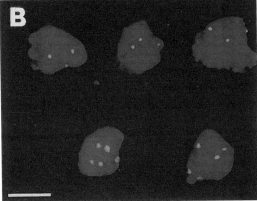

Fig. 1. A DAPI-stained nuclei released from a crushed spore of *S. castanea* and observed under UV light: the nuclei fluoresce bright (*right side*) whilst the spore wall appears dark (*left side*). (*bar* 50 μm). **B** Immunofluorescence in situ hybridization of interphasic nuclei of *G. mosseae* with a digoxigenin-labelled 25S rDNA probe: hybridisation signals are localized by the bright spots of the fluorescent probe (Alexa 488) on the background chromatin (*gray*) stained with propidium iodide (*bar* 5 μm). (Original photographs S. Trouvelot, INRA-CMSE, Dijon, France)

(*A. scrobiculata, A. laevis, A. longula*) vary between 0.33 and 0.54 pg DNA/nucleus. In contrast, values for *Gigaspora rosea* and *Gig. margarita* appear to be very similar (0.65–0.77). The Glomales, consequently, have large genomes as compared to other members of the Zygomycota, where DNA contents of 0.03–0.04 pg/nucleus have been reported (Dusenberry 1975; Dutta 1974; Wöstenmeyer and Burmester 1986). Interestingly, the obligate, biotrophic fungal pathogens *Erysiphe graminis* and *E. chorocearum* have comparably high C values (0.52 and 1.5 pg DNA/ nucleus respectively) (Cavalier-Smith 1985). As haploid organisms, the genome size of these and glomalean fungi would be around 10^8 to 10^9 bp.

B. Repeated DNA Sequences

Assuming that the C values of glomalean species correspond directly to their genome size, the differences within a same genus could result from changes in ploidy levels and/or localised increases in copy numbers of different sequences. Chromosomes have not yet been visualised in the Glomales so that their number is not known. However, the development of an efficient procedure for cloning genomic DNA from spore nuclei has offered possibilities for investigating genome structure and provided first evidence for the existence of highly repeated DNA sequences (Zézé et al. 1994, 1996, 1999). Large genome variations in eukaryotic organisms have generally been attributed to the presence of variable amounts of noncoding repeated DNA sequences (Bennett 1987). Nine genomic clones isolated from *S. castanea* gave high signal intensity with spore DNA, suggesting a high copy number of the corresponding sequences in nuclei. One complete non-encoding sequence (SC1), which is considered to be a member of a family of tandem repeats and is specific to *S. castanea*, represents about 0.24% of the fungal genomic DNA. The calculated corresponding copy number of SC1 is approximately 2600, which classifies this sequence as highly repeated DNA (Zézé et al. 1996). Estimates for another of the isolated repeated sequences (*Mycdire*), which is scattered throughout the genome of different genera, give 0.04, 0.045, 0.039 and 0.107% of the genomic DNA for *S. castanea, A. laevis, Gig. rosea* and *G. caledonium*, respectively (Zézé et al. 1999). This makes it relatively

males vary between species and depending on the analytical methods used. Closest estimates come from microfluorimetry and flow cytometry of DAPI-stained nuclei from spores (Hosny et al. 1998a; Bianciotto and Bonfante 1992). C values which have been obtained for species from four genera range from 0.13 to 1.08 pg DNA/ nucleus (Hosny et al. 1998a). This eight fold variation occurs in the genus *Scutellospora* (*S. pellucida, S. heterogama, S. gregaria*) and it is the greatest found so far between species. *Glomus geosporum* and *G. caledonium* have C values of 0.18 and 0.38 pg, englobing the value of 0.27 pg for *G. versiforme*, whilst *Acaulospora* species

abundant, and copy numbers, calculated from DNA contents published by Hosny et al. (1998a), vary from around 100 to 300 between the species. Sequence analyses of this conserved *Mycdire* element have indicated the presence of autonomously replicating AT-rich regions, homologous to autonomously replicating sequence (ARS) elements in other fungi (Theis and Newlon 1996), and it has been speculated that these may reflect the existence of transposable elements at some time in the history of the Glomales (Zézé et al. 1999).

Another example of repetitive DNA sequences that have recently been found in the genome of glomalean fungi are minisatellites (20-bp repeats) and microsatellites (up to 8-bp repeats). These are hypervariable regions organised in tandem repeats and characterised by an exceptionally high mutation rate in repeats number (Tautz 1993), making them very informative about genetic variation between and within species. Microsatellites have been found in species from the Glomaceae, Acaulosporaceae and Gigasporaceae. Assays for genetic variability using microsatellite-primed PCR gave species-specific DNA fingerprints for seven fungi belonging to the genera *Glomus*, *Acaulospora*, *Gigaspora* and *Scutellospora*. This technique also revealed that intraspecific divergence had developed within an isolate of *G. mosseae* after it had been propagated for more than 12 years in three different culture collections (Longato and Bonfante 1997). In a detailed minisatellite-PCR fingerprinting study of the spore population arising from a single-spore inoculation of *Gig. margarita* onto clover, a high level of genetic diversity rapidly appeared in the first generation of spores (Zézé et al. 1997). Such variation within progeny of an asexual organism could arise from the dissimilar sorting, during the life cycle of the fungus, of different nuclear types present in the original spore.

C. Base Composition

Recent investigations of base composition (% GC or GC content) in glomalean fungi have provided further information about their genome structure (Hosny et al. 1998b). HPLC analysis of nucleotides from spore DNA has given values of 30 to 35% GC for nine species from the genera *Scutellospora*, *Gigaspora*, *Glomus* and *Acaulospora*. In spite of this narrow range, some species differ significantly in their GC contents but variations are not taxon-related. The GC values are relatively low as compared to most fungal taxa but fall within the range reported for other members of the Zygomycota (Storck and Alexopoulos 1970). Particularly low GC (high AT) contents have been observed in a rumen fungus (Brownlee 1989) and soil bacteria under conditions where UV and visible light, which may be instrumental in selecting against mutations to AT, are generally absent (Brownlee 1989; Li and Grauer 1991). It has been proposed that a mutational pressure from GC to AT may exist in the Glomales and that, since these fungi also proliferate under light-deprived conditions in soil and roots, their environment has not exerted any significant counterselection against AT-rich sequences so that GC contents have become low (Hosny et al. 1998b).

Modified DNA bases occur in most eukaryotic genomes but the most frequent, 5-methylcytosine (mC), has only been reported in low amounts in fungi with a maximum of 0.5% of the total genome for *Phycomyces blakesleeanus* (Antequera et al. 1984). Contrastingly high amounts of mC have been detected in the DNA of nine different glomalean species from four genera (Hosny et al. 1998b). Values range from 2.23% of total nucleotides in *Gig. margarita* to 4.26% in *A. laevis*, placing them well above those for other fungi. In addition, from 12.36 to 24.85% of cytosine residues are methylated in glomalean isolates as compared to only 1–2% in *Neurospora crassa* (Russell et al. 1987). This high proportion of mC provides further indirect evidence for the existence of repetitive DNA in the Glomales, since methylation of cytosine is frequent in repeated sequences (Radman 1991). Mutation of mC to T requires only C-4 deamination so that random mC→T transitions should be possible at high frequencies, which will induce rapid divergence in mC-rich repeated sequences and tend to raise the AT content of the genome (Sueoka 1988). Furthermore, nucleotide divergence will reduce recombination events between repeats and may lead to genomes accumulating non-coding "junk" DNA, thus increasing their size and complexity without changing their coding content (Bennett and Smith 1991). Such processes may offer some explanation for the large, AT-rich genomes which have evolved in the Glomales.

III. Ribosomal Genes

Ribosomal genes are by far the most studied multicopy, tandemly repeated genes in glomalean fungi. Ribosomal RNA (rRNA) is the main class of RNA and can represent up to 90% of all RNAs in living cells. Ribosomes being the structural backbone for processes in protein synthesis, the encoding genes have been largely conserved throughout evolution. In eukaryotes, the ribosomal genes code for a pre-RNA of 45S which, after processing, gives the small ribosomal subunit (18S), the 5.8S subunit and the large ribosomal subunit (25S) (Perry 1976). The small, 5.8S and large ribosomal subunit coding regions (rDNA) are separated from each other by an internal transcribed spacer (ITS), and the ribosomal genes by an intergenic nontranscribed spacer (IGS). In fungi, the fourth rRNA involved in ribosome structure (5S rRNA) is encoded within the IGS or in another part of the genome (Garber et al. 1988). Sequence variations are not evenly distributed throughout the ribosomal genes and the three regions evolve at different rates. ITS and IGS are variable regions which mutate more frequently than the three conserved coding subunit regions (18S, 5.8S, 25S) (Mitchell et al. 1995). This generally makes the former more informative for analyses of closely related genomes, whereas the coding regions of the small and the large ribosomal subunit are considered to be more useful for understanding more distant relationships at the species/order level. The various characteristics of rRNA and rDNA have made them a choice target for phylogenetic and taxonomic studies, and comparative studies of the nucleotide sequences in ribosomal genes have provided data for the analysis of phylogenetic relationships over a wide taxonomic range of organisms (Barouin et al. 1988; Bruns et al. 1991).

In the Glomales, the 18S, ITS and 5.8S rDNA of *G. mosseae* (BEG12) and of *S. castanea* (BEG1) have been entirely sequenced from genomic clones (Franken and Gianinazzi-Pearson 1996), and the number of copies per genome has been estimated to be 75 ± 10 for *S. castanea* (Hosny et al. 1999). This copy number is relatively low for ribosomal genes when compared to other fungi and is not in proportion to the large size of the glomalean genome. For example, ribosomal gene copies are 130 to 140 in *Saccharomyces cerevisiae*

(Schweizer et al. 1969), which has a genome about 60 times smaller than that of *S. castanea*, and in *Cochliobolus heterostrophus*, with a genome some 30 times smaller, this number is about 130 (Garber et al. 1988). These observations tend towards the hypothesis that glomalean fungi may have accumulated important amounts of "junk" DNA during their evolution.

A. 18S Subunit

The small rDNA subunit (18S) was the first target region of the glomalean genome to be sequenced for phylogenetic studies based on molecular data (Simon et al. 1992a, 1993a). This initial work represented an important step forward and it has pioneered much of the molecular research on the genomics of these unculturable fungi. By comparing regions in the 18S rDNA from *G. intraradices* and *Gig. margarita* with sequences from other fungal taxon, a PCR primer was designed which, in combination with a universal eukaryotic primer, exclusively amplified glomalean rDNA (Simon et al. 1992a). Phylogenetic data collected for 18S rDNA sequences from 12 fungi were used to date their origin back to a *Glomus*-like fungus 353–462 Ma ago, which corresponds to the period when primitive plants started to colonise terrestrial ecosystems (Simon et al. 1993a). Sequence data were also interpreted as indicating an evolutionary split of fungi in the Acaulosporaceae and Gigasporaceae from those in the Glomaceae, estimated to have occurred some 250 Ma years ago. This relationship, where the Acaulosporaceae clusters with the Gigasporaceae rather than the Glomaceae (Simon 1996), is in contradiction with morphogenic-based taxonomy, which groups the Acaulosporaceae and Glomaceae into the same suborder Glomineae, and separate from the Gigasporaceae, which are placed in the suborder Gigasporineae (Morton 1993). This discrepancy underlines the need for (1) extensive sequencing of clones of the same 18S sequence to evaluate intraspecies divergence and representativity per species, (2) analyses of other regions of the genome across a wide range of Glomales and (3) to interpret such data in the context of analyses based on other taxon-linked characters (morphology, spore ontogeny, proteome, isozymes, etc.).

Variability in small ribosomal subunit sequences has also been sufficient to generate

PCR primers that discriminate between glomalean fungi at the family level (Simon et al. 1993b). Although these primers have been successful for detection of single fungi in plant roots (Simon et al. 1992b; Di Bonito et al. 1995), their sensitivity and specificity was much improved when combined with a selective enrichment of amplified DNA (SEAD) technique (Clapp et al. 1995). Several distinct Glomales were detected within the root system of one and the same plant from field conditions and the makeup of the fungal community in roots was found to differ considerably from that in the soil, estimated from spore frequencies.

B. ITS Regions

The variability observed in the two ITS regions separating the 5.8S from the small and the large ribosomal subunit has also been used to target and distinguish between different glomalean fungi. These regions are more polymorphic than the two large subunits, and have previously been useful in identifying ectomycorrhizal fungi (Gardes and Bruns 1993). However, Glomales spores collected either from the field with a similar phenotype or from pot cultures of a well-defined species have been found to harbour different ITS sequences (Lloyd-MacGilp et al. 1996; Sanders et al. 1995). Nevertheless, interspore polymorphism appears less important within a same isolate from one pot culture than between different isolates of the same fungal species. In contrast, the interspecific diversity, for example between *G. dimorphicum* and *G. fasciculatum*, is not greater than the intraspecific diversity within *G. mosseae* (Lloyd-MacGilp et al. 1996). Although restriction enzyme analysis of the ITS region has been used to identify Glomales at the species level (Redecker et al. 1997), the high inter- and intraspecific variability observed in these fungi greatly restricts the use of these ribosomal regions for genetic and evolutionary studies. Recently, Hosny et al. (1999) isolated 13 full-length rDNA clones from spores of *S. castanea* which separated into nine different types of restriction enzyme profiles. Direct sequencing of the ITS regions and 18S genes from the clones gave five significantly different sequences, one of which was highly divergent from the others. Amplification of single-spore DNA using specific primers for the divergent clone showed that this sequence segregates among spores. Substantial interspore genetic variability has also been found in *Gig. margarita*, with up to 9% sequence divergence in the ribosomal ITS region comprising the 5.8S gene and the flanking ITS regions (Lanfranco et al. 1999a). Furthermore, there is now evidence from analyses of ITS polymorphism in *S. castanea* and *Gig. margarita* that not only nuclei between spores but also those within a single spore are not identical (see below). Heterogeneities in the ribosomal genes have been observed in other organisms (O'Donnell and Cigelnik 1997); polymorphism seems to be the rule in the Glomales and a single glomalean spore probably contains a heterokaryotic "population" of nuclei, and not multicopies of the same nuclei.

C. 25S Subunit

The large ribosomal subunit (25S) is also an interesting polymorphic region of the ribosomal gene. All large rDNAs possess a common core of secondary structure (De Rijk et al. 1998) but large variations in the size of the molecule have arisen during evolution and are accommodated over a dozen rapidly evolving domains (Michot and Bachellerie 1987). Most of the variation is restricted to the two domains D2 and D8. Only the 5′ end of the large ribosomal subunit, covering the two domains D1 and D2, has been studied up to now in the Glomales. As in other organisms, the D2 domain is polymorphic in length and sequence (Fig. 2), and this has enabled generation of taxon-discriminating PCR primers for glomalean fungi (van Tuinen et al. 1994, 1998a,b). When used in a nested-PCR reaction, these easily distinguish between mycelium of different fungi in distinct compartments of cultures, ie bulk soil, mycorrhizosphere and mycorrhiza (S. Trouvelot, E. Jacquot, unpubl. data; Fig. 3). This approach is sufficiently sensitive to analyse single spores or individual small root fragments, and has shown that the latter are frequently colonised by more than one fungus (van Tuinen et al. 1998a). Corresponding PCR fragments, used in fluorescence in situ hybridisation (FISH), have proved useful probes to localise 25S regions in individual nuclei from glomalean spores and to detect a possible nuclear heterogeneity in the number of rDNA sites within *G. mosseae*, *G. intraradices*, *Gig. rosea* and *S. castanea* (Fig. 1B; Trouvelot et al. 1999).

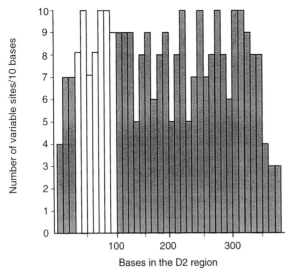

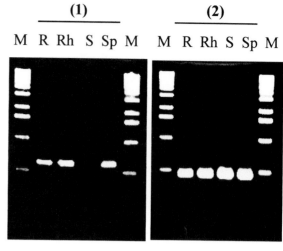

Fig. 2. Histogram comparing the number of variable sites per 10 base pairs in the D2 region of the large ribosomal subunit between spore DNA of 20 isolates from 17 species of Glomales. *White bars* represent the insert that is specific to the genus *Glomus*. As an example, the D2 region of *G. mosseae* spans base 394 to 767 starting from the LR1 primer in the large ribosomal subunit

Fig. 3 1,2. Amplification products from nested PCR of DNA extracted from arbuscular mycorrhiza (*R*), rhizosphere hyphae (*Rh*) and soil mycelium (*S*) of *Scutellospora castanea* (**1**) and *Glomus mosseae* (**2**) after inoculation of micropropagated grapevine plants with a mixture of the two fungi. *G. mosseae* is detected in all the fungal compartments (primer pair 5.25-FLR2) whilst *S. castanea* has not developed in the soil (primer pair 4.24-LR1); *Sp* spores isolated from the inoculum. (S. Trouvelot, D. van Tuinen, V. Gianinazzi-Pearson, unpubl. data)

Table 1. Taxon-discriminating levels of PCR primers designed from sequences in the D2 domain of 25S rDNA

Level of discrimination			
Kingdom	Fungus[a]		
Family	Glomaceae[a]	Acaulosporaceae[a]	Gigasporaceae[b]
Species	*G. mosseae*[b]	*A. laevis*[a]	*Gig. margarita*[a]
	G. versiforme[a]		*Gig. rosea*[c]
	G. caledonium[a]		*Gig. candida*[a]
	G. coronatum[a]		*Scutellospora castanea*[c]
	G. intraradices[c]		
	G. clarum[a]		
Isolate	*G. mosseae*[a] BEG12		

[a] In preparation. [b] van Tuinen et al. (1998b). [c] van Tuinen et al. (1998a).

The polymorphism present in the D2 domain of the 25S rDNA region is such that it has given PCR-primers with different levels of specificity (Table 1). This is important since, depending on the question investigated, molecular tools with different levels of specificity have to be used. The level of specificity which can be obtained when using the different ribosomal regions ranges from nuclei within or between spores (Hijri et al. 1999) to the isolate, species or family level (Table 1), and up to the order (Simon et al. 1992a). Discrimination between nuclei is essential in order to study the genetics of these fungi and to understand the spatial distribution and/or segregation of the nuclei through the fungal mycelium and during spore formation or germination. On the other hand, studies of interactions between a host plant and different fungi require probes that discriminate at the isolate or species level. The use of ribosomal genes has a great advantage over other genomic regions in that more and more sequences are becoming available, providing a broad database for the fine tuning of specific molecular probes for monitoring glomalean fungi in the field.

IV. Protein-Encoding Genes

Relatively few protein-encoding genes have been identified in the glomalean genome. They mainly come from cDNA or genomic libraries, of mycorrhizal roots or spores, probed by differential screening or using PCR primers derived from conserved regions of corresponding genes in other organisms (for more detail see Franken and Requena, Chap. 2, this Vol.). Genomic sequences have been obtained from *G. versiforme* which code for a transmembrane phosphate transporter, a cruciform DNA-binding like protein and class IV chitin synthase (Harrison and van Buuren 1995; Lanfranco et al. 1999b; Burleigh and Harrison 1998). One *Glomus* species has been shown to possess an assimilatory nitrate reductase gene (Kaldorf et al. 1994) and a β-tubulin gene fragment isolated from *G. mosseae* corroborates observations of β-tubulin components in hyphae of this fungus (Aström et al. 1994; Bütehorn et al. 1999).

A. Enzyme-Active Proteins

Indirect evidence for other protein encoding genes comes from enzyme-active fungal gene products. Early research focused on enzymes possibly involved in symbiotic processes in AM and because of the role of the symbiotic fungi in translocating soil phosphate to host cells, investigations began with those linked to phosphate metabolism. The first fungal enzymes to be identified electrophoretically were alkaline phosphatases (EC 3.1.3.2) (reviewed in Gianinazzi et al. 1992) and their origin was deduced from detection in extraradical mycelium and in intraradical mycelium extracted from enzyme-digested mycorrhiza (Gianinazzi-Pearson and Gianinazzi 1986; Tisserant et al. 1998). Spores and external mycelium of glomalean fungi also possess a range of proteins with pectolytic activities which may participate in the process of penetration and colonisation of roots by the symbiotic fungus: pectin esterase (EC 3.1.1.11), endo-polygalacturonase (EC 3.2.1.15),

exo-polygalacturonase (EC 3.2.1.67), pectate (EC 4.2.2.9) and pectin lyases (EC 4.2.2.10) (Garcia-Romera et al. 1991, 1992). Xyloglucan is a major structural component of the primary cell walls of many dicotyledons and monocotyledons, and xyloglucan-hydrolysing glucanases [including endoglucanase (EC 3.2.1.4) and exoglucanase (EC 3.2.1.91)] have been reported to be active in extracts of spores and external mycelium of *G. mosseae* (Garcia-Garrido et al. 1992; Rejon-Palomares et al. 1996). Furthermore, a cellulase has been reported in spores and extraradical hyphae (Garcia-Garrido et al. 1992, 1996). Glomalean cell walls contain chitin (Bonfante-Fasolo et al. 1990), and some of them glucans (Gianinazzi-Pearson et al. 1994), so that the fungi should have their own enzymes for fungal cell wall morphogenesis. Surprisingly, no active isoforms of chitinase (EC 3.2.1.14), chitosanase (EC 3.2.1.99) or β-1,3-glucanase (EC 3.2.1.39) have been detected in extracts of ungerminated spores or mycelium from germinated spores of different glomalean species (Dumas-Gaudot et al. 1994a; Slezack et al. 1996), although additional isoforms of these enzymes are induced in various AM symbioses (Dassi et al. 1996; Dumas-Gaudot et al. 1996; Pozo et al. 1998).

B. Isozyme Polymorphism

Isozymes, which are proteins with the same enzyme activity but encoded by separate loci, can provide information on the genetic and nuclear condition of a fungal isolate. Alkaline phosphatase isozymes have been shown in a range of glomalean fungi but the enzyme does not show a high degree of polymorphism between different species (Gianinazzi et al. 1992). In contrast, other enzymic proteins from glomalean spores display clear variations in electrophoretic banding patterns between species and amongst geographically different isolates. These include esterase (EST, EC 3.1.1.1), glucose-6-phosphate dehydrogenase (G6PDH, EC 2.7.1.4), glutamate oxaloacetate transaminase (GOT, EC 2.6.1.1), hexokinase (HK, EC 2.7.1.1), malate dehydrogenase (MDH, EC 1.1.1.37), malic enzyme (ME, EC 1.1.1.40), peptidase (PEP, EC 3.4.1.1) and phosphoglucomutase (PGM, EC 2.7.5.1) (Hepper et al. 1988b; Sen and Hepper 1986). The presumption that glomalean fungi are haploid comes from their life cycle but also from the banding patterns of some of these isozyme systems (Rosendahl and Sen 1992). The specificity of certain isozymes (EST, GOT and PEP) is such that they can provide diagnostic molecular markers to identify individual fungal species in colonised host roots and to investigate competition between different species (Hepper et al. 1988a). However, isozyme patterns have been reported to vary between intraradical mycelium and spores in *Glomus* species, indicating changes in nuclear conditions and differential expression of genes at different developmental stages of the fungi (Hepper et al. 1986).

Comparisons of isozyme polymorphism between spore cluster-forming *Glomus* isolates have provided evidence for extensive differentiation at the genetic level between morphologically similar fungi and underline the limitations of this approach for taxonomical studies (Rosendahl 1989). In fact, all the reported studies suffer from failure to identify putative loci and are thus difficult to use for comparing genetic distances (Rosendahl and Sen 1992). However, intra- and interspecific isozyme pattern analyses of MDH and EST in various isolates of the morphologically similar fungi *G. mosseae* and *G. coronatum* from different parts of the world confirmed the grouping, based on morphological characters, of these fungi into two species (Dodd et al. 1996). Furthermore, the recent combining of isozyme profiles (MDH) with 18S rDNA analyses for spores of *Gigaspora* species illustrates the usefulness of integrating genetically linked molecular characters into morphological taxonomy for defining groups in this genus (Bago et al. 1998).

C. Fungal Proteome

The terms proteome and proteome analysis first appeared in scientific literature in 1995 (Kahn 1995) to define large-scale studies of translated gene products (proteins) aimed at elucidating gene expression and function in biological systems. Attempts have been made to identify the symbiotic proteome of AM in various plant species (Dumas et al. 1989; Wyss et al. 1990; Schellenbaum et al. 1992; Arines et al. 1993; Garcia-Garrido et al. 1993; Dumas-Gaudot et al. 1994b; Simoneau et al. 1994; Samra et al. 1997). Much less is known about the fungal proteome itself, partly because of the fungi being obligate symbionts with a very time-limited growth phase without their hosts and consequently not easily accessible for biochemical analyses. The fragmentary information available

about the protein makeup of glomalean fungi comes from analyses of isolated spores, or of fungal structures associated with root tissues where the fungus is influenced by symbiotic interactions with the plant.

Preliminary investigations using native PAGE first indicated species differences in spore protein profiles from *A. laevis, G. fasciculatum* and *G. mosseae* (Schellenbaum et al. 1992). Detection and profile resolution was much improved by SDS-PAGE and distinct protein profiles have subsequently been established at the genus, species and isolate level for different Glomales (Dodd et al. 1996; Avio and Giovannetti 1998). The feasibility of using taxon-discriminating fungal protein profiles as a support for taxonomic studies of these fungi has been illustrated in the detailed study on different isolates of *G. mosseae* and *G. coronatum* by Dodd et al. (1996). Although identification of taxon-specific fungal proteins by 1D-PAGE is still restricted by the analytical limits of the method, the existence of differences has prompted the use of protein fractions from spores as effective antigens to produce antibodies against various glomalean species (Sanders et al. 1992; Cordier et al. 1996).

Substantial progress has recently been made in proteome analysis of glomalean fungi by applying high resolution 2D-PAGE analysis to the characterisation of fungal polypeptides (Samra et al. 1996). By miniaturising the procedure for extracting proteins from spores of *S. castanea, A. laevis, Gig. margarita* and *G. mosseae*, between 80 and 200 polypeptides were resolved from no more than $25\,\mu g$ of soluble protein. Taxon-specific polypeptides were distinguished, with some polypeptides being common to the four species and others only being found in one of them. Furthermore, preliminary analyses of subcellular fractions of spore material have given discriminating polypeptide patterns from soluble and cell wall extracts (Fig. 4). Although such research is at a preliminary stage, it undoubtedly opens new perspectives for proteome characterisastion in the Glomales and for producing specific tools to study these fungi. Identification of taxon-specific proteins at the species or isolate level, their purification and microsequencing should provide the opportunity for producing either specific antibodies or oligonucleotide probes to monitor individual fungi in soil and in planta.

2D-PAGE analyses have also been applied to investigate fungal protein expression linked to dif-

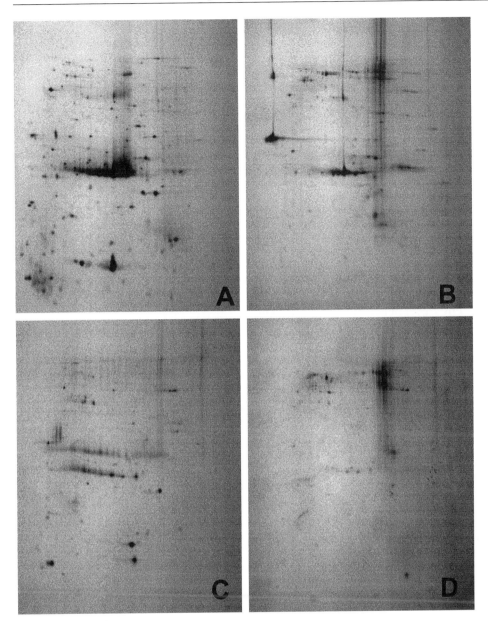

Fig. 4A–D. Two-dimensional PAGE polypeptide patterns of soluble (**A**, **C**) and cell wall (**B**, **D**) extracts of spores of *Gigaspora margarita* (**A**, **B**) and *Scutellospora castanea* (**C**, **D**). Soluble proteins from spores were phenol-extracted as described by Samra et al. (1996), and after centrifugation, cell wall proteins were extracted from the pellet according to Wessels et al. (1991). Samples were separated on immobilized pH gradient three to ten strips (Pharmacia) for the first dimension, on 12% SDS-PAGE for the second dimension and then silver-stained. (G. Bestel-Corre, K. Labour, E. Dumas-Gaudot, unpubl. results)

ferent stages of the life cycle of glomalean fungi during interactions with host plants. For example, a strong increase in the number of polypeptides synthesised by *G. mosseae* was found to occur with spore germination in vitro (Samra et al. 1996). However, in this same study, no qualitative or quantitative modifications in polypeptide expression during germination of spores were elicited by root exudates from pea genotypes differing in their ability to form AM. In attempts to discriminate fungal from plant polypeptides in AM of tomato, polypeptide profiles of mycorrhiza have been compared to those of extraradical hyphae, intraradical mycelium consisting mainly of intraradical vesicles and arbuscules from enzymically digested mycorrhiza, or a mix of germinating spores and hyphae (Dassi et al. 1999; Simoneau et al. 1994). These studies indicate that the additional polypeptides detected in AM are most likely of plant origin. With the development of techniques like mass spectrometry for protein characterisation (Wilkins et al. 1997), it will be possible to work with much smaller amounts of purified protein and envisage more detailed characterisation of the fungal proteome in both presymbiotic (spores, germinating spores and hyphae) and symbiotic phases (intraradical mycelial structures).

V. Genetic Diversity: Possible Mechanisms

The molecular polymorphism in the Glomales indicates that they present a level of genetic diversity that is much greater than that expected from morphological characters alone. Their life cycle is characterised by coenocytic hyphae which contain large numbers of nuclei, and the absence of a uninucleate stage during their development cycle no doubt makes them an exceptional group in the Mycota. If hyphae harbour heterogeneous nuclei and there is no sexual or parasexual reproduction in glomalean fungi, then the nuclei of a same hyphal system could represent a population with little or no exchange between them. Subsequently, the expected genetic drift in the fungal population will not be the same if a spore originates from a small number of these nuclei which then divide, or from an entire population entering as it forms (van Tuinen et al. 1994). Moreover, if glomalean fungi are clonal and each nucleus in a spore undergoes mutation to form a new lineage, genetic divergence within a spore should be less than between spores for a given isolate. By analysing the frequence of polymorphic sequences in the fungal genome, it should be possible to structure the level of diversity, i.e. between spores, between nuclei in a same spore, between regions containing repeated elements within one nucleus (non-allelic genes), or between repeated units of a same region.

A. Inter- and Intraspore Polymorphism

Genetic heterogeneity between spores has been shown in different glomalean species and by different approaches. Sanders et al. (1995) reported different PCR-RFLP profiles for ITS sequences from ten morphologically identical spores collected from a grassland ecosystem. Amplified fragment length polymorphism (AFLP) analysis, which combines targeted restriction digestion of genomic DNA with selective PCR amplification, makes it possible to monitor multiple genetic markers. When individual spores of either *G. mosseae* or *G. caledonium* were analysed by this technique, those originating from the same pot culture gave very different amplification patterns, implying a high polymorphism within each isolate (Rosendahl and Taylor 1997). However, two tests for reproductive mode indicated that this variability is not due to recombination events, suggesting that in glomalean fungi no or very little genetic information is exchanged in this way, which is typical of organisms that reproduce clonally.

It is more difficult to show whether this interspore genetic heterogeneity results from a polymorphism between different nuclei within spores. Intraspore ITS polymorphism was first observed in *G. mosseae* by cloning and sequencing PCR products obtained with universal primers from individual spores (Lloyd-MacGilp et al. 1996; Sanders et al. 1995). At least two ITS sequences appeared to be present in the same spore and these resulted from substitutions or insertions/deletions. More detailed investigations have recently been made on glomalean nuclei by (1) genotyping spores with ITS-type specific primers and analysing the frequency of ITS-types, and (2) dilution of nuclear fractions from single spores to determine segregation of ITS length variants (Hijri et al. 1999). PCR amplification of multisporal DNA from *S. castanea* with universal ITS primers gave several fragments of different lengths, which were cloned and grouped into five types by PCR-RFLP analysis. A high diversity, due to substitutions or insertions/deletions, was observed between some of the sequences. When sequences were amplified from individual spores of *S. castanea*, using ITS type-specific primers, five ITS length variants were found to be present or absent in different frequencies, implying that each spore must be genetically different from another. Furthermore, the absence of each of these five ITS length variants from spores tended to be independent one from the other, providing evidence that there is no genetic linkage between them and that ITS polymorphism is organised mainly between nuclei. Different sequences from the ribosomal ITS region have also been identified within single a spore of *Gig. margarita* (Lanfranco et al. 1999a).

B. Inter- and/or Intranuclear Diversity

Genetic heterogeneity within multinucleated spores will have an important biological impact. If glomalean spores are heterokaryotic, then it is of fundamental interest to understand whether intraspore diversity results from the existence of different nuclear genotypes within individual spores or if it is due to intranuclear variability. When a nuclear suspension from single spores of *S. castanea* was diluted and dilutions amplified using ITS1-type and ITS4-type primers, the ITS length variants from *S. castanea* segregated among the different fractions, indicating that they can be harboured in different nuclei within individual spores (Hirji et al. 1999). The localisation by FISH in individual nuclei of different numbers of rDNA loci in *S. castanea* using ITS probes (M. Hirji, M. Darmency, H. Dulieu, 2nd Int Conf Mycorrhizas 1998), as well as in *G. mosseae*, *G. intraradices*, *Gig. rosea* or *S. castanea* with 25S probes (Trouvelot et al. 1999), lends further support to the conclusion that internuclear variability exists in the Glomales. Although there is no information concerning possible variations in sequences between tandemly repeated rDNA clusters in one nucleus, preliminary evidence for the presence of two ITS types at different loci in a single nucleus of *S. catanea* has been obtained by double colour FISH of spore nuclei using ITS type-specific probes (M. Hirji, M. Darmency, H. Dulieu, 2nd Int Conf Mycorrhizas 1998).

VI. Evolutionary Questions

The existence of variations in sequences in the 18S rDNA region and of several ITS types seems to suggest at first sight that glomalean fungi are an exception to the theory of "concerted evolution" which tends to homogenise repeated elements (Arnheim 1983). However, the fact that there are different absence/presence frequencies in spores of the ITS length variants, whilst there is no genetical linkage between them, means that the loss of each type should result from the random drift of nuclei from hyphae into spores during their for-

mation. This indirectly suggests that the ITS types are not scattered amongst ribosomal gene clusters (loci) but that they are homogeneously organised within all repeats at a given locus. If this is the case, there could have been concerted evolution inside each ribosomal locus whilst different loci progressed towards having different ITS sequences. Different hypotheses can be evoked to explain why glomalean fungi are composed of populations of different nuclear genotypes. One would be molecular divergence through random mutations with no genetic exchange between nuclei since the origin of the Glomales, whilst an alternative is the exchange of genetically different nuclei between hyphae from different spores with the possibility of nuclear fusion and recombination. Less frequent or rare ITS types in *S. castanea*, for example, could only be maintained if there is exchange of nuclei between individuals. Anastomosis between hyphae of geographically close populations has been shown for several glomalean species (Tommerup 1988), so that heterokaryosis is possible through somatic exchange of different nuclei by hyphal fusion. In order to demonstrate that this mechanism is active, it will be necessary to select fungal isolates with two exclusively different DNA polymorphisms and analyse their progeny after inoculation onto a same host plant.

In phylogenetic analyses, where 18S polymorphic sequences in different genomic rDNA clones of *S. castanea* were compared to 18S sequences representative of other glomalean genera, 2 *S. castanea* sequences clustered with *Glomus* and *Acaulospora* whilst 11 grouped with *Scutellospora* and *Gigaspora* (Hosny et al. 1999). Such a divergence within one species isolate suggests that the phylogenetic tree of sequences must have preceded that of the species and that a mode of evolution without genetical exchange may have existed within *S. castanea*. This, which has been considered as theoretically possible in other organisms (Nei 1987), would offer an alternative explanation for polymorphism in the Glomales. It is possible to imagine that if molecular evolution was slow in these fungi due to a lack of genetic exchange, morphological speciation may have been relatively recent once molecular diversification was sufficiently important. The absence of a relationship between genetic divergence and spore morphology concluded by Lloyd-MacGilp et al. (1996) could also be interpreted in this way.

VII. Conclusions

The results from recent investigations of the glo-malean genome appear coherent with the theory that evolution of the Glomales could have pro-ceeded by clonal fragmentation due to geograph-ical isolation in different environments, during which there was random divergence and adapta-tion. There is converging evidence that the genomes of glomalean fungi are unusual in both their complexity and structure, and that these organisms cannot be classed as either sexual or clonal sensu stricto. If sexuality is defined in the very general terms of vertical or horizontal gene transfer between individuals (Cuthill 1991), the Glomales could possess a genetic aptitude for recombination, even at low frequencies, between homologous loci. This would explain the presence of identical ribosomal sequences in different nuclei as well as that of different ones in some nuclei. Nevertheless, the absence of any significant linkage between ITS types in *S. castanea* (Hirji et al. 1999) favours the hypothesis that ability for recombination is restricted within nuclei. Further-more, there could be a primitive form of sexuality (or parasexuality) in glomalean fungi which does not imply nuclear fusion but involves nuclear exchange through anastomosis between hyphae from different spores of a species so that rare sequences persist in populations. If the Glomales are clonal organisms, then the nuclear lineages could be considered as being clones, and multinu-cleate hyphae or spores will be populated by a mixture of several nuclear lineages. In this case, the network of more or less connected hyphae in the soil, rhizosphere and root populations will guarantee that species polymorphism and adaptibility is maintained. Research into the genomics of the Glomales is still very much at its beginning. There remains a lot to learn in order not only to progress in an academic pursuit of the genetics of these fungi but also to more efficiently exploit them in plant production systems.

Acknowledgments. We wish to express our sincere appreciation to Dr. Silvio Gianinazzi for com-ments on the manuscript and for encouraging the work from the authors' laboratory. The latter was partly supported by the French Ministry of the Environment (grant no. 94161), the Burgundy Regional Council (contract 95/5112/1312/462) and an EU Biotechnology Project (BIO-CT97-2225).

References

Antequera F, Tamame M, Villanueva JR, Santos T (1984) DNA methylation in fungi. J Biol Chem 259: 8033–8036

Arines J, Palma JM, Vilarino A (1993) Comparison of protein patterns in non-mycorrhizal and vesicular-arbuscular mycorrhizal roots of red clover. New Phytol 123:763–768

Arnheim N (1983) Concerted evolution of multigene families. In: Nei M, Koehn RK (eds) Evolution of genes and proteins. Sinauer, Sunderland, MA, pp 38–61

Aström H, Giovannetti M, Raudaskoski M (1994) Cytoskeleton components in the arbuscular mycor-rhizal fungus *Glomus mosseae*. Mol Plant Microbe Interact 7:309–312

Avio L, Giovannetti M (1998) The protein pattern of spores of arbuscular mycorrhizal fungi: a comparison of species, isolates and physiological stages. Mycol Res 102:985–990

Bago B, Bentivenga SP, Brenac V, Dodd JC, Piché Y, Simon L (1998) Molecular analysis of *Gigaspora* (Glomales, Gigasporaceae). New Phytol 139:581–588

Barouin A, Perasso R, Qu LH, Brugerolle G, Bachellerie JP, Adoutte A (1988) Partial phylogeny of the unicel-lular eukaryotes based on rapid sequencing of a portion of 28S ribosomal RNA. Proc Natl Acad Sci USA 85:3474–3478

Bennett MD (1987) Variation in genomic form in plants and its ecological implications. New Phytol 106 (Suppl):177–200

Bennett MD, Smith JB (1991) Nuclear DNA amounts in angiosperms. Philos Trans R Soc (Lond) 334:309–345

Bianciotto V, Bonfante P (1992) Quantification of the nuclear DNA content of two arbuscular mycorrhizal fungi. Mycol Res 96:1071–1076

Bonfante-Fasolo P, Faccio A, Perotto S, Schubert A (1990) Correlation between chitin distribution and cell wall morphology in the mycorrhizal fungus *Glomus versi-forme*. Mycol Res 94:157–165

Brownlee AG (1989) Remarkably AT-rich genomic DNA from the anaerobic fungus *Neocallimastix*. Nucleic Acids Res 17:1327–1335

Bruns TD, White TJ, Taylor JW (1991) Fungal molecular systematics. Annu Rev Ecol Syst 22:525–564

Burleigh SH, Harrison MJ (1998) A cDNA from the arbus-cular mycorrhizal fungus *Glomus versiforme* with homology to a cruciform DNA-binding protein from *Ustilago maydis*. Mycorrhiza 7:301–307

Bütehorn B, Gianinazzi-Pearson V, Franken P (1999) Quantification of β-tubulin expression during asymbi-otic and symbiotic development of the arbuscular mycorrhizal fungus *Glomus mosseae*. Mycol Res 103:360–364

Casana MC, Bonfante-Fasolo P (1988) Intercellular hyphae and arbuscules of *Glomus fasciculatum* (Thaxter) Gerd. et Trappe isolated after enzymic digestion. Allionia 25:17–25

Cavalier-Smith T (1985) The evolution of genome size. Wiley, New York

Clapp JP, Young JPW, Merryweather JW, Fitter AH (1995) Diversity of the fungal symbionts in arbuscular myc-orrhizas from a natural community. New Phytol 130:259–265

Cordier C, Gianinazzi S, Gianinazzi-Pearson V (1996) An immunological approach for the study of spatial relationships between arbuscular mycorrhizal fungi in planta. In: Azcon-Aguilar C, Barea JM (eds) Mycorrhizas in Integrated Systems: from genes to plant development. European Commission, EUR 16728, Luxembourg, pp 189–194

Cuthill IC (1991) Sexual semantics. Nature 353:309

Dassi B, Dumas-Gaudot E, Asselin A, Richard C, Gianinazzi S (1996) Chitinase and β-1,3-glucanase isoforms expressed in pea roots inoculated with arbuscular mycorrhizal or pathogenic fungi. Eur J Plant Pathol 102:105–108

Dassi B, Samra A, Dumas-Gaudot E, Gianinazzi S (1999) Different polypeptide profiles from tomato roots following interactions with arbuscular mycorrhizal (*Glomus mosseae*) or pathogenic (*Phytophthora parasitica*) fungi. Symbiosis 26:65–77

De Rijk P, Caers A, van de Peer Y, De Wachter R (1998) Database on the structure of large ribosomal subunit RNA. Nucleic Acids Res 26:183–186

Di Bonito R, Elliot ML, Des Jardin EA (1995) Detection of an arbuscular mycorrhizal fungus in roots of different plant species with the PCR. Appl Environ Microbiol 61:2809–2810

Dodd JC, Rosendahl S, Giovannetti M, Broome A, Lanfranco L, Walker C (1996) Inter- and intraspecific variation within the morphologically similar arbuscular mycorrhizal fungi *Glomus mosseae* and *Glomus coronatum*. New Phytol 133:113–122

Dumas E, Gianinazzi-Pearson V, Gianinazzi S (1989) Production of new soluble proteins during VA endomycorrhiza formation. Agric Ecosyst Environ 29:111–114

Dumas-Gaudot E, Asselin A, Gianinazzi-Pearson V, Gollotte A, Gianinazzi S (1994a) Chitinase isoforms in roots of various pea genotypes infected with arbuscular mycorrhizal fungi. Plant Sci 99:27–37

Dumas-Gaudot E, Guillaume P, Tahiri-Alaoui A, Gianinazzi-Pearson V, Gianinazzi S (1994b) Changes in polypeptide patterns in tobacco roots colonized by two *Glomus* species. Mycorrhiza 4:215–221

Dumas-Gaudot E, Slezack S, Dassi B, Pozo MJ, Gianinazzi-Pearson V, Gianinazzi S (1996) Plant hydrolytic enzymes (chitinases and β-1,3-glucanases) in root reactions to pathogenic and symbiotic microorganisms. Plant Soil 185:211–221

Dusenberry RL (1975) Characterization of the genome of *Phycomyces blakesleeanus*. Biochem Biophys Acta 378:363–377

Dutta SK (1974) Repeated DNA sequences in fungi, Nucleic Acids Res 1:1441–1419

Franken P, Gianinazzi-Pearson V (1996) Construction of genomic phage libraries of the arbuscular mycorrhizal fungi *Glomus mosseae* and *Scutellospora castanea* and isolation of ribosomal genes. Mycorrhiza 6:167–173

Garber RC, Turgeon GB, Selker EU, Yoder OC (1988) Organization of ribosomal RNA genes in the fungus *Cochliobolus heterostrophus*. Curr Genet 14:573–582

Garcia-Garrido JM, Garcia-Romera I, Ocampo JA (1992) Cellulase production by the vesicular-arbuscular mycorrhizal fungus *Glomus mosseae* (Nicol. & Gerd.) Gerd. and Trappe. New Phytol 121:221–226

Garcia-Garrido JM, Toro N, Ocampo JA (1993) Presence of specific polypeptides in onion roots colonized by *Glomus mosseae*. Mycorrhiza 2:175–177

Garcia-Garrido JM, Garcia-Romera I, Parra-Garcia MD, Ocampo JA (1996) Purification of an arbuscular mycorrhizal endoglucanase from onion roots colonized by *Glomus mosseae*. Soil Biol Biochem 28:1443–1449

Garcia-Romera I, Garcia-Garrido JM, Ocampo JA (1991) Pectolytic enzymes in the vesicular arbuscular mycorrhizal fungus *Glomus mosseae*. FEMS Microbiol Lett 78:343–346

Garcia-Romera I, Garcia-Garrido JM, Ocampo JA (1992) Pectinase activity in vesicular arbuscular mycorrhizal colonization of lettuce. Symbiosis 12:189–198

Gardes M, Bruns D (1993) ITS primers with enhanced specificity for basidiomycetes – application to the identification of mycorrhizae and rust. Mol Ecol 2:113–118

Gianinazzi S, Gianinazzi-Pearson V, Tisserant B, Lemoine MC (1992) Protein activities as potential markers of functional endomycorrhizas in plants. In: Read DJ, Lewis DH, Fitter AH, Alexander DJ (eds) Mycorrhizas in ecosystems. CAB International, Oxon, pp 333–339

Gianinazzi-Pearson V, Gianinazzi S (1986) Connaissances actuelles des bases physiologiques et biochimiques des effects des endomycorhizes sur le comportement des plantes. Physiol Vég 24:253–262

Gianinazzi-Pearson V, Lemoine MC, Arnould C, Gollotte A, Morton JB (1994) Localization of β-1,3-glucans in spore and hyphal walls of fungi in the Glomales. Mycologia 86:477–484

Giovannetti M, Azzolini D, Citernesi AS (1999) Anastomosis formation and nuclear and protoplasmic exchange in arbuscular fungi. Appl Environ Microbiol 65:5571–5575

Giovannetti M, Gianinazzi-Pearson V (1994) Biodiversity in arbuscular mycorrhizal fungi. Mycol Res 98:705–715

Harrier LA, Wright JF, Hooker JE (1998) Isolation of the 3-phosphoglycerate kinase gene from the arbuscular mycorrhizal fungus *Glomus mosseae* (Nicol. & Gerd.) Gerdemann & Trappe. Curr Genet 34:386–392

Harrison MJ, van Buuren ML (1995) A phosphate transporter from the mycorrhizal fungus *Glomus versiforme*. Nature 378:626–629

Hepper CM, Mosse B (1975) Techniques used to study the interaction between Endogone and plant roots. In: Sanders FE, Mosse B, Tinker PB (eds) Endomycorrhizas. Academic Press, London, pp 65–75

Hepper CM, Sen R, Maskall CS (1986) Identification of vesicular-arbuscular mycorrhizal fungi in roots of leek (*Allium porrum* L.) and maize (*Zea mays* L.) on the basis of enzyme mobility during polyacrylamide gel electrophoresis. New Phytol 102:529–539

Hepper CM, Azcon-Aguilar C, Rosendhal S, Sen R (1998a) Competition between three species of *Glomus* used as spatially separated introduced and indigenous mycorrhizal inocula for leek (*Allium porrum* L.). New Phytol 110:207–215

Hepper CM, Sen R, Azcon-Aguilar C, Grace C (1988b) Variation in certain isoenzymes amongst different geographical isolates of the vesicular-arbuscular mycorrhizal fungi *Glomus clarum*, *Glomus monosporum* and *Glomus mosseae*. Soil Biol Biochem 20:51–59

Hijri M, Hosny M, van Tuinen D, Dulieu H (1999) Intraspecific ITS polymorphism in *Scutellospora castanea* (Glomales, Zygomycetes) is structured within multinucleate spores. Fungal Gen Biol 26:141–151

Hosny M, Gianinazzi-Pearson V, Dulieu H (1998a) Nuclear contents of 11 fungal species in Glomales. Genome 41:422–428

Hosny M, Païs de Barros, Gianinazzi-Pearson V, Dulieu H (1998b) Base composition of DNA from glomalean fungi: high amounts of methylated cytosine. Fungal Gen Biol 22:103–111

Hosny M, Hijri M, Passerieux E, Dulieu H (1999) The ribosomal units are highly polymorphic in Scutellospora castanea (Glomales, Zygomycota) as revealed by cloning and sequencing. Gene 226:61–71

Hosny M, van Tuinen D, Jacquin F, Füller P, Franken P (1999) Arbuscular mycorrhizal fungi and bacteria: how to construct prokaryotic DNA-free genomic libraries from the Glomales. FEMS Microbiol Lett 170:425–430

Kahn P (1995) From genome to proteome. Science 270: 369–371

Kaldorf M, Zimmer W, Bothe H (1994) Genetic evidence for the occurrence of assimilatory nitrate reductase in arbuscular mycorrhizal and other fungi. Mycorrhiza 5:23–28

Lanfranco L, Delpero M, Bonfante P (1999a) Intrasporal variability of ribosomal sequences in the endomycorrhizal fungus Gigaspora margarita. Mol Ecol 8:37–45

Lanfranco L, Garnero L, Bonfante P (1999b) Chitin synthase genes in the arbuscular mycorrhizal fungus Glomus versiforme: full sequence of a gene encoding a class IV chitin synthase. FEMS Microbiol Lett 170: 59–67

Li W-H, Grauer D (1991) Fundamentals of molecular evolution. Sinauer, Sunderland, MA

Lloyd-MacGilp SA, Chambers SM, Dodd JC, Fitter AH, Walker C, Young JPW (1996) Diversity of the ribosomal internal transcribed spacers within and among isolates of Glomus mosseae and related mycorrhizal fungi. New Phytol 133:103–111

Longato S, Bonfante P (1997) Molecular identification of mycorrhizal fungi by direct amplification of microsatellite regions. Mycol Res 101:425–432

Michot B, Bachellerie JP (1987) Comparisons of large subunit rRNA reveal some eukaryotic-specific elements of secondary structure. Biochimie 69:11–23

Mitchell JI, Roberts PJ, Moss ST (1995) Sequence or structure? Mycologist 9:67–75

Morton JB (1993) Problems and solutions for the integration of glomalean taxonomy, systematics biology, and the study of endomycorrhizal phenomena. Mycorrhiza 2:97–109

Nei M (1987) Molecular evolutionary genetics. Columbia University Press, New York

O'Donnell K, Cigelnik E (1997) Two divergent intragenomic rDNA ITS2 types within a monophyletic lineage of the fungus Fusarium are nonorthologous. Mol Phylogenet Evol 7:103–116

Perry RP (1976) Processing of RNA. Annu Rev Biochem 45:605–629

Pozo MJ, Azcon-Aguilar C, Dumas-Gaudot E, Barea JM (1998) Chitosanase and chitinase activities in tomato roots during interactions with arbuscular mycorrhizal fungi or Phytophthora parasitica. J Exp Bot 49: 1729–1739

Radman M (1991) Avoidance of inter-repeat recombination by sequence divergence and a mechanism of neutral evolution. Biochimie 73:357–361

Redecker D, Thierfelder H, Walker C, Werner D (1997) Restriction analysis of PCR-amplified internal transcribed spacers of ribosomal DNA as a tool for species identification in different genera of the order Glomales. Appl Environ Microbiol 63:1756–1761

Rejon-Palomares A, Garcia-Garrido JM, Ocampo JA, Garcia-Romera I (1996) Presence of xyloglucanhydrolyzing glucanases (xyloglucanases) in arbuscular mycorrhizal symbiosis. Symbiosis 21:249–261

Requena N, Füller P, Franken P (1999) Molecular characterization of GmFOX, an evolutionary highly conserved gene from the mycorrhizal fungus downregulated during interaction with Rhizobacteria. Mol Plant-Microbe Interact 12:934–942

Rosendahl S (1989) Comparisons of spore-cluster forming Glomus species (Endogonaceae) based on morphological characteristics and isozyme banding patterns. Opera Bot 100:215–223

Rosendahl S, Sen R (1992) Isozyme analysis of mycorrhizal fungi and their mycorrhiza. In: Norris JR, Read DJ, Varma AK (eds) Methods in microbiology, vol 24. Academic Press, New York, pp 169–194

Rosendahl S, Taylor JW (1997) Development of multiple genetic markers for studies of genetic variation in arbuscular mycorrhizal fungi using AFLP™. Mol Ecol 6:821–829

Russell PJ, Rodland KD, Rachlin EM, McCloskey JA (1987) Differential DNA methylation during the vegetative life cycle of Neurospora crassa. J Bacteriol 169:2902–2905

Samra A, Dumas-Gaudot E, Gianinazzi-Pearson V, Gianinazzi S (1996) Soluble proteins and polypeptide profiles of spores of arbuscular mycorrhizal fungi. Interspecific variability and effects of host (myc$^+$) and non-host (myc$^-$) Pisum sativum root exudates. Agronomie 16:709–719

Samra A, Dumas-Gaudot E, Gianinazzi S (1997) Detection of symbiosis-related polypeptides during the early stages of the establishment of arbuscular mycorrhiza between Glomus mosseae and Pisum sativum roots. New Phytol 135:711–722

Sanders IR, Ravolanirina F, Gianinazzi-Pearson V, Gianinazzi S, Lemoine MC (1992) Detection of specific antigens in the vesicular-arbuscular mycorrhizal fungi Gigaspora margarita and Acaulospora laevis using polyclonal antibodies to soluble spore fractions. Mycol Res 96:477–480

Sanders IR, Alt M, Groppe K, Boller T, Wiemken A (1995) Identification of ribosomal DNA polymorphisms among and within spores of the Glomales: application to studies on the genetic diversity of arbuscular mycorrhizal fungal communities. New Phytol 130: 419–427

Schellenbaum L, Gianinazzi S, Gianinazzi-Pearson V (1992) Comparison of acid soluble protein synthesis in roots of endomycorrhizal wild type Pisum sativum and corresponding isogenic mutants. J Plant Physiol 141:2–6

Schweizer E, MacKechnie C, Halvorson HO (1969) The redundancy of ribosomal and transfer RNA genes in Saccharomyces cerevisiae. J Mol Biol 40:261–277

Sen R, Hepper CM (1986) Characterization of vesiculararbuscular mycorrhizal fungi (Glomus spp.) by selective enzyme staining following polyacrylamide gel electrophoresis. Soil Biol Biochem 18:29–34

Simon L (1996) Phylogeny of the Glomales: deciphering the past to understand the present. New Phytol 133:95–101

Simon L, Lalonde M, Bruns TD (1992a) Specific amplification of the 18S fungal ribosomal genes from vesicular-arbuscular endomycorrhizal fungi colonizing roots. Appl Environ Microbiol 58:291–295

Simon L, Lévesque RC, Lalonde M (1992b) Rapid quantification by PCR of endomycorrhizal fungi colonizing roots. PCR Methods Appl 2:76–80

Simon L, Bousquet J, Lévesque RC, Lalonde M (1993a) Origin and diversification of endomycorrhizal fungi and coincidence with vascular land plants. Nature 363:67–69

Simon L, Lévesque RC, Lalonde M (1993b) Identification of endomycorrhizal fungi colonizing roots by fluorescent single-strand conformation polymorphism-polymerase chain reaction. Appl Environ Microbiol 59:4211–4215

Simoneau P, Louisy-Lois N, Plenchette C, Strullu DG (1994) Accumulation of new polypeptides in Ri-T-DNA-transformed roots of tomato (*Lycopersicon esculentum*) during the development of vesicular-arbuscular mycorrhizae. Appl Environ Microbiol 60:1810–1813

Slezack S, Dassi B, Dumas-Gaudot E (1996) Arbuscular mycorrhiza-induced chitinase isoforms. In: Muzarelli RAA (ed) Chitin enzymology. Atec Edizioni, Senigallia, pp 339–347

Smith SE, Read DJ (1997) Mycorrhizal symbiosis. Academic Press, London, New York

Storck R, Alexopoulos CJ (1970) Desoxyribonucleic acid of fungi. Bacteriol Rev 34:126–154

Sueoka N (1988) Directional mutation pressure and neutral molecular evolution. Proc Natl Acad Sci USA 85:2653–2657

Swift H (1950) The constancy of desoxyribose nucleic acid in plant nuclei. Proc Natl Acad Sci USA 85:2653–2657

Tautz D (1993) Notes on the definition and nomenclature of tandemly repeated repetitive DNA sequences. In: Pena SDJ, Chakraborty R, Epplen T, Jeffreys AJ (eds) DNA fingerprinting: state of science. Birkhäuser, Basel, pp 21–28

Theis JF, Newlon CS (1996) The replication of yeast chromosomes. In: Brambl R, Marzluf GA (eds) The mycota III Biochemistry and molecular biology. Springer, Berlin Heidelberg New York, pp 3–28

Tisserant B, Brenac V, Requena N, Jeffries P, Dodd JC (1998) The detection of *Glomus* spp. (arbuscular mycorrhizal fungi) forming mycorrhizas in three plants, at different stages of seedling development, using mycorrhiza-specific isozymes. New Phytol 138: 225–239

Tommerup IC (1988) Genetics of vesicular-arbuscular mycorrhizas. In: Sidhu GS (ed) Genetics of pathogenic fungi, Academic Press, New York, pp 81–91

Trouvelot S, van Tuinen D, Hirji M, Gianinazzi-Pearson V (1999) Visualization of ribosomal DNA loci in spore interphasic nuclei of glomalean fungi by fluorescence in situ hybridization. Mycorrhiza 8:203–206

van Tuinen D, Dulieu H, Zézé A, Gianinazzi-Pearson V (1994) Biodiversity and molecular characterization of arbuscular mycorrhizal fungi at the molecular level. In: Gianinazzi S, Schüepp H (eds) Sustainable agriculture and natural ecosystems. Birkhäuser, Basel, pp 13–23

van Tuinen D, Jacquot E, Zhao B, Gollotte A, Gianinazzi-Pearson V (1998a) Characterization of root colonization profiles by a microcosm community of arbuscular mycorrhizal fungi using 25S rDNA-targeted nested PCR. Mol Ecol 7:879–887

van Tuinen D, Zhao B, Gianinazzi-Pearson V (1998b) PCR in studies of AM fungi: from primers to application. In: Varma AK (ed) Mycorrhiza manual. Springer, Berlin Heidelberg New York, pp 387–399

Wessels JGH, De Vrie OMH, Asgeirdottir SA, Schuren FHJ (1991) Hydrophobin genes involved in formation of aerial hyphae and fruit bodies in *Schizophyllum*. Plant Cell 3:793–799

Wilkins MR, Williams KL, Appel RD, Hochstrasser DF (1997) Proteome research: new frontiers in functional genomics. Springer, Berlin Heidelberg New York

Wöstenmeyer J, Burmester A (1986) Structural organisation of the genome of the Zygomycete *Absidia glauca*: evidence for high repetitive DNA content. Curr Genet 10:903–907

Wyss P, Mellor RB, Wiemken A (1990) Vesicular-arbuscular mycorrhizas of wild-type soybean and non-nodulating mutants with *Glomus mosseae* contain symbiosis-specific polypeptides (mycorrhizins), immunologically cross-reactive with nodulins. Planta 182:22–26

Zézé A, Dulieu H, Gianinazzi-Pearson V (1994) DNA cloning and screening of a partial genomic library from an arbuscular mycorrhizal fungus, *Scutellospora castanea*. Mycorrhiza 4:251–254

Zézé A, Hosny M, Gianinazzi-Pearson V, Dulieu H (1996) Characterization of a highly repeated DNA sequence (SC1) from the arbuscular mycorrhizal fungus *Scutellospora castanea* and its detection in planta. Appl Environ Microbiol 62:2443–2448

Zézé A, Sulistyowati E, Ophel-Keller K, Barker S, Smith S (1997) Intrasporal genetic variation of *Gigaspora margarita*, a vesicular arbuscular mycorrhizal fungus, revealed by M13 minisatellite-primed PCR. Appl Environ Microbiol 63:676–678

Zézé A, Hosny M, van Tuinen D, Gianinazzi-Pearson V, Dulieu H (1999) *Mycdire*, a dispersed repetitive DNA element in arbuscular mycorrhizal fungi. Mycol Res 103:572–576

2 Molecular Approaches to Arbuscular Mycorrhiza Functioning

P. Franken and N. Requena

CONTENTS

I. Introduction

Fungi of the order Glomales (Zygomycota) form with the roots of most land plants a mutualistic symbiosis named arbuscular mycorrhiza (Newman and Reddell 1987; Morton and Benny 1990). Fossil and molecular data indicate that these organisms evolved about 400 Ma ago (Pirozynski and Dalpé 1989; Simon et al. 1993) and might have been an important factor for the colonisation process of the land by ancient plants (Pirozynski and Malloch 1975). Nowadays, arbuscular mycorrhizal (AM) fungi are worldwide distributed and a key component of most terrestrial ecosystems (Read et al. 1992). The symbiosis between the two biotrophic organisms is mainly characterised by bidirectional transfer of nutrients which gives access for the plant to low mobile elements like phosphorus (Smith and Gianinazzi-Pearson 1988). However, AM fungi may also influence plant health and development through other non-nutritional mechanisms such as the production of growth-regulating substances (Barea and Azcon-Aguilar 1982) or by increasing the resistance to root pathogens (Azcon-Aguilar and Barea 1996). As a result, the structure of plant

Max-Planck-Institut für terrestrische Mikrobiologie and Laboratorium für Mikrobiologie des Fachbereichs Biologie der Philipps-Universität, Karl-von-Frisch-Straße, 35043 Marburg, Germany

communities can be importantly modified (Allen and Allen 1992), because of the competitive advantage of plants harbouring a mycorrhizal symbiosis compared to non-mycorrhizal plants. In addition, AM symbiosis represents a potential tool to be used in plant production such as agriculture, micropropagation or horticulture (Gianinazzi et al. 1995; Lovato et al. 1996; Azcon-Aguilar and Barea 1997) and for the management and conservation of natural ecosystems (Requena et al. 1996).

The AM symbiosis is the result of a very dynamic process, in both space and time. Key developmental steps are the germination of the spores and limited hyphal development in the absence of the host (Hepper 1983), extensive hyphal branching and subsequent formation of appressoria in the presence of roots (Giovannetti and Citernesi 1993; Giovannetti et al. 1993), colonisation of the cortex in the *Arum* or the *Paris* mode depending on the host (Smith and Smith 1997) and the development of arbuscules, the suggested main site of nutrient exchange (Smith and Smith 1990). Following the development of the intraradical structures, the fungus spread out into the soil by the production of a mycelial network able to infect other roots (Francis and Read 1984). At this mycelium, new asexual spores are formed which are very large, rich in nutrients and which can contain up to 4000 nuclei (Cooke et al. 1987). It is therefore not surprising that many different saprophytic bacteria and fungi are found in the vicinity of these spores (Lee and Koske 1994). These microorganisms might behave as pathogens for AM fungi (Rousseau et al. 1996), or as beneficial saprophytes supporting the growth of hyphae and the formation of mycorrhiza (von Alten et al. 1993; Gryndler et al. 1995; Perotto and Bonfante 1997; Requena et al. 1997). In addition, it has long been known that the cytoplasm of the spores contains structures resembling prokaryotes (Mosse 1970). Recently, these so-called bacteria-like organisms (BLOs) were characterised as

The Mycota IX
Fungal Associations
Hock (Ed.)
© Springer-Verlag Berlin Heidelberg 2001

belonging to the genus *Burkoldheria* and were shown to be present through the whole life cycle of some AM fungal species (Bianciotto et al. 1996). Interestingly, a sequence cloned from such an endosymbiont showed similarity to *nif*D, which belongs to the nitrogen-fixing gene cluster of Rhizobium (Minerdi et al. 1998). Moreover, a putative P-transporter operon was identified in the genome of *Burkholderia*, suggesting that this BLO has the potential for P-uptake (Ruiz-Lozano and Bonfante 1999). Research on AM functioning can be targeted to different levels of symbiotic appearance. Molecular biology is mainly concerned with genes, their regulation and their expression (Fig. 1). This also includes the identification of signals exchanged between plant and fungus, their reception and their transduction. In addition, gene products have to be analysed to elucidate their

direct function in the metabolism of one of the two partners and their role in the symbiosis. Integration of the results will provide insights into the molecular cross-talk between the two symbionts and will help to understand how they guide each other through the different stages of their interaction. This chapter will concentrate on the fungal site of the symbiosis and summarises the current knowledge of the activity and localisation of AM fungal gene products and of plant signals which influence AM fungal development. The main part is, however, focused on the different strategies to clone protein-encoding genes.

II. Protein Activities and Localization

The molecular-oriented research of AM functioning had its origin in the investigation of enzymatic activities related to the symbiosis (Gianinazzi-Pearson and Gianinazzi 1976). Since phosphate metabolism seems to play an important role in the interaction, first analyses were concentrated on the detection and localisation of fungal phosphatases involved in this process (MacDonald and Lewis 1978; Gianinazzi et al. 1979). It was thus possible to correlate alkaline phosphatase activity with symbiotic functioning of the AM fungi (Gianinazzi-Pearson and Gianinazzi 1983). The method was later improved by Tisserant et al. (1993), who developed a technique to estimate efficient mycorrhizal formation by combining in planta immunohistochemical staining for fungal alkaline phosphatase and succinate dehydrogenase activity (Smith and Gianinazzi-Pearson 1990; Smith and Dickson 1991). Alkaline phosphatase activity was also detected in excised intraradical hyphae of AM fungi (Ezawa et al. 1995) and the enzyme was partially purified from the fungus *Gigaspora margarita* (Kojima et al. 1998).

Cox and Tinker suggested already in 1976 that active mechanisms might be involved in the nutrient exchange of the AM symbiosis, implicating the involvement of energy-generating ATPases. Marx et al. (1982) detected ATPase activity at the plasmalemma and in the vacuoles of the fungal arbuscules. This activity was later shown to be due to a proton-pumping H^+-ATPase also present in extraradical hyphae (Gianinazzi-Pearson et al. 1991). In addition, an ATPase activity could be detected during the presymbiotic phase in hyphae of germinating spores of *Gigaspora margarita*,

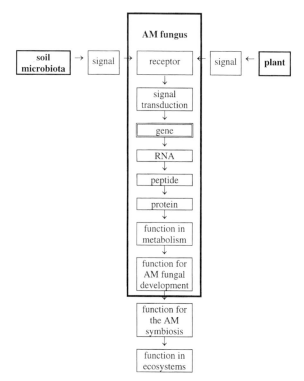

Fig. 1. Different levels of symbiotic appearance of an AM fungus. Signals produced by the plant host or by other soil microorganisms are recognised by specific AM fungal receptors. This leads via certain signal transduction chains to the regulation of specific genes. Those genes express RNA and, in turn, peptides which fold and aggregate to functional proteins. Proteins (or the corresponding genes) exhibit their function on different levels, directly in metabolism (in the broad sense) and indirectly in AM fungal development which, in turn, determines symbiotic functioning and on the highest level, the role of AM fungi in ecosystems

which was, interestingly, stimulated by the presence of root exudates (Lei et al. 1991).

Other targets were enzymes with lytic activities which could be involved in the colonisation of the root (Garcia-Romera et al. 1990). Different proteins were identified in the fungus *Glomus mosseae*, such as pectinases, cellulases or xyloglucanases (Garcia-Romera et al. 1991, 1992; Rejon-Palomares et al. 1998), and a polygalacturonase was localised in the cytoplasm of intracellular hyphae of *Glomus versiforme* (Peretto et al. 1995). However, the relevance of these enzymes for the function of the symbiosis is still not clear. More recently, an endoglucanase (Garcia-Garrido et al. 1996), as well as a protein with lipolytic activity (Gaspar et al. 1997), were purified and characterised.

Additional interesting targets could be cytoskeletal components, since changes in the pattern of these elements have been shown on the plant site during mycorrhizal colonisation (Bonfante et al. 1996; Genre and Bonfante 1997, 1998) and might also occur in AM fungi while colonising the root. Tubulin and actin could already be detected in germinating spores of *Glomus mosseae* by Western blot analyses (Aström et al. 1994).

Many other enzymes have been detected in AM fungi, but mainly used as isozyme patterns for the identification and monitoring of specific isolates (Rosendahl et al. 1994; Tisserant et al. 1998). It might be interesting in the future to use similar methods to look for enzyme activities which are differentially regulated during AM formation. The characterisation of the corresponding proteins could lead to the isolation of their encoding genes for further molecular analysis (see below).

III. Gene Cloning and Analysis

The first gene cloned from an AM fungus was the 18S rRNA gene (Simon et al. 1992). Today, a considerable amount of data about the rRNA gene cluster is available, but together with different repetitive elements and sequences obtained by random approaches, this information is mainly applied to research concerning phylogeny and diversity of AM fungi (Franken 1998). First results in the isolation of genes which may have a function for AM fungal development are summarised in the following paragraphs.

A. Targeted Approaches

Strategies to clone genes can be targeted to different levels of their expression. The classical approach is the mutagenesis of the organism under investigation and the search for interesting new phenotypes. The responsible gene can be subsequently cloned by transforming the mutant with a genomic library or by mapping and chromosome walking. Plant mutants unable to form mycorrhiza are already available for different species like *Pisum sativum* (Duc et al. 1989), *Medicago sativa* (Bradbury et al. 1991), *Medicago truncatula* (Calantzis et al. 1998), *Lotus japonicum* (Wegel et al. 1998) or *Lycopersicum esculentum* (Barker et al. 1998). This kind of approach will however, be, very difficult for the fungal part, since all their developmental stages are coenocytic, and different nuclei in one cytoplasm might easily complement each other. An alternative approach is to target the function of a gene in a certain metabolic pathway. Following this type of strategy, genes could be cloned by complementing the corresponding mutants in a heterologous organism. This has been shown to be possible for AM fungi, as was demonstrated by Harrison and van Buuren (1995), who proved the activity of a phosphate transporter gene from *Glomus versiforme* in the *pho84* mutant of yeast.

Other targeted approaches are always based on the assumption that a gene or a gene product, respectively, plays a role in a certain developmental process. Proteins can be purified as mentioned above and peptide sequencing can lead to the design of oligonucleotides which serve as primers for the PCR amplification of the corresponding DNA fragment. Oligonucleotides might also be designed according to conserved domains of the gene product of interest. Such domains can be found by comparison of the homologous genes or proteins from a set of various organisms. Using RNA from spores of the AM fungus *Gigaspora rosea* as template, this approach was used to amplify fragments of genes, encoding glyceraldehyde-3-phosphate dehydrogenase (GAPDH), β-tubulin and two different ATPases (Franken et al. 1997). In the framework of a European program, the BEG-net (wwwbio.ukc.ac.uk/beg), β-tubulin gene fragments were cloned from many different AM fungal isolates by RT-PCR with spore RNA as template (for a detailed protocol see Franken et al. 1998a). RNA accumulation can be measured by Northern blot analysis. However, this technique can be applied only in the case of highly expressed genes (such as the GAPDH gene), since the fungal material from structures outside the root is scarce and inside a mycorrhizal root substantially diluted by plant material (Franken et al. 1997). An alternative approach to analyse RNA accumulation in different stages of fungal development is the quantitative RT-PCR with an internal standard (Bütehorn et al. 1999).

PCR amplification with oligonucleotides according to conserved domains but using genomic DNA as template has been applied to the isolation of different fragments such as from a nitrate reductase gene of a *Glomus* ssp. (Kaldorf et al. 1994), or as from several chitin synthase genes of *Gigaspora margarita* and *Glomus versiforme* (Lanfranco et al. 1996) and from two H^+-ATPase genes of *G. mosseae* (Ferrol et al. 1998). RNA accumulation of the chitin synthase genes from *Gig. margarita* and of the H^+-ATPase genes from *G. mosseae* was studied by RT-PCR. Differential expression was detected in both cases. RNA of two of the chitin synthase genes was detected only during colonisation, but not in germinating spores (Lanfranco et al. 1999a), and while one H^+-ATPase gene seemed to be expressed in extraradical hyphae, the other showed highest RNA accumulation inside the plant (Ferrol et al. 1998). In situ hybridisation technique was used to show enhanced RNA accumulation of the nitrate reductase gene from *Glomus intraradices* in the arbuscule formed by the fungus in maize roots (Kaldorf et al. 1998).

Most of these approaches, however, render uncompleted sequences of the genes. In order to obtain more information about the structure of a gene, cDNA or genomic libraries can be screened, using as probes either the PCR fragment previously obtained or the corresponding gene from a different organism. One example is the phosphate transporter gene *Gvgpt* from *Glomus versiforme*, which was obtained by screening a cDNA library from mycorrhiza with a heterologous probe from yeast (Harrison and van Buuren 1995). This gene showed specific RNA accumulation in extraradical hyphae and may be involved in phosphate uptake from the soil. The second example is a chitin synthase gene which was isolated from a genomic library constructed with spore DNA of *Glomus versiforme* using as homologous probe a PCR fragment (Lanfranco et al. 1999b). The analysis of this latter clone also gave, in addition to the coding region, information about exon-intron boundaries and the promoter structure which is

the basis for further studies concerning gene regulation. An alternative technique for obtaining more sequences without screening a library is the RACE (rapid amplification of cDNA ends). Using this method, Harrier et al. (1998) could clone the whole coding region of a phosphoglycerate kinase gene from *Glomus mosseae* starting from a PCR fragment obtained by differential RNA display (see next paragraph).

B. Non-Targeted Approaches

Developmental processes are accompanied by changes in gene expression patterns. The up- or downregulation of certain genes is thought to play an important role in the transition from one stage to the other. It might therefore be useful to isolate such genes, using a non-targeted approach without knowing their function. These studies can be carried out at RNA or at protein accumulation level. Protein accumulation patterns have been analysed by comparing extracts from mycorrhizal and non-mycorrhizal plants on 2D-gel electrophoresis (Garcia-Garrido et al. 1993; Dumas-Gaudot et al. 1994; Samra et al. 1997). As a result, novel plant genes can be identified, but also, by comparing the same fungus on two different hosts, fungal-specific polypeptides are detectable. Such peptides can be purified from the gels and the corresponding genes can be cloned as mentioned above. At the RNA level, accumulation patterns of mycorrhiza and control roots have been compared by means of two different techniques, the differential RNA display (Martin-Laurent et al. 1995) and the differential screening of cDNA libraries (Burleigh and Harrison 1996; Murphy et al. 1997). In both cases, interesting new mycorrhiza-regulated plant genes have been cloned (Burleigh and Harrison 1997; Martin-Laurent et al. 1997; Murphy et al. 1997; Krajinski et al. 1998). Fungal genes have been also identified as, for instance, a gene from *Glomus versiforme* encoding amino acid sequence with similarities to a cruciform-binding protein of *Ustilago mayidis* (Burleigh and Harrison 1998) or three different genes from *Glomus intraradices* which might be involved in regulation processes (Delp et al. 1998). However, the presence of the fungal transcripts in mycorrhiza does not indicate that these genes are differentially expressed, since they might be expressed during all stages of fungal development. It is, therefore, necessary to quantitatively compare RNA accumulation patterns in tissues or stages which both contain the AM fungus. One possibility to look for fungal genes without the interference of the plant is to concentrate on the presymbiotic phase of AM development. Germination of AM fungal spores can be investigated on water agar and enough peptides can be labelled for 2D-gel electrophoresis or RNA extracted for differential display analysis. With these techniques, changes in protein and transcript patterns during germination of several AM fungi have been analysed (Franken et al. 1998b; Samra et al. 1996). Moreover, the presence of a bacterial strain which enhances hyphal development of *Glomus mosseae* during this stage of development was shown to induce specific changes in fungal gene expression (Requena and Franken 1998). A differentially displayed cDNA fragment of 375 bp was isolated, and the corresponding gene *GmFox2* was proved by quantitative RT-PCR to be down-regulated in response to rhizobacteria. The identification of the full genomic sequence allowed the protein structure to be deduced. The multidomain protein has homologues in *Neurospora* and human involved in the β-oxidation of fatty acids in peroxisomes. The first domain of the protein, which is also similar to NODG from *Rhizobium*, plays an additional role in humans in the inactivation of estrogens. Since estrogens have been shown to influence the presymbiotic growth of AM fungi (Poulin et al. 1997), a possible involvement of this protein in the signaling in the rhizosphere was proposed (Requena et al. 1999).

The study of RNA accumulation patterns in plant mutants unable to form mycorrhiza is another good approach to tag genes involved in the symbiosis. Comparing mycorrhiza of a *Pisum sativum* wild-type line and of its isogenic mutant where arbuscule development is aborted lead, for instance, to the identification of several genes which are regulated during these late stages of symbiosis formation (Lapopin et al. 1999). One of the fragments turned out to be of fungal origin and might belong to a gene from *Glomus mosseae* which is highly expressed in functional arbuscules.

IV. Signalling Between the Two Symbiotic Partners

The interaction between an AM fungus and the plant root starts even before both partners physically contact each other. Separating the two sym-

bionts by a nitrocellulose membrane, Giovannetti and Citernesi (1993) detected an increased hyphal growth and branching when the fungus was in close proximity to the root. These findings suggest the presence of compounds released by the plant. Analysis of the effect of root exudates on the fungus showed that it was most pronounced when these exudates were derived from phosphate-starved plants (Elias and Safir 1987; Tawaraya et al. 1995; Nagahashi et al. 1996a). This might be one of the reasons why such plants are better colonised than their phosphate-fertilised counterparts, as has long been well known (Mosse 1973). Because of the similarities between the nodule symbiosis and mycorrhiza (Hirsch and Kapulnik 1998), a considerable amount of research regarding root exudates has been concentrated on the flavonoid compounds, which are known to modify bacterial *nod*-gene expression (Philips and Tsai 1992). Several compounds like quercetin or naringenin revealed an effect on presymbiotic development of *Gigaspora* species (e.g., Gianinazzi-Pearson et al. 1989; Becard et al. 1992; Baptista and Sequeira 1994), and it could be shown in one example that the positive influence on fungal growth of biochaninA or the related estradiol in *Glomus intraradices* could be suppressed by an antiestrogen compound (Poulin et al. 1997). However, maize mutants completely lacking flavonoid compounds showed the same degree of colonisation as their corresponding wild-type plants (Becard et al. 1995). Additional compounds must, therefore, exist which influence the development of the mycosymbiont and which were able to complement the lack of flavonoids in the maize mutant. Such compounds could be polyamins or salicylic acid derivatives, which were shown to positively influence the presymbiotic development of *Glomus mosseae* (El Ghashtouli et al. 1996) or *Gig. rosea* (Franken et al. 1998b), respectively. Nagahashi et al. (1996b) developed a method where by they injected root exudates close to developing hyphae into the water agar. At these points, the fungi showed very localised branching 6 to 12 h after injection. Using this method, root exudates and fractions of exudates have been analysed for their influence on hyphal morphology (Buee et al. 1998; Nagahashi et al. 1998). Comparing root exudates of plants differentially fertilised, a particular compound in exudates of phosphate-starved *Petroselinum crispum* was detected by thin layer chromatography (Franken and Gnädinger 1994). Adding this compound to in vitro germination assays of *Scutellospora castanea* resulted in en-hanced hyphal branching (Table 1). In all the different analyses, the final identification of the responsive substances has still to be ascertained.

Table 1. Influence of an unknown compound from root exudates on presymbiotic stages of *Scutellospora castanea*

Order of branches	No. of branches per germinated spore	
	+[a]	−
1	6	4
2	4.5	2
3	1	0.4
4	0.9	0.1
5	0.6	0.6
6	0.2	0.01
7	0.1	0
Total branches	13.3	7.11

[a] Root exudates of phosphate-starved parsley plants were collected and extracted with CH_2Cl_2. A UV-fluorescing compound was isolated from TLC plates and added in 0.1% MeOH to water agar plates (+). 0.1% MeOH alone served as control (−).

V. Future Directions

It is clear from this summary of results obtained so far that molecular analysis of AM fungi is at the very beginning. First genes are available (Table 2), but they seem to be only a very few pieces in a giant and complicated puzzle. The strategies and methods, how to identify and clone more functional genes, are obvious and, when genomic and cDNA clones of such genes are available, studies on their regulation and their role for symbiosis can follow. For these analyses, the development of transformation systems is necessary. Transgenic AM fungi containing antisense constructs which downregulate the expression of the corresponding gene would answer the question whether this gene is important for the developmental stage where it is induced. On the other hand, promoter derivatives driving certain reporter genes could identify regulative sequences which, in turn, can be used to isolate the corresponding regulator genes. Forbes et al. (1998a) showed that transformation of AM fungi is, in principle, possible. Moreover, evidence was presented that the constructs they used were

Table 2. Cloned genes and gene fragments of AM fungi (references see text)

AM fungus	Clones	RNA accumulation[a]	Gene products[b]
Gigaspora margarita	2 PCR fragments	Only in mycorrhiza	Chitin synthases
	1 PCR fragment	Not detectable	Chitin synthase
Gigaspora rosea	4 RT-PCR fragments	In activated spores	GAPDH[c] beta-tubulin ATPase ATPase
Glomus intraradices	3 DD[d] fragments	In mycorrhiza	O-GlcNAc-transferase 'transcriptional regulator' homeobox-containing
	PCR fragment	In arbuscules, not in vesicles[e]	Nitrate reductase
Glomus mosseae	2 PCR fragments	In mycorrhiza	H+-ATPases[f]
	PCR fragment	In all stages	Beta-tubulin
	DD fragment Genomic clone	During germination Downregulated by rhizobacteria	Protein kinase, Fatty acid oxygenase
	cDNA	In mycorrhiza	Phosphoglycerate kinase
	DD fragment	In all stages, enhanced in the symbiotic stage	Unknown
Glomus versiforme	cDNA	Only in extraradical hyphae	Phosphate transporter
	cDNA	Outside and inside the root[g]	Cruciform DNA-binding
	2 PCR fragments	In spores and mycorrhiza	Chitin synthases
	Genomic clone	Not detectable	Chitin synthase

[a] Measured by RT-PCR except.
[b] Deduced from similarities except.
[c] Glyceraldehyde-3-phosphate dehydrogenase.
[d] Differential RNA display.
[e] (By in situ hybridisation).
[f] Shown by complementation of a yeast mutant.
[g] (By Northern blot and RT-PCR).

stably integrated in the genome of the AM fungus (Forbes et al. 1998b).

It is clear that results concerning the molecular biology of AM fungi are very complicated and time-consuming to obtain. There are certainly more appropriate model systems to understand gene regulation, expression and functioning in plant-microbe interactions. The particular characteristics of these biotrophic symbionts and their impact on plant communities and on plant production systems implicate, however, the meaning of this field of research. The first task to achieve will be to understand in more detail the molecular cross-talk of the AM fungi with plant roots and with other soil microorganisms. In the long term, it will then be necessary to use the molecular probes obtained from pot cultures in field experiments in order to investigate their expression and relevance in terrestrial ecosystems. The results from such experiments will help to reach the final goal of applied AM research: the design of tools for formulating inocula for their use in horti- and agriculture, forestry, revegetation of degraded soils, and in the acclimatisation of micropropagated plants.

References

Allen MF, Allen EB (1992) Mycorrhizae and plant community development: mechanisms and pattern. In: Carrol JC, Wick-Low DT (eds) The fungal community: its organization and role in ecosystems. Dekker, New York, pp 455–479

Aström H, Giovannetti M, Raudaskoski M (1994) Cytoskeletal components in the arbuscular mycorrhizal fungus *Glomus mosseae*. Mol Plant-Microbe Interact 7:309–312

Azcon-Aguilar C, Barea JM (1996) Arbuscular mycorrhizas and biological control of soil-borne plant pathogens – an overview of the mechanisms involved. Mycorrhiza 6:457–464

Azcon-Aguilar C, Barea JM (1997) Applying mycorrhiza biotechnology to horticulture – significance and potential. Sci Hortic 68:1–24

Baptista MJ, Siqueira JO (1994) Effect of flavonoids on spore germination and asymbiotic growth of the arbuscular mycorrhizal fungus *Gigaspora gigantea*. Rev Plant Pathol 76:10

Barea JM, Azcon-Aguilar C (1982) Production of plant growth-regulating substances by the vesicular-arbuscular mycorrhizal fungus *Glomus mosseae*. Appl Environ Microbiol 43:810–813

Barker SJ, Stummer B, Gao L, Dispain I, O'Connor P, Smith SE (1998) A mutant in *Lycopersicum esculentum* Mill. with highly reduced VA mycorrhizal colonisation. Isolation and preliminary characterisation. Plant J 15:791–797

Becard G, Douds DD, Pfeffer PE (1992) Extensive in vitro hyphal growth of vesicular-arbuscular mycorrhizal fungi in the presence of CO_2 and flavonols. Appl Environ Microbiol 58:821–825

Becard G, Taylor LP, Douds DD, Pfeffer PE, Doner LW (1995) Flavonoids are not necessary plant signal compounds in arbuscular mycorrhizal symbioses. Mol Plant-Microbe Interact 8:252–258

Bianciotto V, Bandi C, Minerdi D, Sironi M, Tichy HV, Bonfante P (1996) An obligately endosymbiotic mycorrhizal fungus itself harbors obligately intracellular bacteria. Appl Environ Microbiol 62:3005–3010

Bonfante P, Bergero R, Uribe X, Romera C, Rigau J, Puigdomenech P (1996) Transcriptional activation of a maize alpha-tubulin gene in mycorrhizal maize and transgenic tobacco plants. Plant J 9:737–743

Bradbury SM, Peterson RL, Bowley SR (1991) Interactions between three alfalfa nodulation genotypes and two *Glomus* species. New Phytol 119:115–120

Buee M, Nagahashi G, Douds DD, Becard G (1998) Branching signal or growth-promoting factor. In: Ahonen-Jonnarth U, Danell E, Fransson P, Karen O, Lindahl B, Rangel I, Finlay R (eds) Abstr 2nd Int Conf on Mycorrhizae. SLU Service/Repro, Uppsala, 1998, 36 pp

Burleigh S, Harrison M (1996) The cloning of two genes involved in the *Medicago truncatula/Glomus versiforme* mycorrhizal symbiosis. In: Szaro TM, Bruns TD (eds) Abstr 1st Int Conf on Mycorrhizae, Berkeley, 1996, 32 pp

Burleigh SH, Harrison MJ (1997) A novel gene whose expression in *Medicago truncatula* is suppressed in response to colonization by vesicular-arbuscular mycorrhizal fungi and to phosphate nutrition. Plant Mol Biol 34:199–208

Burleigh SH, Harrison MJ (1998) A cDNA from the arbuscular mycorrhizal fungus *Glomus versiforme* with homology to a cruciform DNA binding protein from *Ustilago maydis*. Mycorrhiza 7:301–306

Bütehorn B, Gianinazzi-Pearson V, Franken P (1999) Quantification of β-tubulin RNA expression during asymbiotic and symbiotic development of the arbuscular mycorrhizal fungus *Glomus mosseae*. Mycol Res 103:360–364

Calantzis C, Morandi D, Gianinazzi-Pearson V (1998) Cellular interactions between *G. mosseae* and a myc-1 nod- mutant in *Medicago truncatula*. In: Ahonen-Jonnarth U, Danell E, Fransson P, Karen O, Lindahl B, Rangel I, Finlay R (eds) Abstr 2nd Int Conf on Mycorrhizae. SLU Service/Repro, Upsalla, 1998, 38 pp

Cooke JC, Gemma JN, Koske RE (1987) Observations of nuclei in vesicular-arbuscular mycorrhizal fungi. Mycologia 79:331–333

Cox G, Tinker PB (1976) Translocation and transfer of nutrients in vesicular-arbuscular mycorrhizas. I. The arbuscule and phosphorus transfer: a quantitative ultrastructural study. New Phytol 7:371–378

Delp G, Barker SJ, Smith SE (1998) Isolation by differential display of three cDNAs coding for proteins from the VA mycorrhizal fungus *G. intraradices*. In: Ahonen-Jonnarth U, Danell E, Fransson P, Karen O, Lindahl B, Rangel I, Finlay R (eds) Abstr 2nd Int Conf on Mycorrhizae. SLU Service/Repro, Upsalla, 1998, 51 pp

Duc G, Trouvelot A, Gianinazzi-Pearson V, Gianinazzi S (1989) First report of non-mycorrhizal mutants (myc⁻) obtained in pea (*Pisum sativum* L.) and faba bean (*Vicia faba* L.). Plant Sci 60:215–222

Dumas-Gaudot E, Guillaume P, Tahiri-Alaoui A, Gianinazzi-Pearson V, Gianinazzi S (1994) Changes in polypeptide patterns in tobacco roots colonized by two *Glomus* species. Mycorrhiza 4:215–221

El Ghachtouli N, Martin-Tanguy J, Paynot M, Morandi D, Gianinazzi S (1996) Effect of polyamines and polyamine biosynthesis inhibitors on spore germination and hyphal growth of *Glomus mosseae*. Mycol Res 100:597–600

Elias KS, Safir GR (1987) Hyphal elongation of *Glomus fasciculatus* in response to root exudates. Appl Environ Microbiol 53:1928–1933

Ezawa T, Saito M, Yoshida T (1995) Comparison of phosphatase localization in the intraradical hyphae of arbuscular mycorrhizal fungi, *Glomus* spp. and *Gigaspora* spp. Plant Soil 176:57–63

Ferrol N, Barea JM, Azcon-Aguilar C (1998) Cloning and expression analysis of P-type H⁺-ATPase genes in the arbuscular mycorrhizal fungus *Glomus mosseae*. In: Ahonen-Jonnarth U, Danell E, Fransson P, Karen O, Lindahl B, Rangel I, Finlay R (eds) Abstr 2nd Int Conf on Mycorrhizae. SLU Service/Repro, Uppsala, 1998, 62 pp

Forbes PJ, Millam S, Hooker JE, Harrier LA (1998a) Transformation of the arbuscular mycorrhiza *Gigaspora rosea* by particle bombardment. Mycol Res 102:497–501

Forbes P, Millam S, Harrier L, Gollotte A, Hooker J (1998b) Transformation of the arbuscular mycorrhizal fungus *Gigaspora rosea* Nicolson & Schenck by particle bombardment. In: Ahonen-Jonnarth U, Danell E, Fransson P, Karen O, Lindahl B, Rangel I, Finlay R (eds) Abstr 2nd Int Conf on Mycorrhizae. SLU Service/Repro, Uppsala, 63 pp

Francis R, Read DJ (1984) Direct transfer of carbon between plants connected by vesicular-arbuscular mycorrhizal mycelium. Nature 307:53–56

Franken P (1998) Trends in molecular studies of AM fungi. In: Varma A, Hock B (eds) Mycorrhiza. Springer, Berlin Heidelberg New York, pp 37–49

Franken P, Gnädinger F (1994) Analysis of parsley arbuscular endomycorrhiza: infection development and mRNA levels of defense-related genes. Mol Plant-Microbe Interact 7:612–620

Franken P, Lapopin L, Meyer-Gauen G, Gianinazzi-Pearson V (1997) RNA accumulation and genes expressed in spores of the arbuscular mycorrhizal fungus *Gigaspora rosea*. Mycologia 89:295–299

Franken P, Ressin B, Lapopin L, Gianinazzi-Pearson V, Gianinazzi S (1998a) PCR cloning of gene from arbuscular mycorrhizal fungi. In: Varma A, Hock B (eds) The mycorrhizal manual. Springer, Berlin Heidelberg New York, pp 401–412

Franken P, Bütehorn B, Kuhn G, Lapopin L, Roussel H, Requena N, Gianinazzi-Pearson V (1998b) RNA accumulation patterns in AM fungi. In: Ahonen-Jonnarth U, Danell E, Fransson P, Karen O, Lindahl B, Rangel I, Finlay R (eds) Abstr 2nd Int Conf on Mycorrhizae. SLU Service/Repro, Uppsala, 65 pp

Garcia-Garrido JM, Toro N, Ocampo JA (1993) Presence of specific polypeptides in onion roots colonized by *Glomus mosseae*. Mycorrhiza 2:175–177

Garcia-Garrido JM, Garcia-Romero I, Parra-Garcia MD, Ocampo JA (1996) Purification of an arbuscular mycorrhizal endoglucanase from onion roots colonized by *Glomus mosseae*. Soil Biol Biochem 28:1443–1449

Garcia-Romera I, Garcia-Garrido JM, Martinez-Molina E, Ocampo JA (1990) Possible influence of hydrolytic enzymes on vesicular arbuscular mycorrhizal infection of alfalfa. Soil Biol Biochem 22:149–152

Garcia-Romera I, Garcia-Garrido JM, Ocampo JA (1991) Pectolytic enzymes in the vesicular-arbuscular mycorrhizal fungus *Glomus mosseae*. FEBS Lett 78:343–346

Garcia-Romera I, Garcia-Garrido JM, Ocampo JA (1992) Cellulase production by the vesicular-arbuscular mycorrhizal fungus *Glomus mosseae* (Nicol. & Gerd.) Gerd. and Trappe. New Phytol 121:221–226

Gaspar ML, Pollero R, Cabello M (1997) Partial purification and characterization of a lipolytic enzyme from spores of the arbuscular mycorrhizal fungus *Glomus versiforme*. Mycologia 89:610–614

Genre A, Bonfante P (1997) A mycorrhizal fungus changes microtubule orientation in tobacco root cells. Protoplasma 199:30–38

Genre A, Bonfante P (1998) Actin versus tubulin configuration in arbuscule containing cells from mycorrhizal tobacco roots. New Phytol 140:745–752

Gianinazzi S, Gianinazzi-Pearson V, Dexheimer J (1979) Enzymatic studies on the metabolism of vesicular arbuscular mycorrhiza. III. Ultrastructural localization of acid and alkaline phosphatase in onion roots infected by *Glomus mosseae* (Nicol. & Gerd.). New Phytol 82:127–132

Gianinazzi S, Trouvelot A, Lovato P, Van Tuinen D, Franken P, Gianinazzi-Pearson V (1995) Arbuscular mycorrhizal fungi in plant production of temperate agroecosystems. Crit Rev Biotechnol 15:305–311

Gianinazzi-Pearson V, Gianinazzi S (1976) Enzymatic studies on the metabolism of vesicular-arbuscular mycorrhiza. Physiol Vég 14:833–841

Gianinazzi-Pearson V, Gianinazzi S (1983) The physiology of vesicular-arbuscular mycorrhizal roots. Plant Soil 71:197–209

Gianinazzi-Pearson V, Branzanti B, Gianinazzi S (1989) In vitro enhancement of spore germination and early hyphal growth of a vesicular-arbuscular mycorrhizal fungus by host root exudates and plant flavonoids. Symbiosis 7:243–255

Gianinazzi-Pearson V, Smith SE, Gianinazzi S, Smith FA (1991) Enzymatic studies on the metabolism of vesicular-arbuscular mycorrhizas V. Is H⁺-ATPase a component of ATP-hydrolysing enzyme activities in plant-fungus interfaces? New Phytol 117:61–74

Giovannetti M, Citernesi S (1993) Time-course of appressorium formation on host plants by arbuscular mycorrhizal fungi. Mycol Res 97:1140–1142

Giovannetti M, Avio L, Sbrana C, Citernesi AS (1993) Factors affecting appressorium development in the vesicular-arbuscular mycorrhizal fungus *Glomus mosseae*. New Phytol 123:115–122

Gryndler M, Vejsadova H, Vosátka M, Catska V (1995) Influence of bacteria on vesicular-arbuscular mycorrhizal infection of maize. Fol Microbiol 40:95–99

Harrier L, Wright F, Hooker J (1998) Isolation of the 3-phosphoglycerate kinase gene of mRNA transcript from the arbuscular mycorrhizal fungus *Glomus mosseae* (Nicol. & Gerd.) Gerdemann & Trappe. Curr Genet 34:386–392

Harrison MJ, Van Buuren ML (1995) A phosphate transporter from the mycorrhizal fungus *Glomus versiforme*. Nature 378:626–629

Hepper CM (1983) Limited independent growth of a vesicular-arbuscular mycorrhizal fungus in vitro. New Phytol 93:537–542

Hirsch A, Kapulnik Y (1998) Signal transduction pathways in mycorrhizal associations: comparisons with the *Rhizobium*-legume symbiosis. Fung Genet Biol 23:205–212

Kaldorf M, Zimmer W, Bothe H (1994) Genetic evidence for the occurrence of assimilatory nitrate reductase in arbuscular mycorrhizal and other fungi. Mycorrhiza 5:23–28

Kaldorf M, Schmelzer E, Bothe H (1998) Expression of maize and fungal nitrate reductase genes in arbuscular mycorrhiza. Mol Plant-Microbe Interact 11:439–448

Kojima T, Hayatsu M, Saito M (1998) Characterization and partial purification of a mycorrhiza-specific phosphatase from *Gigaspora margarita* – *Allium cepa* symbiosis. In: Ahonen-Jonnarth U, Danell E, Fransson P, Karen O, Lindahl B, Rangel I, Finlay R (eds) Abstr 2nd Int Conf on Mycorrhizae. SLU Service/Repro, Uppsala, 100 pp

Krajinski F, Martin-Laurent F, Gianinazzi S, Gianinazzi-Pearson V, Franken P (1998) Cloning and analysis of *psam2*, a gene from *Pisum sativum* L. regulated in symbiotic arbuscular mycorrhiza and pathogenic root-fungus interactions. Mol Physiol Plant Pathol 52:297–307

Lanfranco L, Van Buuren M, Longato L, Garnero L, Harrison MJ, Bonfante P (1996) Chitin synthase genes in arbuscular mycorrhizal fungi (*Glomus versiforme* and *Gigaspora margarita*). In: Szaro TM, Bruns TD (eds) Abstr 1st Int Conf on Mycorrhizae, Berkeley, 74 pp

Lanfranco L, Vallino M, Bonfante P (1999a) Expression of chitin synthase genes in the arbuscular mycorrhizal fungus *Gigaspora margarita*. New Phytol 142:347–354

Lanfranco L, Garnero L, Bonfante P (1999b) Chitin synthase genes in the arbuscular mycorrhizal fungus *Glomus versiforme*: full sequence of a gene encoding a class IV chitin synthase. FEMS Microbiol Lett 170:59–76

Lapopin L, Gianinazzi-Pearson V, Franken P (1999) Comparative differential display analysis of arbuscular mycorrhiza in *Pisum sativum* and a mutant defective in late stage development. Plant Mol Biol:669–677

Lee P-J, Koske RE (1994) *Gigaspora gigantea*: parasitism of spores by fungi and actinomycetes. Mycol Res 98:458–466

Lei J, Bécard G, Catford JG, Piché Y (1991) Root factor stimulate ^{32}P uptake and plasmalemma ATPase activity in vesicular-arbuscular mycorrhizal fungus, *Gigaspora margarita*. New Phytol 118:289–294

Lovato P, Gianinazzi-Pearson V, Trouvelot A, Gianinazzi S (1996) The state of art of mycorrhiza and micropropagation. Adv Hortic Sci 10:46–52

MacDonald RM, Lewis M (1978) The occurrence of some acid phosphatases and dehydrogenases in the in the vesicular-arbuscular mycorrhizal fungus *Glomus mosseae*. New Phytol 80:135–141

Martin-Laurent FA, Franken P, Gianinazzi S (1995) Screening of cDNA fragments generated by differential RNA display. Anal Biochem 228:182–184

Martin-Laurent F, Van Tuinen D, Dumas-Gaudot E, Gianinazzi-Pearson V, Gianinazzi S, Franken P (1997) Differential display analysis of RNA accumulation in arbuscular mycorrhiza of pea and isolation of a novel symbiosis-regulated plant gene. Mol Gen Genet 256:37–44

Marx C, Dexheimer J, Gianinazzi-Pearson V, Gianinazzi S (1982) Enzymatic studies on the metabolism of vesicular-arbuscular mycorrhizas IV. Ultracytoenzymological evidence (ATPase) for active transfer processes in the host-arbuscular interface. New Phytol 90:37–43

Minerdi D, Fani R, Gallo R, Bonfante P (1998) Identification of nitrogen fixation genes in *Burkholderia* endosymbionts of arbuscular mycorrhizal fungi. In: Ahonen-Jonnarth U, Danell E, Fransson P, Karen O, Lindahl B, Rangel I, Finlay R (eds) Abstr 2nd Int Conf on Mycorrhizae. SLU Service/Repro, Uppsala, 1998, 120 pp

Morton JB, Benny GL (1990) Revised classification of arbuscular mycorrhizal fungi (Zygomycetes): a new order, Glomales, two new suborders, Glomineae and Gigasporineae, and two new families, Acaulosporaceae and Gigasporaceae, with an emendation of Glomaceae. Mycotaxon 37:471–491

Mosse B (1970) Honey-coloured, sessile endogone spores: II. Changes in fine structure during spore development. Arch Microbiol 74:129–145

Mosse B (1973) Plant growth responses to vesicular-arbuscular mycorrhiza IV. In soil given additional phosphate. New Phytol 72:127–136

Murphy PJ, Langride P, Smith SE (1997) Cloning plant genes differentially expressed during colonization of roots of *Hordeum vulgare* by the vesicular-arbuscular mycorrhizal fungus *Glomus intraradices*. New Phytol 135:291–301

Nagahashi G, Douds DD Jr, Abney GD (1996a) Phosphorus amendment inhibits hyphal branching of the VAM fungus *Gigaspora margarita* directly and indirectly through its effect on root exudation. Mycorrhiza 6:403–408

Nagahashi G, Douds DD, Abney G (1996b) A rapid microinjection technique allows for the sensitive detection of root exudate signals which stimulate the branching and growth of germinated VAM fungus spores. In: Szaro TM, Bruns TD (eds) Abstr 1st Int Conf on Mycorrhizae, Berkeley, 1996, 91 pp

Nagahashi G, Douds DD, O'Connor J (1998) Fractioning of AM fungal branching signals from aqueous exudates of Ri T-DNA transformed carrot roots. In: Ahonen-Jonnarth U, Danell E, Fransson P, Karen O, Lindahl B, Rangel I, Finlay R (eds) Abstr 2nd Int Conf on Mycorrhizae. SLU Service/Repro, Uppsala, 1998, 125 pp

Newman EI, Reddell P (1987) The distribution of mycorrhizas among families of vascular plants. New Phytol 106:745–751

Peretto R, Bettini V, Favaron F, Alghisi P, Bonfante P (1995) Polygalacturonase activity and location in arbuscular mycorrhizal roots of *Allium porrum* L. Mycorrhiza 5:157–163

Perotto R, Bonfante P (1997) Bacterial associations with mycorrhizal fungi: close and distant friends in the rhizosphere. Trends Microbiol 5:496–501

Philips DA, Tsai SM (1992) Flavonoids as plant signals to rhizosphere microbes. Mycorrhiza 1:55–58

Pirozynski KA, Dalpé Y (1989) Geological history of the Glomaceae with particular reference to mycorrhizal symbiosis. Symbiosis 7:1–36

Pirozynski KA, Malloch DW (1975) The origin of land plants: a matter of mycotrophism. Biosystems 6:153–164

Poulin M-J, Simard J, Catford J-G, Labrie F, Piche Y (1997) Responses of symbiotic endomycorrhizal fungi to estrogens and antiestrogens. Mol Plant-Microbe Interact 10:481–487

Read DJ, Lewis DH, Fitter AH, Alexander IJ (1992) Mycorrhizas in ecosystems. CAB International, Oxford

Rejon-Palomares A, Ocampo JA, Garcia-Romera I (1998) Production of xyloglucanase enzyme by different Glomus strains. In: Ahonen-Jonnarth U, Danell E, Fransson P, Karen O, Lindahl B, Rangel I, Finlay R (eds) Abstr 2nd Int Conf on Mycorrhizae. SLU Service/Repro, Uppsala, 1998, 144 pp

Requena N, Franken P (1998) Pre-symbiotic gene expression in AM fungi. In: Ahonen-Jonnarth U, Danell E, Fransson P, Karen O, Lindahl B, Rangel I, Finlay R (eds) Abstr 2nd Int Conf on Mycorrhizae. SLU Service/Repro, Uppsala, 1998, 144 pp

Requena N, Füller P, Franken (1999) Molecular characterisation of GmFOX2, an evolutionary highly conserved gene from the mycorrhizal fungus Glomus mosseae, downregulated during interaction with rhizobacteria. Mol Plant-Microbe Interact 12:934–942

Requena N, Jeffries P, Barea JM (1996) Assessment of natural mycorrhizal potential in a desertified semiarid ecosystem. Appl Environ Microbiol 62:842–847

Requena N, Jiminez I, Toro M, Barea JM (1997) Interactions between plant-growth-promoting rhizobacteria (PGPR), arbuscular mycorrhizal fungi and Rhizobium spp. in the rhizosphere of Anthyllis cytisoides, a model legume for revegetation in mediterranean semi-arid ecosystems. New Phytol 136:667–677

Rosendahl S, Dodd J, Walker C (1994) Taxonomy and phylogeny of the Glomales. In: Gianinazzi S, Schüepp H (eds) Impact of arbuscular mycorrhizas on sustainable agriculture and natural ecosystems. Birkhäuser, Basel, pp 1–12

Rousseau A, Benhamou N, Chet I, Piché Y (1996) Mycoparasitism of the extramatrical phase of Glomus intraradices by Trichoderma harzianum. Phytopathology 86:434–443

Ruiz-Lozano JM, Bonfante P (1999) Identification of a putative P-transporter operon in the genome of a Burkholderia strain living inside the arbuscular mycorrhizal fungus Gigaspora margarita. J Bacteriol 181:4106–4109

Samra A, Dumas-Gaudot E, Gianinazzi-Pearson V, Gianinazzi S (1996) Soluble proteins and polypeptides profiles of spores of arbuscular mycorrhizal fungi. Interspecific variability and effects of host (myc+) and non-host (myc−) Pisum sativum root exudates. Agronomie 16:709–719

Samra A, Dumas-Gaudot E, Gianinazzi S (1997) Detection of symbiosis-related polypeptides during the early stages of the establishment of arbuscular mycorrhiza between Glomus mosseae and Pisum sativum roots. New Phytol 135:711–722

Simon L, Lalonde M, Bruns TD (1992) Specific amplification of 18S fungal ribosomal genes from vesicular-arbuscular endomycorrhizal fungi colonizing roots. Appl Environ Microbiol 58:291–295

Simon L, Bousquet J, Lévesque RC, Lalonde M (1993) Origin and diversification of endomycorrhizal fungi and coincidence with vascular land plants. Nature 363:67–69

Smith FA, Smith SE (1997) Tansley review no. 96: structural diversity in (vesicular)-arbuscular mycorrhizal symbioses. New Phytol 137:373–388

Smith SE, Dickson S (1991) Quantification of active vesicular-arbuscular mycorrhizal infection using image analysis and other techniques. Aust J Plant Physiol 18:637–648

Smith SE, Gianinazzi-Pearson V (1988) Physiological interactions between symbionts in vesicular-arbuscular mycorrhizal plants. Annu Rev Plant Physiol Plant Mol Biol 39:221–244

Smith SE, Gianinazzi-Pearson V (1990) Phosphate uptake and vesicular-arbuscular activity in mycorrhizal Allium cepa L.: effect of photon irradiance and phosphate nutrition. Aust J Plant Physiol 17:177–188

Smith SE, Smith FA (1990) Structure and function of the interfaces in biotrophic symbioses as they relate to nutrient transport. New Phytol 114:1–38

Tawaraya K, Watanabe S, Yoshida E, Wagatsuma T (1995) Effect of onion (Allium cepa) root exudates on the hyphal growth of Gigaspora margarita. Mycorrhiza 6:57–59

Tisserant B, Gianinazzi-Pearson V, Gianinazzi S, Gollote A (1993) In planta histochemical staining of fungal alkaline phosphatase activity for analysis of efficient arbuscular endomycorrhizal infections. Mycol Res 97:245–250

Tisserant B, Brenac V, Requena N, Jeffries P, Dodd JC (1998) The detection of Glomus spp. (arbuscular mycorrhizal fungi) forming mycorrhizas in three plants, at different stages of seedling development, using mycorrhiza-specific isozymes. New Phytol 138: 225–239

Von Alten H, Linderman A, Schönbeck F (1993) Stimulation of vesicular-arbuscular mycorrhiza by fungicides or rhizosphere bacteria. Mycorrhiza 2:167–173

Wegel E, Schauser L, Sandal N, Stougaard J, Parniske M (1998) Mycorrhiza mutants of Lotus japonicus define genetically independent steps during symbiotic infection. Mol Plant-Microbe Interact 11:933–936

3 Immunochemical Characterization of Mycorrhizal Fungi

A. Hahn[1], S. Wright[2], and B. Hock[1]

CONTENTS

I. Introduction

The highly selective binding properties of antibodies (Abs) have been well documented in many different contexts. Detection and identification of fungal antigens were early applications of immunochemical methods. Pioneering work in use of antibodies against soil microorganisms was done by Schmidt and Bankole (1962), who used fluorescent Abs to detect fungi (*Aspergillus flavus*). Schmidt et al. (1974) also were the first to use antisera against mycorrhizal antigens that discriminated between hyphae of *Pisolithus tinctorius* and *Telephora terrestris*. Since the introduction of the hybridoma technology by Köhler and Milstein (1975), immunochemistry has advanced considerably, due to the uniformity and unlimited supply of the monoclonal antibodies (mAbs) produced from hybridoma cell culture.

In this chapter, we will review the work performed to date with mAbs and mycorrhizal fungi as well as give an overview of the production of mAbs. However, it should be mentioned in advance, that, depending on the nature of the research, the polyclonal approach to immunochemistry has some merits not to be overlooked. Properties of each type of antibody are presented as a guide for comparison of polyclonal and mAbs for specific use.

Antibodies are glycoproteins that are produced by the vertebrate immune system upon introduction (i.e., immunization) of a foreign substance (the antibody generator = antigen). They selectively bind to a restricted surface area of their antigen, the epitope, which usually has a size of about five to six amino acids or nucleic acids or monosaccharides. Other antigens may, or may not, have similar or identical epitopes. Multiple epitopes on the antigen induce the proliferation of several responding clones of Ab-producing cells. Consequently, a polyclonal antiserum is unique in its specific composition of polyclonal Abs (pAbs) and, therefore, in binding properties and performance. Different antisera (even from the same animal) must be separately assessed for their suitability in any particular immunochemical assay.

Each antibody-producing lymphocyte cell line in the organism, however, produces one and only one discrete type of antibody. The hybridoma technology, introduced by Köhler and Milstein (1975), provides the means to immortalize and single out that clone of a cell line which produces the antibody with the desired properties (the monoclonal Ab = mAb). This mAb is then directed against one single epitope and characterized by a uniform specificity and affinity towards an antigen. In this way, mAbs are able to identify single molecules as well as individual regions in the same molecule.

The best starting point for a selective polyclonal antiserum is usually a pure antigen. Use of crude preparations of an antigen for immunization will most probably result in pAbs recognizing many different antigens. Antigen-enriched

[1] Technical University of München at Weihenstephan, Department of Botany, Alte Akademie 12, 85350 Freising, Germany
[2] USDA-ARS Soil Microbial Systems Laboratory, 10300 Baltimore Ave., Beltsville, Maryland 20705-2350, USA

The Mycota IX
Fungal Associations
Hock (Ed.)
© Springer-Verlag Berlin Heidelberg 2001

fractions of crude preparations are usually adequate for immunizations to obtain mAbs, provided that subsequent screening permits the selection of the mAbs of interest (Dewey et al. 1997). In general, purification of the antigen is more important when immunizing for pAbs, while for mAbs, purification or some kind of identification is later required for the screening process. The production of polyclonal antibodies (pAb) is a well-developed technique which has been in use for around 50 years. The adaptation of this technique for the production of pAb against mycorrhizal antigens has recently been summarized (Hahn et al. 1998).

The advantages of polyclonal antisera over mAbs include their easy production with considerably less cost as well as a higher possible titer and higher affinity for most antigens. This often results in an immunochemical signal stronger than with mAbs due to the greater number of epitopes recognized by an antiserum.

The problems encountered with pAb are linked in particular to the possibility of cross-reactivities due to their variable specificity. Their limited and variable supply and the changing properties with each bleeding and each animal used are additional drawbacks. These can be overcome by the use of hybridoma technology. The benefits derived from this technology include the virtually unlimited supply of specified Abs, fewer problems with unspecific binding and a system that, within limits, allows a selective screening for those mAbs that have the desired specificities and affinities towards a given antigen.

The affinity of a mAb can be selected in the screening process, just like the isotype and other desired properties of the immunoglobulin. Impure antigens or mixtures can still be used to raise highly specific antibodies; this is especially relevant when antigens are to be detected in complex matrices such as soil or plant tissue.

II. Production of Monoclonal Antibodies Against Mycorrhizal Antigens

The technical details of the production of monoclonal antibodies (mAb) have been described before and a review pertaining to the prospects and limitations of this technique for research on arbuscular mycorrhizal (AM) fungi was published by Perotto et al. (1992). We will discuss here only those aspects of the method that have implications for studies of fungal symbioses.

Substances of microbial origin that may act as antigens can be extracellular or intracellular and have a structural, metabolic, or regulatory function. They are usually proteins, glycoproteins, polysaccharides, lipopolysaccharides, or lipids (Mernaugh et al. 1990). The usefulness of an antigen for immunochemistry depends on its biochemical and immunological properties as well as on the test system that is later to be employed for its detection. Other important properties of the antigen include immunogenicity, location, and abundance in the organism, solubility, and stability. Before the start of an experiment to produce mAbs, it is advisable to take as many of these points into account as possible. The procedure for mAb production, although well standardized, may need to be adapted in some cases.

Hybridomas are obtained through the fusion of a myeloma cell line that contributes genetic material for continuous cell growth and a spleen cell (or lymph node cell) from an immunized animal that contributes the genetic information for an antibody. Hybridoma production is based upon genetic and chemical manipulation of nucleic acid synthesis pathways. A screening step must follow in which an immunochemical test system is utilized to find the cell lines producing mAb which recognize the desired antigen. Detailed descriptions of the procedures used by the authors have been published (Dazzo and Wright 1996; Göbel et al. 1998).

Immunochemical tests are firmly established as analytical methods and there are a number of standard handbooks for their application (Tijssen 1985; Harlow and Lane 1988). In general, it is advisable to choose a screening system for the mAbs, which most closely reflects their intended use. Since mAbs directed against soluble antigens will not necessarily be useful for immunolabeling structural antigens and vice versa, the screening procedure will not single out every mAb that is produced by the hybridomas, but rather only those mAbs which are detected by the assay system. In the course of hybridoma culture, a large number of cultures need to be tested for mAb production within a short time. This must be possible with the chosen test format.

If the purpose of the desired mAbs is the localization of structural antigens or bound residues in complex matrices such as plant tissue or fixed soil samples, then the procedure for screening will, in principle, be an indirect immuno-labeling protocol, where the supernatants from the hybridoma cell cultures are incubated with the fixed antigen-containing matrix. Then potential antibody binding is detected by a secondary molecule, usually an antibody directed against murine immunoglobulins, although protein A is sometimes used. This detector molecule is coupled to a signal molecule such as a fluorescent stain or a gold particle. These may then be observed microscopically. The necessary control experiments include the use of serum from a nonimmunized mouse or any other unspecific murine antibody as well as (if possible) antigen-free matrix material.

When those hybridoma cell cultures which produce mAbs with the required specifications against the antigen have been identified, a cloning procedure follows. The cell populations are suspended and reseeded at a dilution which is statistically designed to result in a density of one cell per vessel. The cell cultures derived from these clones consistently produce a single type of antibody and secrete it into their medium.

Antibody-antigen reactions are known for their high degree of specificity. Nevertheless, Abs can cross-react with other, usually related, antigens under certain circumstances. These cross-reactions constitute, along with the isotype and the concentration, the most important quality of the mAb. When the cell lines are monoclonal, these qualities must be determined.

The problem associated with mAb appears to be the relatively statistically low chances of raising them against AM fungal antigens. Although a number of mAb have been obtained by this method (Wright et al. 1987; Hahn et al. 1993; Göbel et al. 1995), low fusion rates of spleen cells and myelomas have been experienced, and low percentages (<20%) of Ab-producing hybridomas obtained. Out of six fusions of murine spleen cells immunized with different isolates of *Glomus*, only four cell lines of hybridomas expressed mAb with a variety of specificities against the antigens (Hahn et al. 1993). Similarly, Wright et al. (1987) obtained only two mAb-secreting cell lines from two spleens using soluble AM fungal antigen.

This difficulty in obtaining mAb has also often been described for nonmycorrhizal fungal antigens. For example, out of 16 fusions of murine spleens immunized with the pathogen *Pseudocer-*

cosporella herpotrichoides, only 3 cell lines could be derived that expressed mAb which specifically recognized the antigen (Dewey et al. 1991). Schulze and Bahnweg (1998) describe poor immunogenic responses against fungal antigens in the production of mAbs for the identification and detection of *Armillaria* spp. and *Heterobasidion annosum*, whereas Zollfrank et al. (1987) encountered no difficulties with pAbs. Dewey et al. (1997) recommend the use of coimmunization techniques, where the formation of nonspecific antibodies is blocked by the coinjection of fungal target antigens and antisera previously raised against the same or a closely related fungus, to overcome this problem of low specific responses.

III. Immunochemical Test Systems

Basically, two types of test systems can be employed for the detection of antigens in mycorrhizal research. Immunoassay formats (e.g., ELISAs) measure the concentration of antigens in solution near the equilibrium and give rise to a concentration-dependent signal. Immunolabeling methods utilize the high selectivity of Abs to locate an antigen in a structured matrix, working with an excess of antibody. Immunoblotting is a form of an assay where soluble or unsoluble antigens are bound to a membrane. We summarize here a protocol from each test type.

A. Immunoassay

A very sensitive enzyme-linked immunoassay (ELISA) against a glycoprotein produced by AM fungi has been developed (Wright and Upadhyaya 1996). The mAb resulted from use of crushed fresh spores of *Glomus intraradices* as the immunogen in an attempt to produce a species-specific probe for this fungus. During screening of mAbs, it was noted that one mAb (32B11) was generally cross-reactive with all AM fungi tested. The mAb cell line was saved and later tested to determine the location of reactivity and the extent of cross-reactivity using immunofluorescence (see Sect. III.B). This mAb eventually revealed that AM fungi produce copious amounts of a glycoprotein on hyphae; this protein "glomalin" is sloughed from hyphae to soil and is probably a major contributor to soil stability (Wright et al. 1996,

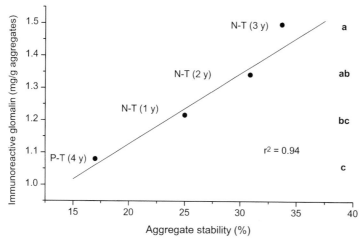

Fig. 1. Soil aggregates (1–2-mm size range) from a soil in transition from plow tillage (*PT*) to no tillage (*NT* after 1, 2, and 3 years) probed by ELISA for the content of glomalin. The glycoprotein is extracted from hyphae or soil using citrate and heat and 0.125 µg of glomalin (based on a Bradford protein assay) in 50 µl of PBS are placed in wells of 96 well polyvinyl chloride microassay plates. Glomalin extracted from hyphae of an AM fungus is used as a control on each plate. Samples are dried at 37 °C. A 2% solution of nonfat dried milk in PBS (w/v) is used to block the remaining reactive sites in wells. Incubation is for 15 min. The blocking solution is poured off, and mAb32Bll diluted in PBS is added. This and subsequent steps are incubated for 1 h at RT. Extensive washing with PBS + Tween 20 (polyoxyethylene sorbitanmonolaurate) is carried out between each step and after the final incubation. A goat antimouse IgM antibody with biotin attached by a long spacer arm (Jackson ImmunoResearch Laboratories, Inc. West Grove, PA, USA) diluted in PBS with 1% BSA is added to wells. The final incubation is with streptavidin peroxidase diluted in PBS with 1% BSA. Peroxidase is detected by ABTS (2,2′-azino-bis(3-ethylbenzthiazoline-6-sulfonic acid) (Tijssen 1985). The immunoreactivity of glomalin is compared to the control and can be related to the original concentration of glomalin in a sample as determined by Bradford protein analysis (i.e., percent immunoreactive glomalin in 0.125 µg is used to calculate concentration of immunoreactive glomalin in total glomalin). A minimum of six replicates was measured for each treatment. Aggregate stability is correlated with immunoreactive glomalin with $r^2 = 0.94$. Identical lower case letters in the right margin denote separation with a least significant difference of means < 0.05

1998). Figure 1 displays aggregate stability and concentrations of immunoreactive glomalin in a soil planted to corn during transition from plow tillage (PT) to no tillage (NT). Both aggregates stability and glomalin showed significant increases after 2 years of NT compared with PT (manuscript submitted). These results were based on the immunoreactivity of 0.125 µg of total glomalin extracted from 1 g of aggregates. A quantitative ELISA, using glomalin as a standard, was developed to simplify estimation of immunoreactive glomalin in soil extracts (Wright and Upadhyaya 1998b).

Indirect immunoassays utilize a secondary molecule to detect the primary antibody bound to the antigen. The immunoassay developed for quantification of glomalin extracted from hyphae and soil is based on an indirect ELISA using a mAb of the IgM class and a series of secondary reactions to overcome the large molecular mass that accumulates when an IgM (900 kDa) is used. The assay utilizes the sensitive biotin-streptavidin reaction to detect the mAb, but overcomes loss of sensitivity due to insufficient space for layering a large molecular mass at the reactive site. Estimates of molecular sizes (kDa) of the molecules used are: IgM = 900; IgG antimouse antibody = 160; biotin = 0.2; streptavidin = 60; peroxidase = 44. Biotin attached to the antimouse secondary antibody by a long spacer arm allows streptavidin peroxidase to attach to biotin. Optical density (405 nm) readings of four replicate samples of 0.04 µg of glomalin were 0.102 ± 0.005, 1.098 ± 0.097, and 1.645 ± 0.037 for antimouse IgM-peroxidase, biotin, and streptavidin-peroxidase, and long spacer arm biotin and streptavidin-peroxidase, respectively. These data show the remarkable increase in sensitivity obtained by the use of spacer arm biotin. Assays must be developed with the class of immunoglobulin and space for layering of reactants in mind. Potentially important mAbs may be discarded at the preliminary screening stage, because the detection assay was not optimized.

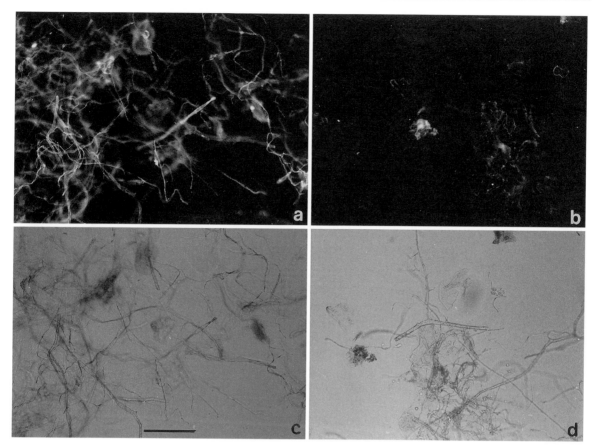

Fig. 2a–d. Immunofluorescent labeling of hyphae of *Glomus mosseae* using a monoclonal antibody (8A7) specific for this species of AM fungi. Hyphae are bound to silanized microscope slides, labeled by a (primary) anti-AM antibody (IgM), which is subsequently labeled by a antimurine IgM antibody conjugated to biotin, and detection takes place by a streptavidin molecule bound to the fluorescent dye phycoerythrin. Unspecific binding is blocked before and after binding of the primary mAb by adding gt-IgG at a concentration of $10\,\mu g\,ml^{-1}$. A fluorescent signal is visible only when the specific mAb is used as primary Ab (**a**), not when unspecific murine IgM is employed (**b**). The bottom row (**c**, **d**) shows the corresponding bright field micrographs. *Bar* $200\,\mu m$

B. Immunolabeling

Fluorescence labeling is the most frequently used form of signal generation in immunohisto- and cytochemistry, although other labels are available, such as gold particles, radioactive labels, or insoluble enzyme substrates.

Washed mycelial fragments were used as immunogens and a number of genus-specific mAb have been raised to date against insoluble surface antigens of AM fungi (Göbel et al. 1995). For the screening process, an indirect immunolabeling protocol is utilized, where hyphae are immobilized on microscope slides, primary (anti-AM) mAbs, and, subsequently, a secondary Ab (antimurine IgG, coupled to biotin) is added. Detection takes place under an epifluorescent microscope after adding streptavidin-phycoerythrin. Since antibody excess concentrations are used, blocking of unspecific binding is a very crucial step.

Figure 2 gives an example of the type of fluorescent signal derived from this procedure. It can clearly be distinguished that only hyphae of the fungal antigen are labeled, while background fluorescence is very low. The control experiment with a supernatant from a cell culture producing unspecific murine Ig gives no rise to any fluorescent signal.

C. Immunoblotting

Wright and Morton (1989) developed a dot blot immunoassay (DIBA) utilizing a mAb that was able to specifically stain fungal matter on a nitrocellulose "print" of the crushed root. In this way, even the nonstainable AM fungus. *G. occultum* could be detected and quantified in whole root systems. This assay was developed to overcome the limitation of having an antibody against an antigen in soluble internal spore material, but requiring an assay for root colonization. The soluble antigen is released from spores by crushing pieces of roots with attached fungus on nitrocellulose. The presence and position of spores are revealed by DIBA. Figure 3 shows a dot-blot of root material from red clover.

This DIBA of whole root systems was adapted in a study by Plattner and Hall (1995), who produced an antiserum against soluble proteins from ascoma of *Tuber melanosporum*. The antigen was probed in roots of weeds growing near trees which were colonized by *T. melanosporum*. The antigen was closely associated with necroses of the weed roots and this indicates that the Perigord black truffle is pathogenic to the weeds. The pAb used in this study was highly specific for its antigen and did not cross-react with 20 other soil fungi, including ectomycorrhizal and pathogenic species. The authors conclude that Ab techniques are more sensitive than isoenzyme analyses and more universally applicable than molecular techniques and that immunochemical methods are appropriate and sufficiently powerful to monitor the presence of a specific fungus such as *T. melanosporum* in soil or on root tips.

IV. Identification of Mycorrhizal Fungi at the Genus and Subgenus Level

The difficult identification of endomycorrhizal fungi, especially within the Glomales, prompted the search for techniques to identify them unequivocally and rapidly. The high specificity of antibodies was utilized in a number of experiments to this end. Polyclonal and monoclonal antibodies have been raised against spore material (both soluble fractions and wall epitopes) and extraradical mycelium of a few species of AM fungi. The Ab are generally checked for cross-reactivity against species of fungi from the Glomales and other fungal taxa, using immunodiffusion, ELISA, DIBA, and immunocytochemical techniques. The latter are the most discriminating since antigen concentration exposed to the Ab is not defined. Unspecificity of pAb is a problem often encountered, but this may

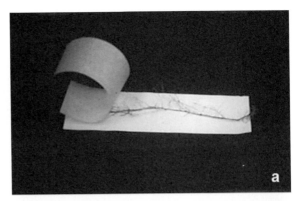

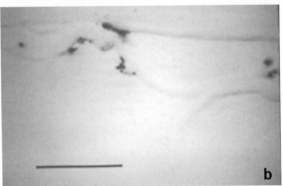

Fig. 3a,b. Dot blot of *Glomus occultum* colonizing roots of red clover. A root section was submerged in water and then floated on to a strip of nitrocellulose membrane. After blotting until roots and membrane were almost dry, waxed paper was placed over the root and taped to hold it in place. A rolling pin was passed over the layers in different directions at least 15 times (**a**). Roots were removed and the nitrocellulose was placed in 15% H_2O_2 for 1h. Strips were washed three times (5min each) in Tris-buffered saline (TBS; 20mM Tris-0.25M NaCl adjusted to pH 7.5). The ELISA was performed as follows: 1-h incubation in 6% nofat milk (w/v in distilled water), 4-h incubation in tissue culture supernatant fluid containing the mAb against *G. occultum*, three washes with TBS, 4h in antimouse IgG-peroxidase, and three washes with TBS followed by exposure to the color developer (0.015g of 4-chloro-1-naphthol, 25ml TBS, and 150µl 3% H_2O_2). Precipitated purple stain reveals where spores were located on the root section, and an imprint of the root remains in the membrane after the dot blot is developed (**b**)

be overcome by using purified fractions of soluble protein from fungal preparations as immunogen (Cordier et al. 1994). In this work, pAbs were described that recognize only *Gigaspora* and that are even capable of discrimination between species in this genus.

Thingstrup et al. (1995) were able to detect the AM fungus *Scutellospora heterogama* in mycorrhizal roots using polyclonal antibodies developed against a soluble fraction of the spores. The cross-reactivities of the antiserum were tested

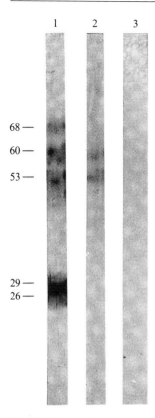

Fig. 4. Western blot of extracts from *Scutellospora heterogama*, separated by SDS-PAGE from: spores of *S. heterogama* (*lane 1*), roots of *Brachiaria dictyoneura* colonized by this AM fungus (*lane 2*), and noncolonized roots of *B. dictyoneura* (*lane 3*). The *numbers on the vertical* axis are molecular weight data in kDa (Thingstrup et al. 1995)

against spores and/or hyphae of eight other AM fungal species from five genera, and no reaction was detected with any of them in a Ouchterlony double diffusion assay. Figure 4 shows results from the subsequently developed Western Blot, utilizing antigens that are shared between spores (lane 1) and intraradical hyphae (lane 2). The similarity of antigens in spores and hyphae of AM fungi plays a role in the study of gene expression in different stages of the fungal life cycle, and the findings from this report suggest that the spectrum of molecules produced by the fungus changes during its symbiotic association with the host plant. Some antigens are formed in the spore stage, others only in the intraradical hyphae during the symbiotic phase.

However, Oramas-Shirey and Morton (1990) found significant differences in absorbance values in an ELISA with an mAb raised against *G. occultum* when the same amount of antigen from different isolates of the AM species was employed.

This indicates that antigenic properties of AM fungi may vary even at the strain level.

Because of the selective binding of Abs to species-specific antigens, immunochemistry can contribute to the qualitative and quantitative differentiation of fungal isolates. The Glomales can serve as examples for the potential of immunochemical methods in taxonomic applications. This group of fungi can be differentiated only with considerable difficulty on the basis of the morphological characteristics of their chlamydospores. In addition, Morton and Bentivenga (1994) illustrate the variation in spore morphology within individual species. The microscopic appearance of the hyphae is even less reliable as a taxonomic criterion. Here. the immunochemical approach can take advantage of submicroscopic biochemical structures, preferably on the surface of spores and hyphae. When this is correlated with at least one other method of differentiation, such as morphological appearance, isozyme analysis (Tisserant et al. 1998), or DNA fragment analysis (Van Tuinen et al. 1998), the Ab technique can be a valuable tool in classifying the AM fungi.

V. Serological Relations of AM Fungi

In the above-mentioned report by Thingstrup et al. (1995), species-specific antigens were subsequently identified by molecular weight. Even without knowledge of the target molecules, however, antisera may be raised against purified extracts, and this can improve specificity considerably. The antisera raised by Sanders et al. (1992) against soluble protein extracts of spores of *Acaulospora laevis* and *Gigaspora margarita* recognized untreated fungal extracts of their homologous antigens in a DIBA (cf. Fig. 5). Cross-reactivities were determined and the antiserum raised against *G. margarita* gave a reaction with two isolates of *G. margarita*, and two isolates of *A. laevis* as well as one isolate of *Scutellospora* sp. *S. pellucida* and three species of *Glomus* were not recognized.

These findings would point to a closer serological relationship between *Acaulospora* and *Gigaspora*, which is consistent with earlier findings (Aldwell et al. 1985; Wilson et al. 1983) but not with the current taxonomic concept of the Glomales as classified by spore morphology (Morton 1998). Table 1 gives an overview of the

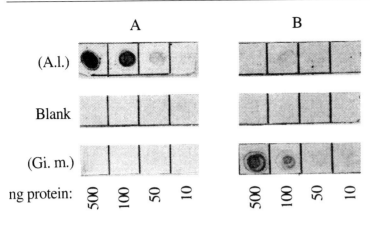

Fig. 5A,B. Dot immunoblot assay with soluble protein fractions of **A** *Acaulospora laevis* and **B** *Gigaspora margarita*, using an antiserum against either *A. laevis* (*A.1.*) or *G. margarita* (*Gi.m.*). The antigen is expressed in ng protein (Sanders et al. 1992)

Table 1. Discrimination of AM fungi by immunochemistry

Reference	Glomales					
	Glomineae				Gigasporinae	
	Glomaceae		Acaulosporaceae		Gigasporaceae	
	Sclerocystis	*Glomus*	*Entrophospora*	*Acaulospora*	*Gigaspora*	*Scutellospora*
Wilson et al. (1983)		−		+	+	
Aldwell et al. (1985)	−	−		+	+	
Sanders et al. (1992)		−		+	+	−
Sanders et al. (1992)		−		+	+	+
Friese and Allen (1991)		−			+	
Cordier et al. (1994)				−	+	+
Thingstrup et al. (1995)		−	−	−	+	+
Kough et al. (1983)		+		−	−	
Aldwell et al. (1985)	+	+		−	−	
Wright et al. (1987)		+	−	−	−	
A. Hahn et al. (unpubl.)		+	−	−	−	
Göbel et al. (1995)		+	−	−	−	−

Test systems include immunofluorescence, ELISA, and DIBA, results are summarized as + for a positive and − for a negative reaction.

immunochemical relationships within the Glomales, as determined from literature on work with pAbs and mAbs. Three studies reported antisera which recognize Acaulosporaceae and Gigasporaceae but not Glomaceae. This result points to functional and phylogenetic similarities between these fungi. The majority of antibodies raised, however, are capable of discriminating between these families. Below the generic level, however, the discrimination by the immunochemical approach gives a rather incomplete picture, due to the fact that most of the studies used heterogenous isolates of AM fungi for cross-reactivity tests. This makes a comparison, e.g., within the Glomaceae, difficult.

Figure 6 shows the results of immunochemical discrimination of taxa in vivo with a genus-specific mAb, F5G5, directed against *G. mosseae* and reacting with a number of other members of the species *Glomus*, but not with *Entrophospora* sp. (Göbel et al. 1995). The fluorescence micrograph in Fig. 6a depicts a mixture of hyphae of these two genera, which were isolated from pure cultures of the AM fungi and subjected to immunolabeling. From a comparison with the corresponding brightfield micrograph (Fig. 6b), it

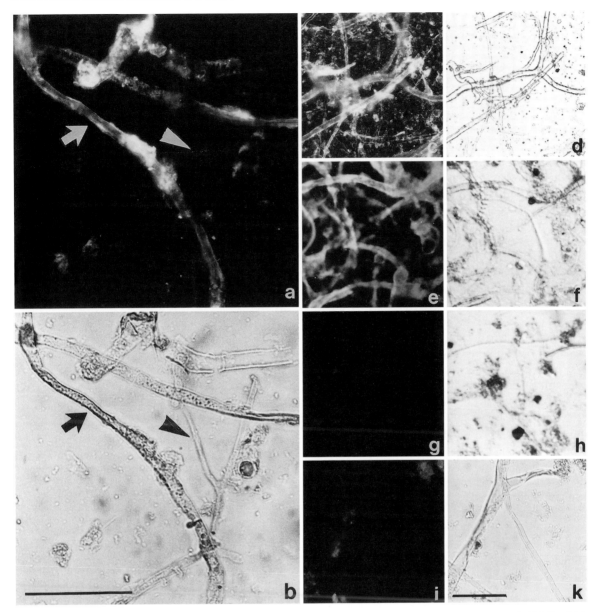

Fig. 6a–k. Immunolabeling of *G. mosseae* and *Entrophospora colombiana* with a mAb (F5G5) capable of discrimination between these two species (**a, b** corresponding bright field micrograph). Only some hyphae are labeled (*arrows*), others are not fluorescing (*arrowheads*). The control experiments are depicted in the third column. **c** Labeling of the hyphal mixture with mAb H8F7, which does not discriminate between the species. **e** Labeling of pure *G. mosseae* with F5G5. **g** Labeling of pure *E. colombiana* with F5G5. **i** Labeling of the hyphal mixture with control IgM as primary mAb. **d–k** corresponding bright field micrographs. *Bars* 50 μm

becomes obvious that only some hyphae are labeled, while others do not raise a fluorescent signal (arrow). The necessary controls include positive labeling of the hyphal mixture with a nondifferentiating mAb (Fig. 6c), positive labeling of pure culture *G. mosseae* with F5G5 (Fig. 6e), negative labeling of *E. colombiana* with F5G5 (Fig. 6g), and negative labeling of the hyphal mixture with control IgM (Fig. 6i). The right column shows the corresponding bright field micrographs to these control experiments.

With this method, it is possible to selectively label hyphae of one AM fungal genus isolated from mixed cultures. Apart from detecting different fungi in natural habitats, this approach allows studies of competition between AM fungi.

VI. Detection and Localization of Compounds Involved in Fungal Symbioses

In addition to the potential of species-specific recognition of antigens in soluble and insoluble fractions of the mycorrhizal fungi, immuno-chemical methods can contribute to a better understanding of the establishment and the physiological processes involved in mycorrhizal symbioses. This is possible if antibodies are provided that recognize relevant compounds such as phytohormones, pathogenesis-related substances or enzymes. One aspect of these studies involves the quantification of these substances and comparison of mycorrhizal and non-mycorrhizal states in order to detect significant changes in the expression levels of these substances that may be due to the presence of the symbiosis partners. The second approach involves the localization of these compounds in various tissues and the deduction of information on their synthesis and their presence both intra- and extraradically.

Glomalin is an abundantly produced glycoprotein, which may have great influence on soil stability and aggregation. It is revealed on hyphae, roots, and plastic hyphae traps in soil by immunofluorescence using mAb32B11 (Wright et al. 1996). Glomalin fluoresces brightly green and is abundantly produced by hyphae of AM fungi (Fig. 7a). Pieces of glomalin often are seen on roots, root hairs, and attached to plastic traps (Wright and Upadhyaya 1998a,b). Autofluorescence of fungal spores is yellow. On some plants root hairs show a green autofluorescence, but the smooth margins are distinctly different from the general unevenness of glomalin deposits. Glomalin was detected in soil using the above-described immunofluorescent labeling procedure. This is shown in Fig. 7b, where the presence of glomalin is revealed on aggregates from a prairie soil that has never been plowed. A simple indirect immunofluorescence assay can be used to reveal glomalin on water-stable aggregates. The same mAb used in the previously described immunoassay (Sect. III.A) was used for this assay. The combined molecular mass of the IgM mAb and goat antimouse IgM tagged with fluoresceine isothiocyanate fits within the reactive site and gives a strong signal.

Hyphae of AM fungi and glomalin sloughed from hyphae adhere to plastic horticultural

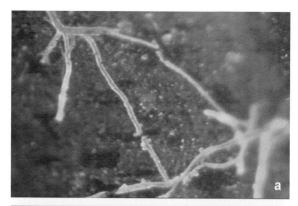

Fig. 7a,b. Immunolocalization of glomalin in situ. Pieces of fresh hyphae, fresh roots with attached hyphae, plastic mesh used to trap hyphae in soil (**a**) (manuscript submitted), or soil aggregates (**b**) are placed in small sieves. Nonfat dry milk in PBS (2% w/v) is added, incubated for 15 min, and poured off. Diluted mAb32B11 is added to cover the test material and incubated for 1 h. Goat-antimouse IgM labeled with fluorescein isothiocynate (FITC) diluted in PBS with 1% BSA is added and incubated for 1 h. Samples are washed three times for 5 min in PBS + Tween 20 between antibody incubation steps and after the final step. Samples are mounted in a medium that quenches photofading, and viewed with an epifluorescence microscope using filters for FITC (490 nm excitation and 510–550 nm emission)

mesh inserted into a pot culture (Wright and Upadhyaya 1998b). A 1-cm-square section of the horticultural mesh from a pot culture of *Glomus caledonium* (Fig. 7a) shows a thick coating of glomalin that was revealed by the same assay as used for aggregates.

The high selectivity of antibodies may also be used to search for substances which have a function in the mycorrhizal symbiosis. They can give clues on spatial and temporal expression, as well as the amount, of relevant molecules.

Wyss et al. (1990) made use of a pAb that was raised against nodule-specific host proteins in the *Rhizobium*-legume symbiosis. The nodule-specific serum showed cross-reactivities to in vitro translated products of polyadenylated RNA from mycorrhizal roots indicating that mycorrhizal infection also induces symbiosis-specific polypeptides, which subsequently were termed mycorrhizins.

The similarity of antigenic components in nodules and mycorrhizal roots of soybean was again utilized by Perotto et al. (1994). They used an mAb against a glycoprotein expressed in the infection thread area of the nodulated legumes. The difference between the two symbiotic processes was revealed by the fact that host defense systems induced by the infection of *Rhizobium* are absent during colonization with AM fungi.

Myc⁻ pea mutants were later also subjected to immunolabeling studies with mAbs raised originally against *Rhizobium*-induced glycoproteins. In this case, the antigens were induced by AM inoculation at the appressorium level as opposed to wild-type *P. sativum*, where they were detected only at some later stages of the infection process (Gollotte et al. 1995).

Together with observations of the presence of molecules accumulating in plant-pathogen interactions, these polysaccharides from the *Rhizobium* symbiosis indicate a weak activation of defense mechanisms during the establishment of AM symbiosis. The myc⁻ pea mutants, on the other hand, did show a high expression of pathogenesis-related defense reactions. This points to complex interactions between defense genes and the mycorrhizal symbiosis.

Balestrini et al. (1996) used mAbs raised against rhamnogalacturonan I for the detection of epitopes of this compound in roots of different mycorrhizal plants, thereby proving that the composition of the interface zone between host and fungus in the AM symbiosis is of plant origin. This is apparently also true for the late stages of orchid mycorrhizae, as Peterson et al. (1996) proved by using mAbs JIM5 and JIM7, directed against pectins. These molecules were predominantly found in the middle lamella of mycorrhizal protocorm cells in early colonization stages, but not in the interfacial material. Only in the late stages did the interfacial matrix material surrounding the collapsing hyphae give a signal with the anti-pectin mAb JIM5. A pAb against β-1,3-glucans, in contrast, labeled the fungal cell wall in early stages and the interfacial material in late stages. This points to a complex process of responses in the orchid mycorrhizal symbiosis.

Perotto et al. (1995) immunocytochemically detected pectin-degrading enzymes during the infection process of ericoid mycorrhizal fungi in host and nonhost plants. They suggest that a crucial step in the establishment of a successful mycorrhiza is the regulation of the production of cell wall-degrading enzymes.

Immunochemical assays of symbiosis-regulated polypeptides were studied in the mycorrhiza of *Pisolithus tinctorius* with *Eucalyptus* ssp. by Martin et al. (1995). These studies showed major alterations of fungal cell walls during the early stages of ectomycorrhizal interaction and emphasized the importance of reciprocal control of genes and proteins in the development of the symbiotic interface.

A number of studies shed light on the role of phytohormones and other messenger molecules in mycorrhizal symbioses. Table 2 gives an overview of the compounds that were measured using ELISAs and a summary of the results pertaining to mycorrhizal regulation through soluble compounds. With the exception of one study (Baas and Kuiper 1989), all the findings presented here indicate a significant role of the plant hormones in the establishment and the physiological processes in mycorrhizal formation.

VII. Conclusions

The highly specific binding abilities of antibodies for a virtually unlimited variety of antigens and haptens have proven useful in the study of mycorrhizal symbioses. Especially the hybridoma technique, making monoclonal antibodies biochemically definable and reproducible reagents, has contributed important insights into the complex physiological and structural conditions of the mycorrhizae (Perotto et al. 1992).

The production of monoclonal antibodies against fungal antigens meets a number of challenges, including the fact that the immunological response to fungal antigens is generally relatively weak. Dewey et al. (1997) summarized the findings for soil fungi pertaining especially to mAbs. Any development of immunodiagnostic methods for plant-invading fungi is impaired by the difficulty

Table 2. Effects of phytohormones and second messengers observed in mycorrhizal symbioses by the use of ELISA techniques

Phytohormone, messenger, enzyme	Mycorrhiza type	Effects observed	Reference
Auxins	Ectomycorrhiza	Mycorrhization upregulates free phytohormone in host	Liebmann and Hock (1989)
Auxins	"	Mycorrhization upregulates adventitious root formation and elongation in host	Karabaghli et al. (1998)
Cytokinins	"	Host root presence upregulates phytohormone production in fungi	Kraigher et al. (1991)
Cytokinins	"	Mycorrhization upregulates phytohormone content in host	Kraigher and Hanke (1996)
Cyclic AMP	"	Addition of C upregulates cAMP, cAMP upregulates glycolysis in fungi	Hoffmann et al. (1997)
Auxins, abscisic acid	Arbuscular mycorrhiza	Fungi control gas exchange of host plant through phytohormone production	Goicoechea et al. (1997)
Auxins, abscisic acid, cytokinins	"	Mycorrhization upregulates ABA, cytokinins, IAA unchanged	Danneberg et al. (1993)
Abscisic acid	"	Hyphae contain high levels of phytohormone	Esch et al. (1994)
Cytokinins	"	Mycorrhization upregulates phytohormone level in host	Druege and Schönbeck (1993)
Cytokinins	"	Phytohormones are not primary growth regulator in host	Baas and Kuiper (1989)
Basic chitinase	"	Mycorrhization downregulates chitinase production	David et al. (1998)
Acid phosphatases	Ericoid mycorrhiza	Host regulates fungal metabolism by enzyme inhibition	Straker et al. (1989)
Auxins	Orchid mycorrhiza	Fungus produces phytohormone, N concentration regulates symbiosis	Beyrle et al. (1991)

in raising specific antisera (Dewey et al. 1991). The antisera to fungi (hyphal fragments, soluble extracts, or culture supernatants) cross-react heavily with related fungi, species from unrelated genera, and host-tissue molecules.

Reasonably high specificities may usually be obtained after immunization with protein precipitates of culture filtrates or mycelial extracts. As stated earlier, the specificity is satisfactory in most cases, if it is possible to use specific fungal fractions such as enzymes, toxins, or soluble carbohydrate fractions as antigens.

The use of immunochemical methods for the study of mycorrhizae has in recent years lost some of its unique advantages in the highly specific recognition and detection of molecules due to the rapid advance in DNA-based techniques; but the two molecular approaches to the biochemical studies of symbiotic processes complement each other very well. Pain et al. (1994) utilized the unique high-affinity binding of mAbs for immunomagnetic separation of intracellular hyphae from plant tissue. This approach could be adopted for studies of AM mycelium.

Where the immunochemical techniques recognize structural and surface components, the molecular genetic techniques have their advantage in the recognition of physiologially relevant processes. Liu et al. (1998) recently cloned two high-affinity phosphate transporters from roots of *Medicago truncatula* and were able to show that the transcript levels of these genes are significantly downregulated upon mycorrhization, as the symbiont acquires the phosphate for the plant. For the localization and detection of functionally important molecules in situ, however, the production of mAbs against them is a feasible and sensible approach. This shows that immunochemical techniques will find many more applications in the future.

References

Aldwell FEB, Hall IR, Smith JMB (1985) Enzyme-linked immunosorbent assay as an aid to taxonomy of the Endogonaceae. Trans Br Mycol Soc 84:399–402

Baas R, Kuiper D (1989) Effects of vesicular-arbuscular mycorrhizal infection and phosphate on *Plantago major* ssp. pleiosperma in relation to internal cytokinin concentrations. Physiol Plant 76:211–215

Balestrini R, Hahn MG, Faccio A, Mendgen K, Bonfante P (1996) Differential location of carbohydrate epitopes in plant cell walls in the presence and absence of arbuscular mycorrhizal fungi. Plant Physiol 111: 203–213

Beyrle H, Penningsfeld F, Hock B (1991) The role of nitrogen concentration in determining the outcome of the interaction between *Dactylorhiza incarnata* (L.) Soó and *Rhizoctonia* sp. New Phytol 117:665–672

Cordier C, Gianinazzi-Pearson V, Gianinazzi S (1994) Immunodétection de champignons endomycorhiziens à arbuscules in planta. Acta Bot Gall 141:465–468

Danneberg G, Latus C, Zimmer W, Hundeshagen B, Schneider-Poetsch H, Bothe H (1993) Influence of vesicular-arbuscular mycorrhiza on phytohormone balances in maize (*Zea mays* L.). J Plant Physiol 141:33–39

David R, Itzhaki H, Ginzberg I, Gafni Y, Galili G, Kapulnik Y (1998) Suppression of tobacco basic chitinase gene expression in response to the colonization by the arbuscular mycorrhizal fungus *Glomus intraradices*. Mol Plant Microbe Interact 11:489–497

Dazzo FB, Wright SF (1996) Production of anti-microbial antibodies and their use in immunofluorescence microscopy. Mol Microb Ecol Manual 4.1.2:1–27

Dewey FM, Evans D, Coleman J, Priestley R, Hull R, Horsley D, Hawes C (1991) Antibodies in plant science. Acta Bot Neerl 40:1–27

Dewey FM, Thornton CR, Gilligan CA (1997) Use of monoclonal antibodies to detect, quantify and visualize fungi in soils. Adv Bot Res 24:275–308

Druege U, Schönbeck F (1993) Effect of vesicular-arbuscular mycorrhizal infection on transpiration, photosynthesis and growth of flax (*Linum usitatissimum* L.) in relation to cytokinin levels. J Plant Physiol 141:40–48

Esch H, Hundeshagen B, Schneider-Poetsch H, Bothe H (1994) Demonstration of abscisic acid in spores and hyphae of the arbuscular mycorrhizal fungus *Glomus* and in the N_2-fixing cyanobacterium *Anabaena variabilis*. Plant Sci (Limerick) 99:9–16

Friese CF, Allen MF (1991) Tracking the fates of exotic and local VA mycorrhizal fungi: methods and patterns. Agric Ecosyst Environ 34:87–96

Göbel C, Hahn A, Hock B (1995) Production of polyclonal and monoclonal antibodies against hyphae from arbuscular mycorrhizal fungi. Crit Rev Biotechnol 15:293–304

Göbel C, Hahn A, Giersch T, Hock B (1998) Monoclonal antibodies for the identification of arbuscular mycorrhizal fungi. In: Varma A (ed) Mycorrhiza manual. Springer, Berlin Heidelberg Now York, pp 271–287

Goicoechea N, Antolin MC, Sanchez-Diaz M (1997) Gas exchange is relatd to the hormone balance in mycorrhizal or nitrogen-fixing alfalfa subjected to drought. Physiol Plant 100:989–997

Gollote A, Gianinazzi-Pearson V, Gianinazzi S (1995) Immunodetection of infection thread glycoprotein and arabinogalactan protein in wild type *Pisum satiuum* (L.) or an isogenic mycorrhiza-resistant mutant interacting with *Glomus mosseae*. Symbiosis 18:69–85

Hahn A, Bonfante P, Horn K, Pausch F, Hock B (1993) Production of monoclonal antibodies against surface antigens of spores from arbuscular-mycorrhizal fungi by an improved immunization and screening procedure. Mycorrhiza 4:69–78

Hahn A, Göbel C, Hock B (1998) Polyclonal antibodies for the detection of arbuscular mycorrhizal fungi. In:

Varma A (ed) Mycorrhiza manual. Springer, Berlin Heidelberg New York, pp 255–270

Harlow E, Lane D (1988) Antibodies. A laboratory manual. Cold Spring Harbor Laboratory Press, Cold Spring Harbor

Hoffmann E, Wallenda T, Schaeffer C, Hampp R (1997) Cyclic AMP, a possible regulator of glycolysis in the ectomycorrhizal fungus Amanita muscaria. New Phytol 137:351–356

Karabaghli C, FreyKlett P, Sotta B, Bonnet M, LeTacon F (1998) In vitro effect of Laccaria bicolor S238N and Pseudomonas fluorescens strain BBc6 on rooting of de-rooted shoot hypocotyls of Norway spruce. Tree Physiol 18:103–111

Köhler G, Milstein C (1975) Continuous cultures of fused cells secreting antibody of predefined specificity. Nature 256:495–497

Kough J, Malajczuk N, Linderman RG (1983) Use of the indirect immunofluorescent technique to study the vesicular-arbuscular mycorrhizal fungus Glomus epigaeum and other Glomus species. New Phytol 94:57–62

Kraigher H, Hanke DE (1996) Cytokinis in Norway spruce seedlings and forest soil pollution Phyton (Horn) 36:57–60

Kraigher H, Grayling A, Wang TL, Hanke DE (1991) Cytokinin production by two ectomycorrhizal fungi in liquid culture. Phytochemistry 30:2249–2254

Liebmann S, Hock B (1989) Auxin concentrations in roots of spruce during in vitro ectomycorrhizal synthesis. Plant Physiol (Life Sci Adv) 8:99–104

Liu H, Trieu AT, Blaylock LA, Harrison MJ (1998) Cloning and characterization of two phosphate transporters from Medicago truncatula roots: regulation in response to phosphate and to colonization by arbuscular mycorrhizal (AM) fungi. Mol Plant-Microbe Interact 11:14–22

Martin F, Laurent P, De-Carvalho D, Burgess T, Murphy P, Nehls U, Tagu D (1995) Fungal gene expression during ectomycorrhiza formation. Can J Bot 73 (Suppl 1): S541–S547

Mernaugh RL, Mernaugh GR, Kovacs GR (1990) The immune response: antigens, antibodies, antigen-antibody interactions. In: Hampton R, Ball E, De Boer S (eds) Serological methods for the detection of viral and bacterial plant pathogens. APS Press, St Paul, pp 3–14

Morton JB (1998) Classification of Glomales. INVAM homepage. http://invam.caf.wvu.edu/classification.html

Morton JB, Bentivenga SP (1994) Levels of diversity in endomycorrhizal fungi (Glomales, Zygomycetes) and their role in defining taxonomic and non-taxonomic groups. Plant Soil 159:47–59

Oramas-Shirey M, Morton JB (1990) Immunological stability among different geographic isolates of the arbuscular mycorrhizal fungus Glomus occultum. Abstr 90th ASM:311

Pain NA, Green JR, Gammie F, O'Connell RJ (1994) Immunomagnetic isolation of viable intracellular hyphae of Colletotrichum lindemuthianum (Sacc. and Magn.) Briosi and Cav. from infected bean leaves using a monoclonal antibody. New Phytol 127:223–232

Perotto S, Malavasi F, Butcher GW (1992) Use of monoclonal antibodies to study mycorrhiza: present applications and perspectives. In: Norris JR, Read DJ,

Varma AK (eds) Methods in microbiology 24. Academic Press, London, pp 221–248

Perotto S, Brewin NJ, Bonfante P (1994) Colonization of pea roots by the mycorrhizal fungus Glomus versiforme and by Rhizobium bacteria: immunological comparison using monoclonal antibodies as probes for plant cell surface components. Mol Plant Microb Interact 7:91–98

Perotto S, Peretto R, Faccio A, Schubert A, Varma A, Bonfante P (1995) Ericoid mycorrhizal fungi: cellular and molecular bases of their interactions with the host plant. Can J Bot 73 (Suppl 1):S557–568

Peterson RL, Bonfane P, Faccio A, Uetake Y (1996) The interface between fungal hyphae and orchid protocorm cells. Can J Bot 74:1861–1870

Plattner I, Hall IR (1995) Parasitism of non-host plants by the mycorrhizal fungus Tuber melanosporum. Mycol Res 99:1367–1370

Sanders IR, Ravolanirina F, Gianinazzi-Pearson V, Gianinazzi S, Lemoine MC (1992) Detection of specific antigens in the vesicular-arbuscular mycorrhizal fungi Gigaspora margarita and Acaulospora laevis using polyclonal antibodies to soluble spore fractions. Mycol Res 96:477–480

Schmidt EL, Bankole RO (1962) Detection of Aspergillus flavus in soil by immunofluorescent staining. Science 136:776–777

Schmidt EL, Biesbrock JA, Bohlool BB, Marx DH (1974) Study of mycorrhizas by means of fluorescent antibody. Can J Microbiol 20:137–139

Schulze P, Bahnweg G (1998) Critical review of identification techniques for Armillaria spp. and Heterobasidion annosum root and buttrot diseases. J Phytopathol 146:61–72

Straker CJ, Gianinazzi-Pearson V, Gianninazzi S, Cleyet-Marel JC, Bousquet N (1989) Electrophoretic and immunological studies on acid phosphatase from a mycorrhizal fungus of Erica hispidula L. New Phytol 111:215–221

Thingstrup I, Rozycka M, Jeffries P, Rosendahl S, Dodd JC (1995) Detection of the arbuscular mycorrhizal fungus Scutellospora heterogama within roots using polyclonal antisera. Mycol Res 99:1225–1232

Tijssen P (1985) Practice and theory of enzyme immunoassays. Laboratory techniques in biochemistry and molecular biology, vol 15. Elsevier, Amsterdam

Tisserant B, Brenac V, Requena N, Jeffries P, Dodd JC (1998) The detection of Glomus spp. (arbuscular mycorrhizal fungi) forming mycorrhizas in three plants, at different stages of seedling development, using mycorrhiza-specific isozymes. New Phytol 138:225–239

Van Tuinen D, Jaquot E, Zhao B, Gollotte A, Gianinazzi-Pearson V (1998) Characterization of root colonization profiles by a microcosm community of arbuscular mycorrhizal fungi using 25S rDNA-targeted nested PCR. Mol Ecol 7:879–887

Wilson JM, Trinick MJ, Parker CA (1983) The identification of vesicular-arbuscular mycorrhizal fungi using immunofluorescence. Soil Biol Biochem 15:439–445

Wright SF, Morton JB (1989) Detection of vesicular-arbuscular mycorrhizal fungus colonization of roots by using a dot-immunoblot assay. Appl Environ Microbiol 55:761–763

Wright SF, Upadhyaya A (1996) Extraction of an abundant and unusual protein from soil and comparison with hyphal protein of arbuscular mycorrhizal fungi. Soil Sci 161:575–586

Wright SF, Upadhyaya A (1999) Quantification of arbuscular mycorrhizal fungi by the concentration on hyphal traps. Mycorrhiza 8:283–285

Wright SF, Morton JB, Sworobuk JE (1987) Identification of a vesicular-arbuscular mycorrhizal fungus by using monoclonal antibodies in an enzyme-linked immunosorbent assay. Appl Environ Microbiol 53:2222–2225

Wright SF, Franke-Snyder M, Morton JB, Upadhyaya A (1996) Time-course study and partial characterization of a protein on hyphae of arbuscular mycorrhizal fungi during active colonization of roots. Plant Soil 181: 193–203

Wright SF, Upadhyaya A, Buyer JB (1998) Comparison of N-linked oligosaccharides of glomalin from arbuscular mycorrhizal fungi and soils by capillary electrophoresis. Soil Biol Biochem 30:1853–1857

Wyss P, Mellor RB, Wiemken A (1990) Vesicular-arbuscular mycorrhizas of wild-type soybean and non-nodulating mutants with *Glomus mosseae* contain symbiosis-specific polypeptides (mycorrhizins), immunologically cross-reactive with nodulins. Planta 182:22–26

Zollfrank U, Sautter C, Hock B (1987) Fluorescence immunohistochemical detection of *Armillaria* and *Heterobasidion* in Norway spruce. Eur J For Pathol 17:230–237

4 At the Interface Between Mycorrhizal Fungi and Plants: the Structural Organization of Cell Wall, Plasma Membrane and Cytoskeleton

P. Bonfante

CONTENTS

I. Introduction

Mycorrhizal fungi are a heterogeneous group of soil fungi that colonize the roots of about 240 000 plant species in nearly all terrestrial ecosystems to form symbiotic associations called mycorrhizas. Mycorrhizas have long been held to be an essential feature of the biology and ecology of most terrestrial plants, since they influence the growth of plants, their water and nutrient absorption, and protect them from root diseases. A remarkable review of current knowledge of these topics has been provided by Smith and Read (1997), who demonstrate how mycorrhizas play a central role in nutrient cycles, in soil stability and in plant health. In the past 10 years a wealth of experimental investigations, together with the development of new technologies, has led to substantial advances in the knowledge of mycorrhizal func-

tioning, mostly in the field of cellular and molecular biology. Here, progress can be very fast, and important achievements have been made in mycorrhizal research as reviewed by Bonfante and Perotto (1995), Gianinazzi-Pearson (1996), Harrison (1999), Martin et al. (1997, 1999), and illustrated in several chapters of this Volume. These papers have reviewed the molecular mechanisms underlying mycorrhizal development. Attention has been mainly directed to the molecular basis of early morphogenetic events, analysis of differential gene expression in the symbionts and signals exchanged between the partners. Other fruitful fields of research have been the comparison between legume/rhizobium symbiosis and arbuscular mycorrhizas (Gianinazzi-Pearson and Denarie 1997; Hirsh and Kapulnik 1998), and the more general aspects of plant-microorganism interactions, leading to the identification of host defence reactions against mycorrhizal fungi (Gianinazzi-Pearson et al. 1996).

This chapter offers a synopsis of the literature on the interactions between mycorrhizal fungi and their host cells and its aim is to demonstrate how a good knowledge of cellular and molecular interactions can provide a unifying key to interpret the many scenarios that lie behind the terms endomycorrhiza, where fungi develop intracellular structures in root cells, or ectomycorrhiza, in which hyphae develop around and among the root cells (Fig. 1). Molecular probes aimed to identify fungal symbionts have demonstrated that the same fungal species may form ecto- or endostructures depending on its host (Zelmer and Currah 1995): the capacity of a fungal symbiont to establish one type of association or the other seems therefore wider than described in the past.

The establishment of compatible contacts between the fungus and the root, their crucial role in the deposition of new molecules, nutrient transfer and signal exchanges will be described. The analysis of cellular interactions identifies cell walls, membranes and cytoskeleton of both

Dipartimento di Biologia Vegetale dell'Università di Torino – Centro di Studio sulla Micologia del Terreno del CNR, Viale Mattioli 25, 10125 Turin, Italy

The Mycota IX
Fungal Associations
Hock (Ed.)
© Springer-Verlag Berlin Heidelberg 2001

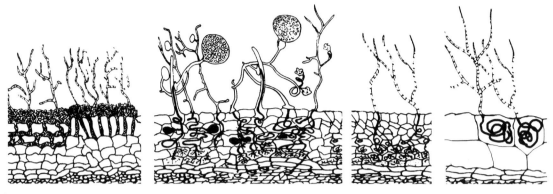

Fig. 1. Diagram of the different mycorrhizal types and of the main colonization structures. *From the left* Ectomycorrhizas, vesicular-arbuscular mycorrhizas, orchid and ericoid mycorrhizas

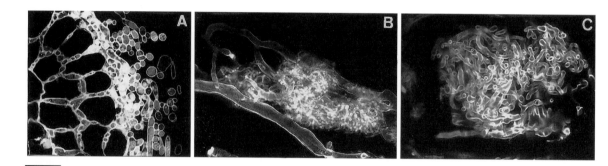

Fig. 2A,B. Micrographs showing the fungal mantle and the hartig net in a *Quercus* ectomycorrhiza (**A**); an arbuscule in the cortical cell of a leek root (**B**); a coil produced by an ericoid fungus in a hair root of *Calluna vulgare* (**C**). *Bar* 50, 16, 15 µm, respectively

partners as important areas of research. In plant cell biology, this area is currently defined as the *cytoskeleton-membrane-wall continuum* and regarded as a crucial compartment for plant cell development (Miller et al. 1997a,b). The major insights derived from cellular, immunobiochemical and molecular studies on the fungal-root cell interface will be summarized and discussed in the light of a more advanced conceptual framework. Gaps in our current knowledge will also be highlighted.

II. Overview of the Colonization Process in Mycorrhizas

The impressive biodiversity of the plant and fungal taxa involved in mycorrhizal symbiosis has resulted in their anatomical description in many hosts since the start of this century (see Smith and

Smith 1997 for references). The main aspects of mycorrhizal colonization are here briefly summarized only to provide the structural background for further discussion. Mycorrhizal fungi produce presymbiotic structures that can also be developed independently of the host plant and consist of propagules (sexual or asexual spores) and hyphae. Fungal hyphae are of paramount importance for survival, completion of the life cycle and mobilization of nutrients, and are sometimes organized in highly structured bundles and rhizomorphs. When they move to the symbiotic phase, mycorrhizal fungi adopt various colonization patterns and different cellular interactions with the host (Fig. 2). Endomycorrhizal fungi penetrate the host root cells and show a great variety of infection patterns. Among them, Glomalean fungi are highly dependent on their hosts and cannot survive for long in their absence. They produce infection units that involve several cells of different root tissues: extraradical hyphae originating from germinating spores form appressoria on the epidermal cells, penetrate the cortical tissue with coils and intercellular hyphae which spread the infection and eventually form intracellular, highly branched structures called arbuscules (Fig. 2B) (Bonfante 1984). The *Arum* and *Paris* terminology first used by Gallaud at the start of the century was proposed again by Smith and Smith (1997) to better differentiate among colonization patterns. However, host cell types and not only plant genotype influence the colonization patterns: the fungus usually produces coils and does not branch in epidermis and exodermis, while it forms arbuscules exclusively in cortical cells (Bonfante and Genre 2000).

Ericoid mycorrhizal fungi also colonize the root cells of their host, but they do so via simple hyphal structures that do not change their morphology during the presymbiotic and the intraradical symbiotic phase. Each hypha produces an infection unit involving a single host cell: after the production of an ill-defined appressorium, the hypha penetrates into the epidermal cell and produces a coil without spreading to the neighboring root cells (Fig. 2C) (Perotto et al. 1995). In orchid mycorrhiza, intracellular coils in cortical cells are also produced by Basidiomycetes during both the protocorm and root colonization, although the infection unit comprises in this case a larger number of host cells (Peterson et al. 1998). Collapsed fungal clumps are seen in the inner part of the protocorm or root tissues.

Ectomycorrhizal fungi do not penetrate the host cell wall, and complete their colonization process through two major events. Having contacted the surface of the root and developed inside its mucilage, hyphae aggregate to produce a tissue-like structure (the mantle) covering the root surface. Then, they progress into the apoplastic region around epidermal (angiosperms) and cortical (gymnosperms) cells where they display a labyrinthine growth and form an extracellular hyphal network within the root tissues, termed the Hartig net (Fig. 2A). Other mycorrhizal morphologies include the ectendomycorrhiza, the arbutoid and monotropoid mycorrhizas, all with distinct structural attributes (Smith and Read 1997). Colonization of mycoheterotrophic plants is of particular ecological interest (Leake 1994), but these types of plant-fungal associations have not been subjected to detailed cellular analysis.

III. The Interface Is a Unifying Feature of the Cellular Interactions in Mycorrhizas

The cell walls of higher plants form a unique extracellular matrix that controls growth and development in the absence of cell migration (Reiter 1998). Similarly, the fungal wall enables fungi to assume a variety of shapes, and protects the protoplast from a range of environmental stresses (Gooday 1995). The notion that cell walls are deeply involved in the molecular talk among plant cells in the same or different tissues, and between plant and microbial cells in heterologous communication, has encouraged investigation of the role of cell walls in mycorrhizas (Bonfante and Scannerini 1992; Tagu and Martin 1996; Martin et al. 1997). Plant-fungal communication requires walls which are permeable to molecules exchanged between the partners. Since plasmodesmata do not occur between plant and fungi, the only working pathway would be across the apoplastic material that separates the plant and fungal plasma membranes, consisting of the fungal wall, the plant wall and/or interfacial material. This compartment, defined by the physical contact between the plant and the fungal cell and including the respective plasma membranes, has been termed interface. Different types of interface have been described in detail by Scannerini and Bonfante (1983), Smith and Smith (1990), and Bonfante and Perotto (1995). Two main types are produced in mycorrhizas. These are known as intracellular and intercellular interfaces, depending on whether or not the fungus penetrates the host cell. However, both types are always composed of the membranes of both partners, separated by an apoplastic region (Fig. 3). Their fundamental role is assumed to be that of allowing a two-way exchange of signal molecules and nutrients. The type of compounds transferred and the transfer mechanisms depend on the type of mycorrhiza and the biological characteristics (transport protein) of the plasma membrane.

Many reports have demonstrated that cytoskeletal proteins are crucial in plant/fungal associations (Kobayashi et al. 1995; McLusky et al. 1999). Due to the close interactions between the plasma membrane and proteins of the cytoskeleton (Miller et al. 1997a,b), these latter will be regarded as components of the interface in mycorrhizas.

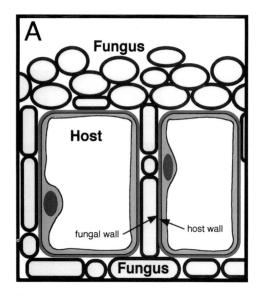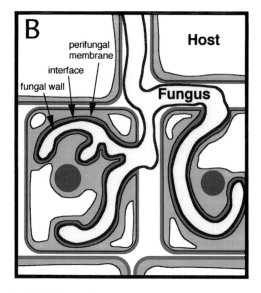

Fig. 3A,B. Diagram of the two major interfaces described in mycorrhizas. **A** Intercellular interfaces (fungal wall/host wall) are found both in ecto- and endomycorrhizas. **B** Intracellular interfaces (fungal wall/interface compartment/perifungal host membrane) are typical of all endomycorrhizal associations

IV. Wall Molecules of Fungal Origin at the Interface

When a hypha comes in touch with the surface of the host plant, a contact area between them is readily established. Molecules typical of the fungal wall (Herrera 1992) have been consistently located by affinity techniques in contacting hyphae, irrespective of their colonization pattern. Chitin and glucans with β-1,3 or β-1,4 linkages have been revealed by specific lectins or antibodies in a range of arbuscular, ericoid, orchid and ectomycorrhizal fungi (Bonfante 1988; Bonfante et al. 1990, 1998; Gollotte et al. 1996, 1997; Peterson et al. 1998).

Whereas β-1,3 glucans have been consistently located in asco- and basidiomycetous symbionts, they have only been detected in some Glomales; e.g., in *Glomus* but not in *Gigaspora* (see Gollotte et al. 1996, 1997 for references). In addition, striking changes in cell wall architecture occur during the life cycle of Glomales and are mirrored by modifications in composition (Bonfante 1988). The main change is in chitin architecture: this structural component of the fungal wall occurs in the spores as arched fibrils with a helicoidal organization, but appears as parallel fibrils in the hyphae, and is found as shorter chains in the arbuscule (Figs. 4, 5). This is accompanied by a decrease in the abundance of β-1,3 glucans, when present, and by a general thinning of the wall concomitant to the progress of root colonization. The arbuscular branches display the thinnest and most simplified wall (Bonfante 1988).

Chitin contributes to the structural rigidity of the fungal wall, but may, in the oligomeric form, be also regarded as a signal molecule that induces host defence responses in plant-fungus interactions, including ectomycorrhizas (Salzer et al. 1997). A chitin-like molecule is also present in the Nod factor, the molecule which triggers nodule morphogenesis in the plant-rhizobium interaction (Long 1996). Plant defence reactions are weak and transient in arbuscular mycorrhizal plants (Gianinazzi-Pearson et al. 1996) and mostly localized in arbuscule-infected cells. The presence of short chitin chains in the thin wall of arbuscules (Fig. 4B) may suggest that this structure is a potential source of elicitors. In the other fungal structures, the longer chitin molecules which are generally found are also masked by different wall components, thus decreasing the potential eliciting effect.

Because of the structural and elicitor functions of chitin, the identification of genes involved in chitin synthesis could help in understanding both the development of mycorrhizal fungi and their relationships with the host. Chitin synthase genes (*chs*) have been investigated by Lanfranco et al. (1998), who demonstrated that a large family of *chs* belonging to different classes is present in the genome of AM, and ectomycorrhizal and

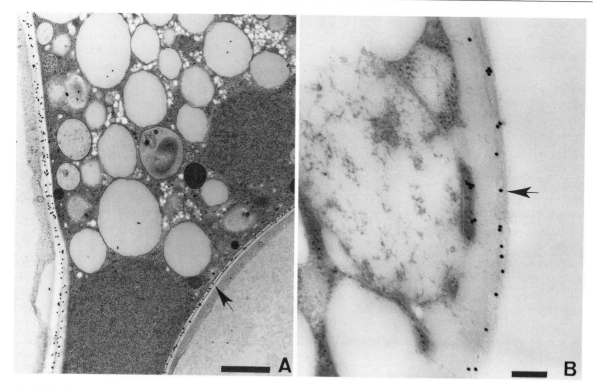

Fig. 4A,B. Localization of chitin in the wall of AM fungi by using wheat germ agglutinin bound to colloidal gold granules (*arrows*). **A** Intracellular hypha of *Gigaspora mar-garita*. **B** Thin arbuscule branch of *Glomus versiforme*. Wall is very thin, but chitin is still present. *Bars* 0.6 and 0.2 µm, respectively

ericoid fungi. For example, *Gvchs*3 from *Glomus versiforme* codes for a predicted protein of 1194 amino acids, with a very close similarity to class IV chitin synthases, while class II and IV *chs* were found in *Gigaspora margarita*. RT-PCR experiments demonstrated that the latter two genes were expressed in mycorrhizal roots, but not during spore germination (Lanfranco et al. 1999). These experiments suggest that the expression of specific *chs* may be regulated during the life cycle of AM fungi, some of them being required solely during the interactions with the host. Specific roles of *chs* in growth and morphogenesis have been already demonstrated in yeasts and filamentous fungi (Borgia et al. 1996).

Striking changes in wall morphology during the colonization process have not been observed in other endo- and ectomycorrhizal fungi. In these fungi, the thickness of the hyphal wall also remains more or less constant during both the saprotrophic/presymbiotic and the symbiotic phase. However, biochemical and molecular analysis of ectomycorrhizal fungi has led to the identification of specific cell wall proteins (Martin et al. 1997, 1999; Tagu and Martin 1996). A novel

class of cell wall polypeptides, defined as symbiosis-regulated acidic polypetides (SRAP) of 31–32-kDa, has been found in *Pisolithus tinctorius* (Laurent et al. 1999). SRAP 32d and its related polypeptides are present on the hyphal surface, mostly when the fungus is attached to the eucalyptus root surface or root hairs. The protein contains the Arg-Gly-Asp motif found in cell-adhesion proteins. Identification of the corresponding cDNA clone (*PtSRAP32*) has allowed analysis of the transcripts by Northern blot analysis. There is a drastic upregulation in its expression 3 days after the contact between the hyphae and the root surface, showing that major changes in fungal wall composition occur during mycorrhizal development.

Hydrophobins are fungal proteins widely distributed in the Mycota kingdom (Wessels 1996), where they play important roles in fungal morphogenesis and pathogenesis. They are secreted in the substrate or inside the wall and produce an amphipathic film at the hydrophilic-hydrophobic interface: this assembly is responsible in vivo for the formation of hydrophobic structures. Two types of hydrophobin mRNAs are produced and highly accumulated by *P. tinctorius* during the formation of ectomycorrhizas (Tagu et al. 1996). Preliminary immunolocalization experiments with antibodies produced against one recombinant hydrophobin polypeptide confirm that the protein is located in the fungal wall: it is present in the hyphae contacting the root surface, but is mostly expressed in the mantle or in fruit bodies where hyphae tightly aggregate (unpubl. results).

One of the most abundant proteins in the fruitbody of *Tuber borchii* (Tb11.9) is a cell wall protein encoded by the gene *tb11.9*. Northern blot analysis revealed that *tb11.9* is highly expressed in immature and mature fruitbodies and was not detectable in cultured mycelium or ectomycorrhizal roots. Immunogold experiments show that the encoded protein may be located in the hyphal wall, but not in ascus or sporal wall (De Bellis et al. 1998).

Taken as a whole, these data indicate that cell walls undergo drastic changes during the life cycle of mycorrhizal fungi, even in the absence of visible changes in wall morphology. These changes may be of great importance in the establishment of mycorrhizas (Martin et al. 1999), and a clearer picture will be obtained with the investigation of related proteins in other mycorrhizal systems and the analysis of their functional significance.

V. The Organization of the Cell Walls in the Early Contacts

Plant cell walls are highly heterogenous in structure and composition (Carpita and Gibeaut 1993). Current models indicate that microheterogeneity also occurs in tissues and individual cells and it is best illustrated by the use of specific antibodies to locate matrix components. Several authors have therefore investigated the occurrence of specific and local changes in the plant cell wall caused by the contact with mycorrhizal fungi.

Unlike pathogenic fungi, mycorrhizal symbionts cause limited modifications to their host's wall at their first contact point. A good example is given by ericoid mycorrhizas. The Ericaceae have a peculiar root anatomy, and the epidermal cells of their very fine hair roots increase in size and separate from each other. The epidermal cell wall is thickened and rich in cellulose organized with a fibrillar helicoidal structure (Perotto et al. 1995). During root colonization, ericoid fungi first develop inside the mucilage material covering the epidermal cells, then attach firmly to the root surface thanks to an extracellular coat where glycoproteins and enzymic activities have been detected (Perotto et al. 1995). Lastly, they produce infection points and cross the thick epidermal wall, but avoid excessive damage to the target cell. In the presence of nonhost plants, they cross the host walls by producing evident damages, including the breakdown of the membrane and necrosis.

The contacts of AM fungi are much more complex than those of other endomycorrhizal fungi and in some cases produce local changes in the host wall. An increasing number of observations suggest that complex morphogenetic events occur in the fungus before host colonization (Bago et al. 1998; Giovannetti et al. 1996; Nagahashi and Douds 1997). After spore germination, the fungus may perceive host-derived signals and produce highly branched hyphae and appressoria within 36 h. However, information on the contacts between appressoria and root epidermal cells is scanty (Bonfante and Genre 2000). In leek and maize, wall thickening is produced probably as a consequence of mechanical pressure caused by *G. versiforme*, whose appressoria are conspicuous structures with thick chitinous walls (Fig. 5). Appressoria produce intercellular penetration hyphae, which grow between the epidermal maize cells, causing their separation. A fungal peg crosses the anticlinal wall at the base of the epidermal cell, initiating intracellular colonisation. In addition to mechanical pressure, the penetration strategy of the epidermal host cells probably involves the release of hydrolytic fungal enzymes (Bonfante and Perotto 1995).

An increase in cell wall thickness characterizes the contact area of AM fungi with mutant pea plants resistant to mycorrhizal infection (Gianinazzi-Pearson 1996). Most of them block fungal development on the root surface at the appressorium level (Myc[-1] phenotype) (Gianinazzi-Pearson 1996). It is not clear whether the wall thickening observed in some plants species and in these mutants derives from a similar plant response. Cytochemical characterization of this type of interface is, in fact, limited to mutant plants. Their ultrastructural analysis has shown local callose and phenol depositions on the epidermal walls in contact with the appressoria of *Glomus mosseae* (Gollotte et al. 1993). This demonstrates that AM fungi may evoke a defence response, as confirmed by upregulation of

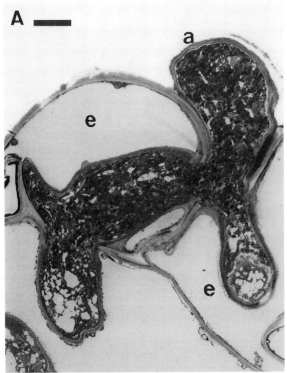

Fig. 5A,B. *Glomus versiforme* and maize roots: electron micrographs. **A** An appressorium (*a*) produces intracellular hyphae which colonize the epidermal layer (*e*). **B** Detail of the appressorium wall: chitin is organized in highly regular fibrils (*arrows*). *Bars* 4 and 0.1 μm, respectively

defence-related genes during the early interactions between the endosymbiont and these mutants (Ruiz-Losano et al. 1999). Cell wall thickenings have also been found in nonmycorrhizal *Lotus* mutants (Wegel et al. 1998) in the presence of *Gigaspora margarita*. Here, the fungus colonizes the epidermal cells which suffered from degradation and accumulation of electron-dense material, indicating cell death. The underlying cortical cells, which were never colonized, exhibited a highly localized wall thickening, rich in esterified and nonesterified pectins (P. Bonfante, J. Webb, M. Parniske, unpubl. observ.).

Hyphae of some ectomycorrhizal fungi (*Pisolithus, Suillus, Laccaria*) cause the breakdown of a thin electron-dense layer lining the cell wall of the root cap cells, which represents a specialized cell wall domain (Kottke 1997; Bonfante et al. 1998). Further, hyphae develop between this layer and make firm contact with the underlying part of the host wall. At this stage, the fungal wall of *S. collinitus* seems to loosen in texture, even though chitin and glucans – the major skeletal molecules in many higher fungi – are still detected (Fig. 6). Interestingly, ultrastructural observations revealed that an *S. collinitus* strain isolated from a Mediterranean area caused changes indicative of a defence reaction in the roots of *Pinus nigra* which grows typically in Alpine areas. Similarly to the pea Myc[-1] mutants, here, too, an amorphous wall thickening containing β-1,3 glucans was detected (Bonfante et al. 1998). Ectomycorrhizal fungi are known to release elicitors of defence reactions in cell cultures (Salzer et al. 1997), though this activity is reduced in planta. Experiments on the elicitors produced by *Hebeloma crustuliniforme* allowed Salzer et al. (1997) to demonstrate that only some reach the plant receptors and initiate hypersensitive responses in the host cells, whereas others are degraded.

Taken as a whole, these results demonstrate that physical contacts between the root surface and the extraradical hyphae depend on both the type of mycorrhiza and the compatibility between plant and fungal genomes. Mycorrhizal fungi generate elicitors of plant defence reactions, leading to typical wall defence reaction in the early stages of noncompatible interactions. By contrast, in fully compatible interactions such responses may be repressed directly by suppressors, or indirectly by the expression of putative plant symbiosis-related genes (Gollotte et al. 1993).

Deeper in the root, the contact with both ecto- and endomycorrhizal fungi causes only limited

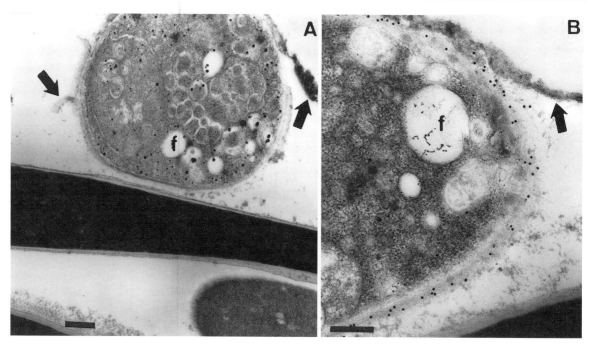

Fig. 6A,B. Rootlets of *Pinus nigra* inoculated with *Suillus collinitus* strain J3-15-31. Electron micrographs of the tip region. **A** First contact between the hyphal wall and the electron dense layer lining the wall of a cap cell (*arrows*).

B The fungus (*f*) develops in the space between the electron-dense layer (*arrow*) and the host cell wall. The fungal wall is labeled by the polyclonal antibody to locate glucans. *Bars* 0.2 μm

changes to the plant cell walls. Limited thinning as well as a more evident fibrillar texture have been described, but without any substantial change in the wall composition (Gea et al. 1994).

It thus seems that, when the barrier of the outer root layers is overcome, the pathway for the colonizing fungus becomes very smooth until the formation of arbuscules, which represents another major checkpoint during fungal colonization. This is also suggested by the analysis of the *Lotus japonicus* mutants described by Wegel et al. (1998), where fungal infection is blocked in the epidermal layer. Here, the AM fungus can occasionally overcome the plant response, then developing a normal colonization pattern.

VI. Intracellular Interfaces: From the Location of Plant Cell Wall Molecules to the Genetic Analysis of Their Biosynthesis

When root cells are colonized by endomycorrhizal fungi that form either intracellular coils, hyphae

or arbuscules, the host plasma membrane invaginates and proliferates around the developing fungus (Fig. 3B). Apoplastic material is laid down between this membrane and the fungal cell surface, and a new interface compartment is created. This is one of the most important events in the successful colonization of plant cells by endomycorrhizal fungi. Irrespective of the mycorrhizal type (orchid, ericoid, arbuscular) the compartment is always composed of the membranes of both partners, separated by apoplastic material (Smith and Smith 1990; Bonfante and Scannerini 1992; Bonfante and Perotto 1995). In high pressure/freeze-substituted samples, the interface is narrow (about 80–100 nm), the host membrane being close to the fungal wall (Fig. 7A). Cellular and molecular probes have led to a better definition of this thin but complex subcellular compartment (Fig. 8).

The interfacial material surrounding intracellular hyphae is topologically continuous with the host wall, but its morphology and composition change during development of the intracellular structures. In AMs, the material is electron-dense at the penetration point; it thins around the arbuscular branches and thickens again around col-

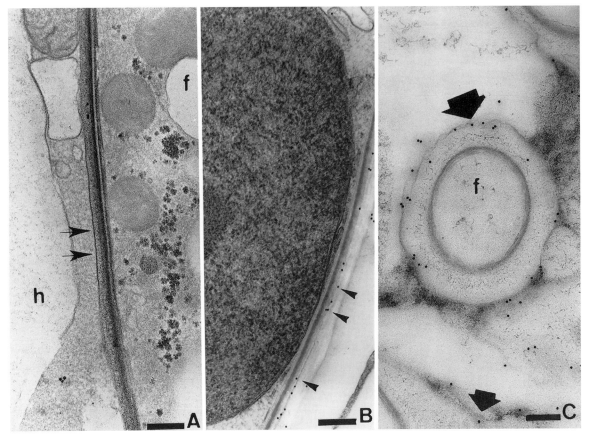

Fig. 7A–C. Ultrastructural organization of the interface compartment in AMs. **A** In a clover root prepared by high pressure freeze substitution, the host (*h*) membrane (*arrow*) surrounding the fungus (*f*) is smooth (*arrow*), the interface material is electron-dense after periodic acid-thiocarbohydrazide-silver proteinate treatment, and the fungal wall is very thin. **B** Pectin residues are present in the interface as revealed by the monoclonal antibody JIM 5, which binds to homogalacturonans (*arrows*). **C** Arabino-galactan proteins are detected on the perifungal membrane in addition to the plasma membrane (*arrowheads*) by the monoclonal antibody CCRC-M7. *Bars* 0.25 μm, 0.4 μm and 0.2 μm, respectively

lapsed branches. In situ use of enzymes, lectins, or antibodies has shown that this compartment is a zone of high molecular complexity (Bonfante and Perotto 1995; Gollotte et al. 1996, 1997). Molecules common to the plant primary wall, such as cellulose, β-1,4 glucans, polygalacturonans, hemicelluloses such as xyloglucans, proteins rich in hydroxyproline (HRGP) and arabinogalactan-proteins have been located on many mycorrhizal plant walls (Figs. 7B,C; 8). In maize, β-1,4 glucans were detected around the large intracellular coils of *G. versiforme*, but not around the thinner arbuscular branches (Balestrini et al. 1994). (β-1,3) (β-1,4) glucans were regularly present on the same maize walls, but never detected in the interface (unpubl. observ.), while β-1,3 glucans were only occasionally revealed (Balestrini et al. 1994; Gollotte et al. 1996). These findings suggest that

morphological changes of the interfacial material are mirrored by changes in its composition and are developmentally regulated. In particular, some matrix components display a highly heterogeneous location, as illustrated by the distribution of xyloglucans and arabinogalactans. Four mycorrhizal plants (leek, maize, clover and tobacco) were investigated with two monoclonal antibodies generated against rhamnogalacturonan I, specific for a terminal fucosyl-containing epitope (CCRC-M1), and for an arabinosylated β-(1,6)- galactan epitope (CCRC-M7) (Balestrini et al. 1996). A specific tissue distribution of immunogold label was found in the root of the different plant species: for example, CCRC-M1 mostly labeled epidermal and hypodermal cells in leek, while CCRC-M7 mainly labeled cortical cells of maize. In the presence of the AM fungus, additional label

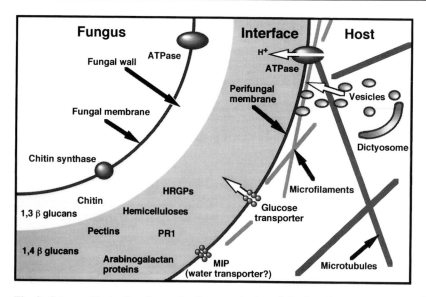

Fig. 8. Scheme illustrating the molecular complexity of the interface compartment, listing the structures involved and many of the molecules so far determined

was found in the interface around the intracellular hyphae, but in the different plant species the pattern mirrored the highly specific distribution of the epitopes in the respective host cell walls (Fig. 8C). The conclusion was drawn that the two probes recognize epitopes whose distribution is highly heterogeneous in a single organ, and that this heterogeneity is maintained when a new deposition of wall molecules occurs following fungal colonization.

Current models assume that plant cell walls consist of three interwoven domains, one network of cellulose and hemicellulose, another of heterogenous pectins, and a third of proteins (Carpita and Gibeaut 1993). The presence in the plant-fungus interface of many molecules typical of the primary plant cell wall suggests that the host perifungal membrane (described in the next section) retains the enzymatic machinery involved in both synthesis (i.e., cellulose) and secretion (i.e., pectins, hemicellulose, HRGPs) of cell wall material. The morphology of the interface material indicates, however, that these components are not assembled into a fully structured wall. This may be the result either of a weak lytic activity by the fungus, which impairs assembly mechanisms by producing pectic enzymes, or of the use of plant sugars as a food source, as suggested by the absence of starch in colonized cells (Bonfante and Perotto 1995 and references).

Molecular biology techniques have been used to investigate the genesis of the interfacial material. The presence of HRGPs in mycorrhizal roots is of particular interest because they are considered to be both developmental and stress markers. They are strongly expressed in the meristem and during the formation of lateral and adventitious roots (Vera et al. 1994), but their expression also increases after pathogenic infection or wounding (Esquerrè-Tugayè et al. 1979). HRGPs have been located in the interface area in maize mycorrhizal roots (Balestrini et al. 1994), and one of the questions there was whether the HRGP located around the fungus is a protein of new synthesis, or is derived from previously laid down cell wall protein. In situ hybridization experiments clearly demonstrate that expression of the HRGP gene is limited to specific cells containing coiled hyphae and arbuscules (Balestrini et al. 1997). In mycorrhizal roots, these wall proteins may be involved in the reprogramming of infected root cells and deposition of the interfacial material. However, a potential role of such proteins in localized plant defence cannot be excluded as other typical components of plant defence, such as PR 1 proteins, have been sometimes detected at the arbuscule interface (Gollotte et al. 1997).

Arabinogalactan proteins (AGPs) are cell wall and membrane proteins that play a crucial role in plant development (Ding and Zhu 1997). They also occur in the interface of many plant/fungal combinations. A recombinant library enriched for mycorrhiza-specific expressed sequences has allowed van Buuren et al. (1999) to identify a cDNA clone encoding a putative AGP. The gene shows only little expression in *Medicago* roots prior to symbiosis, but is induced during mycor-rhizal development. In situ hybridization demon-strates that the transcripts are mostly located in the arbuscule-containing cells, thus indicating a new synthetic activity in mycorrhizal cells at the inter-face area and corroborating previous data (Fig. 8) obtained by immunolocalization experiments (Perotto et al. 1994; Balestrini et al. 1996; Gollotte et al. 1997).

Deposition of cell wall material requires the combined activities of both polysaccharide syn-thases and lytic enzymes. Xyloglucan endotransg-lycolases (XETs) cleave xyloglucan chains and attach the cut ends to new xyloglucan acceptors (Fry et al. 1992). Two XET genes have been isolated from *Medicago truncatula* (Maldonado-Mendoza and Harrison 1998), one being expressed only in mycorrhizal roots. The authors propose that the gene product may be involved either in facilitating hyphae penetration by allowing local-ized cell wall loosening or in modifying the struc-ture of xyloglucans in the interface compartment. Lytic and transglycosylation events during wall deposition may be facilitated by the interface pH that is becoming more acidic owing to an H^+-ATPase activity (Smith and Smith 1990, see next section).

These new findings demonstrate that intracel-lular interfaces are the result of dynamic events (upregulation of some proteins by activation of new genes, changes in structure and composition of the interface material). These processes can probably be extended to the other endomycor-rhizal associations, though much less information is available.

The nature of the interfacial material in orchid mycorrhizas depends on the stage of inter-action between the fungal hyphae and protocorm cells. Hyphae at early stages of colonization and development of pelotons are surrounded by a matrix lacking pectins, cellulose and β-1,3 glucans, as shown by affinity gold methods (Peterson et al. 1996). By contrast, degenerating and collapsed pelotons are surrounded by an interfacial matrix again containing pectins, cellulose and β-1,3 glucans, which are also found in and around the clumps of degenerated hyphae. The absence of several polysaccharides around developing pelo-tons may have different explanations. The fungal species associated with orchid protocorms are effective saprotrophs and can use complex cell wall polysaccharides after the release of cellulases, pectinases and other hydrolytic enzymes. Alterna-tive explanations are that hyphae may be produc-

ing inhibitors of polysaccharide synthesis, or the host plasma membrane may be expanding so rapidly during peloton formation that it is not con-figured for wall synthesis (Peterson et al. 1996). Similar hypotheses can be formulated for ericoid mycorrhizas, where many of the probes used in the other mycorrhizal systems failed to reveal plant wall molecules. Glucans with β-1,3 and β-1,4 linkages are easily located on the fungal or host walls, but they are not detected on the interface material (Perotto et al. 1995).

These investigations raise some novel ques-tions concerning the significance of the interfacial material laid down between plant and fungal membranes in mycorrhizas. Its presence seems to be a unifying feature. However, its composition is strongly dependent on both the plant host and the nutritional strategy of the fungus, since its com-plexity increases when moving from mycorrhizal fungi with saprotrophic capabilities towards the obligate biotrophic AM fungi.

VII. The Interface as a Working Compartment: the Role of Membranes

It is assumed that the fundamental role of plant-fungal interfaces is to allow a two-way exchange of signal molecules and nutrients (Smith and Smith 1990). The fungus can be seen as a sink for organic molecules derived from plant photosynthesis and as a source of mineral nutrients. The chemical nature of the molecules transferred and the mecha-nisms involved are largely hypothetical. Biochemi-cal and molecular methods are, however, offering new approaches to study the basis of nutrient trans-port in mycorrhiza. Harrison's group was the first to investigate by PCR the molecular components of a functioning mycorrhiza. They identified a cDNA coding for a transmembrane phosphate trans-porter from the mycorrhizal fungus *G. versiforme* (Harrison and van Buuren 1995). This cDNA could complement a yeast mutant affected in phosphate uptake. An important discovery was that in *G. ver-siforme* the corresponding gene is transcribed only in the external hyphae, which can thus be regarded as the initial site of P uptake from the soil. How-ever, the molecular mechanisms of its release to the plant are not clear.

Genes connected with plant P metabolism have been identified Rosewarne et al. (1999). A cDNA sharing some homology with a tomato gene inducible by phosphate starvation was identified by differential screening in *Medicago truncatula* (Burleigh and Harrison 1997). Expression was negatively influenced by the presence of *G. versiforme* and mRNA level decreased during mycorrhizal development. A decrease was also observed when the interaction was not successful, as in Myc(−) mutants of *M. sativa*. Interestingly, this expression was strongly affected by P availability, since transcripts were present only in roots grown under P-deficient conditions. AM fungi depend on plant sugar sources for their growth; glucose is preferentially taken up by the fungus and then metabolized to trehalose and mobile lipids (Pfeffer et al. 1999). The molecular basis of this uptake is not known, while it is clear that mycorrhizal plants express transmembrane proteins that transport glucose and fructose (Harrison 1996) and thus make them available to the fungus. The expression of one of the corresponding genes is regulated in response to colonization: transcript levels increased two- to fourfold in *M. truncatula* following colonization, but they did not during the unsuccessful interaction with the *M. sativa* Myc(−) mutant. In situ hybridization suggests that the transcripts were located in the most colonized areas of the root (Harrison 1996).

If these proteins are involved in active sugar transport, other membrane proteins may be involved in passive transport. Roussel et al. (1997) detected a membrane protein that belongs to the large family of membrane intrinsic proteins (MIPs). This family comprises aquaporins, that mediate the transport of water, and channels for amino acids or small peptides. Water uptake from the interface compartment may be crucial for maintenance of the osmotic pressure. The breakdown of this equilibrium could be one reason why arbuscules collapse.

Molecular approaches have also been successful in ectomycorrhizas, where the biochemical sugar exchange pathways were already known (Smith and Read 1997). Nehls et al. (1998) used a PCR approach to identify a monosaccharide transporter from *Amanita muscaria*. Its gene expression was enhanced up to fourfold in symbiotic roots, as well as in the mycelium grown at elevated glucose concentration, suggesting that monosaccharides act as signal molecules regulating gene expression.

Mechanisms of active transport that work against a concentration gradient require energy. The proton driving force created by H^+-ATPase activity could provide the energy required for

P loading from the interface towards the plant cell or for the sugar loading from the root cells to the fungus (Smith and Smith 1990). Different approaches have confirmed that ATPase genes are expressed in mycorrhizal roots. In barley, mycorrhizal roots show increased level of transcripts highly homologous to H^+-ATPase genes from several plants (Murphy et al. 1997). In addition, transgenic plants in which the H^+-ATPase gene promoter has been fused to the GUS gene show a strong GUS expression in colonized cells (Gianinazzi-Pearson 1996).

Of course, the fungus also requires energy for active transport. Five partial gene sequences with high homology to H^+-ATPase from other fungi, e.g., yeasts, were discovered in *Glomus mosseae* (Ferrol et al. 2000). It suggests that these components may play an important role in the bidirectional nutrient exchange, although further studies are needed on gene expression.

It is becoming clear that the interface is a working area where the key events of both ecto- and endomycorrhizas take place. However, many questions are still open: how and where is P transferred across the fungal membrane? Is in situ hybridization the best way to locate the site of such events? How does the extensive biogenesis of the host membranes occur, especially in endomycorrhizas? By contrast with peribacteroid membranes in legume nodules, we do not know anything about perifungal membrane biogenesis. In a nodule cell, there is almost 30 times more peribacteroid than plasma membrane (Verma and Fortin 1989) suggesting that massive membrane synthesis is needed to house the symbiont (Cheon et al. 1993). The mechanisms by which membrane proliferation is initiated, or by which vesicle transport towards the infection thread or the peribacteroid membrane occurs, are still unknown. Cheon et al. (1993) have shown the importance in plant membrane synthesis of plant homologues of Rab1p and Rab7p, small GTP-binding proteins involved in vesicular transport in animal cells.

VIII. The Cytoskeleton as a Cellular Scaffolding

The way in which signals at the cell surface are propagated via signal transduction pathways is a major subject of current investigation (Miller et al.

1997a). It is suggested that the cytoskeleton acts as an integrated lattice that physically connects the nucleus with cytosolic components and with the cell surface (Miller et al. 1997a; Baluska et al. 1998). Some specialized cells (for example root hairs) are excellent models in this respect (Miller et al. 1997b), since they show that polar growth is the result of specific organization of actin and tubulin proteins, location of spectrin-like antigen, changes in cell wall texture, and incorporation of exocitic vesicles into the plasma membrane.

Cytoskeletal elements provide a network of tracks and highways along which molecules and organelles move around the cell (Kobayashi et al. 1995). This function is of particular importance during biotrophic plant/fungus interactions, which require reorganization of the infected cell as well as the development of a contact area for uni- or bidirectional signaling and nutrient exchanges. The deep impact of mycorrhizal fungi on the root cells, from elongation of epidermal cells in ecto-mycorrhizas to the impressive re-building of cortical cells in endomycorrhizas, has promoted investigation on the role of cytoskeleton in such events (Bonfante and Perotto 1995). During the development of *Pisolithus* on eucalyptus roots, Carnero Diaz et al. (1996) observed a substantial increase in the expression of α-tubulin genes in the host, mostly in connection with the development of lateral roots. Variation in the steady-state levels of tubulin isoforms is observed when *Suillus* develops its mantle on pine roots (Niini et al. 1996). The morphological changes in the microfilament pattern of *Suillus* following depolymerization treatments have led Raudaskoski et al. (1998) to suggest that the altered morphology of the hyphae may result from actin reorientation, mediated by protein kinases which could be regulated by root signals.

As previously discussed, establishment of an intracellular interface is only one sign of successful colonization by endomycorrhizal fungi. Other morphological changes involve fragmentation of the central vacuole and movement of the nucleus and other organelles towards the fungal branches. These processes recall events where the cytoskeleton is involved. Tobacco plants transformed with the promoter of an α-tubulin gene fused to the bacterial GUS gene have shown that, in roots, the gene is expressed only in colonized cells and in the meristem (Bonfante et al. 1996). We therefore looked to see whether this new transcription was mirrored by a different cytoskeletal organization.

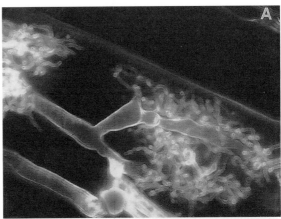

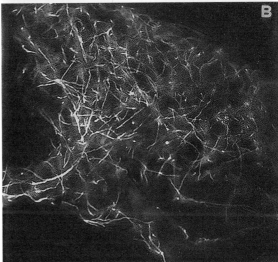

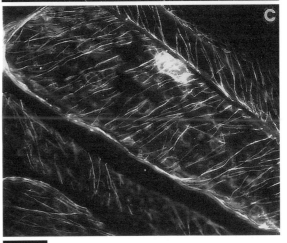

Fig. 9A–C. Microtubule organization in tobacco cell in the presence and the absence of a mycorrhizal fungus. **A** Arbuscule of *Gigaspora margarita*, after WGA-FITC treatment. **B** A tridimensional network of microtubules crosses the arbuscule branches. **C** Microtubules in an uninfected cortical cells. Their pattern is the result of the cortical helical organization, found in the differentiating cells. *Bar* 20 μm

An increase in complexity of the microtubule (MT) arrays was observed in infected cells (Genre and Bonfante 1997), where MTs ran along large intracellular hyphae, connected hyphae with each other and with the nucleus (Fig. 9). Substantial changes were also found in the actin microfilaments: they closely followed the fungal branches and enveloped the whole arbuscule in a dense coating network, supporting the idea that actin proteins are closely related to the perifungal membrane (Genre and Bonfante 1998). Interestingly, these data agree with the observations of Uetake and Peterson (1997a,b), who were the first to investigate cytoskeletal rearrangement in orchid cells following fungal colonization. In addition, antibody against γ-tubulin revealed microtubule organizing centers (MTOC) along the nuclear envelope and along the host membrane that surrounds the plant/fungus interface, while clathrin was observed along the peripheral and perifungal membranes, as well as along a tubular system of endomembranes (Genre and Bonfante 1999).

The similarities between the systems so far investigated indicate that mycorrhizal fungi activate different tubulin and actin genes in their host plants. The new proteins are dynamically organized in new arrays that follow the fungal structures and are therefore responsible for the new make up of the colonized cell.

However, one basic question is still open: is cytoskeleton reorientation the direct consequence of penetration by the fungus, or is this mediated by a more complex mechanism involving signaling systems from both plant and fungus? Cytoskeleton reorients in response to many stimuli, including light quality, plant growth regulators, gravity, pressure, electric fields and several biotic factors (Wymer and Lloyd 1996). Dynamic reorganization of cytoskeletal elements occurs when pathogenic fungi attack their plant hosts (Kobayashi et al. 1995; McLusky et al. 1999). Actin microfilaments are required for the expression of nonhost resistance in higher plants through the polarization of organelles or defense-related compounds at fungal penetration points, and it has been suggested that proteins associated with the cytoskeleton play an important part in signal transduction pathways (Heath 1997; Kobayashi et al. 1997).

More direct evidence of the response of cytoskeleton to signals has been provided for rhizobium/legume symbiosis. Cardenas et al. (1998) have clearly demonstrated that actin microfilaments rearrange in root hairs following the exposure to rhizobium nodulation signals.

Changes in the actin cytoskeleton were observed as soon as 5–10 min after application of the Nod factor. The long filamentous actin became fragmented into short filaments, maybe due to increased calcium. Actin rearrangement could be usefully employed to discover whether signal molecules released by mycorrhizal fungi – such as the chitinous elicitors identified in ectomycorrhizal fungi (Salzer et al. 1997) – induce a visible rerrangement of the cytoskeleton. Myc(–) mutants provide another experimental system with which to look for signal molecules and determine their effect on the plant cytoskeleton.

The cytoskeleton can thus be convincingly postulated as a part of the interface, and as a scaffolding that first bridges the cytoplasm with the membrane/wall complex in both partners, and then accommodates a two-way exchange of information between the cellular exterior and the nucleus. It could also be of structural significance in endomycorrhizas, where the development of new membranes and the deposition of wall molecules require a new assembly. Actin, together with other membrane-bound molecules, may be the first to respond to signaling from the fungus. In a similar way, signals released by the plant may stimulate reorganization in the fungus cytoskeleton and produce the different morphogenetic events previously discussed.

IX. Conclusions

The concept of interface is both useful and meaningful. The impressive biodiversity of mycorrhizas may be reduced to a few types of plant/fungal contacts. The structures involved (cytoskeleton, membranes, cell walls) are highly dynamic cell compartments which change their organization and composition depending on the mycorrhizal association and the stage of interaction.

A combination of in situ, molecular and biochemical investigations has proved very useful in the analysis of specific compartments in the mycorrhizal tissues. This multiple approach allows us to describe how plant and fungi interact at cellular and molecular level and raises new questions on how the partners perceive and respond to signals. The use of plant and fungal mutants, and the setup of fungal transformation, will provide us with powerful tools to dissect such complex events and to establish a function for the genes so far identified.

Acknowledgments. I wish to thank all the colleagues who allowed me to use unpublished results; Silvia Perotto for critical reading of the manuscript; Andrea Genre for his skilful computer assistance in the preparation of the plates; and Antonella Faccio for her help in the reference list. Investigation was supported by the Consiglio Nazionale delle Ricerche, by the Ministero della Ricerca Scientifica (MURST-PRSRN) and by the following European projects: ECC Biotechnology Project (IMPACT Project, Contract No BIO2-CT93-0053); ECC AIR Project (MYCOMED, Contract N. AIR2-CT94-1149).

References

Bago B, Azcon-Aguilar C, Goulet A, Piché Y (1998) Branched absorbing structures (RAS): a feature of the extraradical mycelium of symbiotic arbuscular mycorrhizal fungi. New Phytol 139:375–388

Balestrini R, Romera C, Puigdomenech P, Bonfante P (1994) Location of a cell wall hydroxyproline-rich glycoprotein, cellulose and β-1,3 glucans in apical and differentiated regions of maize mycorrhizal roots. Planta 195:201–209

Balestrini R, Hahn M, Faccio A, Mendgen K, Bonfante P (1996) Differential localization of carbohydrate epitopes in plant cell walls in the presence and absence of arbuscular mycorrhizal fungi. Plant Physiol 111:203–213

Balestrini R, Josè-Estanyol M, Puigdomènech P, Bonfante P (1997) Hydroxyproline-rich glycoprotein mRNA accumulation in maize root cells colonized by an arbuscular mycorrhizal fungus as revealed by in situ hybridization. Protoplasma 198:36–42

Baluska F, Barlow PW, Lichtscheidl KI, Volkmann D (1998) The plant cell body: a cytoskeletal tool for cellular development and morphogenesis. Protoplasma 202:1–10

Bonfante P (1984) Anatomy and morphology of VA mycorrhizae. In: Powell CL, Bagyaraj DJ (eds) VA mycorrhizas. CRC Press, Boca Raton, FL, pp 5–33

Bonfante P (1988) The role of the cell wall as a signal in mycorrhizal associations. In: Scannerini S, Smith D, Bonfante P, Gianinazzi-Pearson V (eds) Cell to cell signals in plant, animal and microbial symbiosis. Springer Berlin Heidelberg New York, pp 219–235

Bonfante P, Genre A (2000) Outside and inside the roots: AM fungi differently interact with epidermal and cortical cells of their host plants. In: de Wit PJGM, Bisseling T, Stiekema WS (eds) 2000 IC-MPMI Proceedings: Biology of Plant-Microbe Interactions, vol 2. (in press)

Bonfante P, Perotto S (1995) Strategies of arbuscular mycorrhizal fungi when infecting host plants. Tansley Review No. 82. New Phytol 130:3–21

Bonfante P, Scannerini S (1992) The cellular basis of plant-fungus interchanges in mycorrhizal associations. In: Allen MG (ed) Functioning mycorrhizae. Chapman & Hall, London, pp 65–101

Bonfante P, Faccio A, Perotto S, Schubert A (1990) Correlation between chitin distribution and cell wall morphology in the mycorrhizal fungus *Glomus versiforme*. Mycol Res 94:157–165

Bonfante P, Bergero R, Uribe X, Romera C, Rigau J, Puigdomenech P (1996) Transcriptional activation of a maize α-tubulin gene in mycorrhizal maize and transgenic tobacco plants. Plant J 9:737–743

Bonfante P, Balestrini R, Martino E, Perotto S, Plassard C, Mousain D (1998) Morphological analysis of early contacts between pine roots and two ectomycorrhizal *Suillus* strains. Mycorrhiza 8:1–10

Borgia PT, Iartchouk N, Riggle PJ, Winter KR, Koltin Y, Bulawa CE (1996) The *chsB* gene of *Aspergillus nidulans* is necessary for normal hyphal growth and development. Fungal Genet Biol 20:193–203

Burleigh SH, Harrison MJ (1997) A novel gene whose expression in *Medicago truncatula* roots is suppressed in response to colonization by vesicular-arbuscular mycorrhizal (VAM) fungi and to phosphate nutrition. Plant Mol Biol 34:199–208

Cardenas L, Vidali L, Dominguez J, Pérez H, Sanchez F, Hepler PK, Quinto C (1998) Rearrangement of actin microfilaments in plant root hair responding to *Rhizobium etli* nodulation signals. Plant Physiol 116:871–877

Carnero-Diaz E, Martin F, Tagu D (1996) Eucalypt α-tubulin: cDNA cloning and increased level of transcripts in ectomycorrhizal root system. Plant Mol Biol 31:905–910

Carpita NC, Gibeaut DM (1993) Structural models of primary cell walls in flowering plants: consistency of molecular structure with the physical properties of the walls during growth. Plant J 3:1–30

Cheon C, Lee NG, Siddique ARM, Bal AK, Verma DPS (1993) Roles of plant homologs of Rab1p and Rab7p in the biogenesis of peribacteroid membrane, a subcellular compartment formed de novo during root nodule symbiosis. EMBO 11:4125–4135

De Bellis R, Agostini D, Piccoli G, Vallorani L, Potenza L, Polidori E, Sisti D, Amoresano A, Pucci P, Arpaia G, Macino G, Balestrini R, Bonfante P, Stocchi V (1998) The *tb11.9* Gene from the white truffle *Tuber borchii* encodes for a structural cell wall protein specifically expressed in fruitbody. Fungal Gen Biol 25:87–99

Ding L, Zhu JK (1997) A role for arabinogalactan-proteins in root epidermal cell expansion. Planta 203:289–294

Esquerrè-Tugaye MT, Lafitte C, Mazau D, Toppan, Touze A (1979) Cell surfaces in plant-microorganisms interactions. II. Evidence for the accumulation of hydroxyproline-rich glycoproteins in the cell wall of diseased plants as a defence mechanism. Plant Physiol 64:320–326

Ferrol N, Barea JM, Azcón-Aguilar C (2000) The plasme membrene H(+)-ATPase gene family in the arbuscular mycorrhizal fungus *Glomus mosseae*. Curr Genet 37:112–118

Fry SC, Smith RC, Renwick KF (1992) Xyloglucan endo-trasnglycosylase, a new wall-loosening enzyme activity from plants. Biochem J 282:821–828

Gea L, Normand L, Vian B, Gay G (1994) Structural aspects of ectomycorrhiza of *Pinus pinaster* (Ait.) Sol. formed by an IAA-overproducer mutant of *Hebeloma cylindrosporum* Romagnési. New Phytol 128:659–670

Genre A, Bonfante P (1997) A mycorrhizal fungus changes microtubule orientation in tobacco root cells. Protoplasma 199:30–38

Genre A, Bonfante P (1998) Actin versus tubulin configuration in arbuscule-containing cells from mycorrhizal tobacco roots. New Phytol 140(4):745–752

Genre A, Bonfante P (1999) Cytoskeleton-related proteins in tobacco mycorrhizal cells: γ-tubulin and clathrin localisation. Eur J Histochem 43:105–111

Gianinazzi-Pearson V (1996) Plant cell responses to arbuscular mycorrhizal fungi: getting to the roots of the symbiosis. Plant Cell 8:871–1883

Gianinazzi-Pearson V, Denarie J (1997) Red carpet genetic programmes for root endosymbioses. Trends Plant Sci 2:371–372

Gianinazzi-Pearson V, Dumas-Gaudot E, Gollotte A, Tahi-Alaoui A, Gianinazzi S (1996) Cellular and molecular defence-related root responses to invasion by arbuscular mycorrhizal fungi. New Phytol 133:45–47

Giovannetti M, Sbrana C, Citernesi AS, Avio L (1996) Analysis of factors involved in fungal recognition responses to host-derived signals by arbuscular mycorrhizal fungi. New Phytol 133:65–71

Gollotte A, Gianinazzi-Pearson V, Giovannetti M, Sbrana C, Avio L, Gianinazzi S (1993) Cellular localization and cytochemical probing of resistance reactions to arbuscular mycorrhizal fungi in a "locus a" myc-mutant of Pisum sativum L. Planta 191:112–122

Gollotte A, Lemoine MC, Gianinazzi-Pearson V (1996) Morphofunctional integration and cellular compatibility between endomycorrhizal symbionts. In: Mukerji KG (ed) Concepts in mycorrhizal research. Kluwer, Dordrecht, pp 91–111

Gollotte A, Cordier C, Lemoine MC, Gianinazzi-Pearson V (1997) Role of fungal wall components in interactions between endomycorrhizal symbionts. In: Schenk HEA, Herrmann R, Jeon KW, Muller NE, Schwemmler W (eds) Eukaryotism and symbiosis. intertaxonomic combination versus symbiotic adaptation. Springer, Berlin Heidelberg New York, pp 412–427

Gooday GW (1995) Cell walls. In: Gow NAR, Gadd GM (eds) The growing fungus. Chapman and Hall, London, pp 43–62

Harrison MJ (1996) A sugar transporter from Medicago truncatula: altered expression pattern in roots during vesicular-arbuscular (VA) mycorrhizal associations. Plant J 9:491–503

Harrison MJ (1999) Molecular and cellular aspects of the arbuscular mycorrhizal symbiosis. Annu Rev Plant Physiol Plant Mol Biol 50:361–389

Harrison MJ, van Buuren ML (1995) A phosphate transporter from the mycorrhizal fungus Glomus versiforme. Nature 378:626–629

Heath CM (1997) Signalling between pathogenic rust fungi and resistant or susceptible host plants. Ann Bot 80:713–720

Herrera JR (1992) Fungal cell wall: structure, synthesis and assembly. CRC Press, Boca Raton

Hirsch AM, Kapulnik Y (1998) Signal trasduction pathways in mycorrhizal associations: comparisons with the Rhizobium-legume symbiosis. Fungal Gen Biol 23: 205–212

Kobayashi I, Murdoch LJ, Kunoh H, Hardham AR (1995) Cell biology of early events in the plant resistance response to infection by pathogenic fungi. Can J Bot 73 (Suppl 1):S418–S425

Kobayashi Y, Kobayashi I, Funaki Y, Fujimoto S, Takemoto T, Kunoh H (1997) Dynamic reorganization of microfilaments and microtubules is necessary for the expression of non-host resistance in barley coleoptile cells. Plant J 11:525–537

Kottke I (1997) Fungal adhesion pad formation and penetration of zoot cuticle in early stage mycorrhizas of Picea abies and Laccaria amethystea. Protoplasma 196:55–64

Lanfranco L, Garnero L, Bonfante P (1998) Chitin synthase genes in the arbuscular mycorrhizal fungus Glomus versiforme: full sequence of a gene encoding a class IV chitin synthase. FEMS Microbiol Lett 170:59–67

Lanfranco L, Vallino M, Bonfante P (1999) Expression of chitin synthase genes in the arbuscular mycorrhizal fungus Gigaspora margarita. New Phytol 142:347–354

Laurent P, Tagu D, De Carvalho D, Nehls U, De Bellis R, Balestrini R, Bauw G, Inz D, Bonfante P, Martin F (1999) A novel class of cell wall polypeptides in Pisolithus tinctorius contain a cell-adhesion RGD motif and are up-regulated during the development of Eucalyptus globulus ectomycorrhiza. Mol Plant-Micr Inter 12:862–871

Leake JR (1994) The biology of myco-heterotrophic ("saprophytic") plants. New Phytol 127:171–216

Long SR (1996) Rhizobium symbiosis: nod factors in perspectives. Plant Cell 8:1885–1898

Maldonado-Mendoza IE, Harrison MJ (1998) A xyloglucan endo-transglycosylase (XET) gene from Medicago truncatula induced in arbuscular mycorrhizae. Abstr 2nd Int Conf on Mycorrhiza, July 5–10, 1998, Uppsala, Sweden

Martin F, Lapeyrie F, Tagu D (1997) Altered gene expression during ectomycorrhizal development. In: Carroll GC, Tudzynky P (eds) The Mycota V Part A. Springer, Berlin Heidelberg New York, pp 223–242

Martin F, Laurent P, De Carvalho D, Voiblet C, Balestrini R, Bonfante P, Tagu D (1999) Cell wall proteins of the ectomycorrhizal basidiomycete Pisolithus tinctorius: identification, function, and expression in symbiosis. Fungal Genet Biol 27:161–174

McLusky SR, Bennett MH, Beale MH, Lewis MJ, Gaskin P, Mansfield JW (1999) Cell wall alterations and localized accumulation of feruloy-3′-methoxytyramine in onion epidermis at sites of attempted penetration by Botrytis allii are associated with actin polarisation, peroxidase activity and suppression of flavonoid biosynthesis. Plant J 17:523–534

Miller DD, Hable W, Gottwald J, Ellard-Ivey M, Demura T, Lomax T, Carpita N (1997a) Connections: the hard wiring of the plant cell for perception, signaling and response. Plant Cell, 12:2105–2117

Miller DD, Deruijter NCA, Emons AMC (1997b) From signal to form: aspects of the cytoskeleton plant membrane cell wall continuum in root hair tips. J Exp Bot 48:1881–1896

Murphy PJ, Langridge P, Smith SE (1997) Cloning plant genes differentially expressed during colonization of roots of Hordeum vulgare by the vesicular-arbuscular mycorrhizal fungus Glomus intraradices. New Phytol 135:291–301

Nagahashi G, Douds DD Jr (1997) Appressorium formation by AM fungi on isolated cell walls of carrot roots. New Phytol 127:703–709

Nehls U, Wiese J, Guttenberger M, Hampp R (1998) Carbon allocation in ectomycorrhizas: identification and expression analysis of an Amanita musaria monosaccharide transporter. Mol Plant Micr Inter 11:167–176

Niini S, Tarkka T, Raudaskoski M (1996) Tubulin and actin protein patterns in Scots pine (*Pinus sylvestris*) roots and developing ectomycorrhiza with *Suillus bovinus*. Physiol Plant 96:186–192

Perotto S, Brewin NJ, Bonfante P (1994) Colonization of pea roots by the mycorrhizal fungus *Glomus versiforme* and by *Rhizobium* bacteria: immunological comparison using monoclonal antibodies as probes for plant cell surface components. Mol Plant Micr Inter 7:91–98

Perotto S, Peretto R, Faccio A, Schubert A, Varma A, Bonfante P (1995) Ericoid mycorrhizal fungi: cellular and molecular bases of their interactions with the host plant. Can J Bot 73 (Suppl 1):S557–S568

Peterson RL, Bonfante P, Faccio A, Uetake Y (1996) The interface between fungal hyphae and orchid protocorm cells. Can J Bot 74:1861–1870

Peterson RL, Uetake Y, Zelmer C (1998) Fungal symbioses with orchid protocorms. Symbiosis 25:29–55

Pfeffer PE, Douds DD Jr, Bécard G, Shachar-Hill Y (1999) Uptake, transport and metabolism of carbon in arbuscular mycorrhizal carrot roots. Plant Physiol 120:587–598

Raudaskoski M, Niini S, Tarkka M (1998) Regulation of hyphal morphology in *Suillus bovinus* during asymbiotic and symbiotic growth. Abstr 2nd Int Conf on Mycorrhiza, July 5–10, 1998, Uppsala, Sweden

Reiter WD (1998) The molecular analysis of cell wall components. Trends Plant Sci 3:27–32

Rosewarne GM, Barker SJ, Smith SE, Smith FA, Schatchtman D (1999) *A Lycopersicon esculentum* phosphate transporter involved in phosphorus uptake from a vesicular-arbuscular mycorrhizal fungus. New Phytologist 144:507–516

Roussel H, Bruns S, Gianinazzi-Pearson V, Hahlbrock K, Franken F (1997) Induction of a membrane intrinsic protein-encoding mRNA in arbuscular mycorrhiza and elicitor-stimulated cell suspension cultures of parsley. Plant Sci 126:203–210

Ruiz-Lozano JM, Roussel H, Gianinazzi S, Gianinazzi-Pearson V (1999) Defence genes are differentially induced by a mycorrhizal fungus and *Rhizobium* in wild type and symbiosis-defective pea genotypes. Mol Plant Micr Inter 12:976–984

Salzer P, Hubner B, Sirrenberg A, Hager A (1997) Differential effect of purified spruce chitinases and β 1,3-glucanases on the activity of elicitors from ectomycorrhizal fungi. Plant Physiol 114:957–968

Scannerini S, Bonfante P (1983) Comparative ultrastructural analysis of mycorrhizal associations. Can J Bot 61:917–943

Smith FA, Smith SE (1997) Structural diversity in (vesicular)-arbuscular mycorrhizal symbioses. Tansley Rev No 96 New Phytol 137:373–388

Smith SE, Read DJ (1997) Mycorrhizal symbiosis, 2nd edn. Academic Press, London

Smith SE, Smith FA (1990) Structure and function of the interfaces in biotrophic symbioses as they relate to nutrient transport. Tansley Rev No 20 New Phytol 114:1–38

Tagu D, Martin F (1996) Molecular analysis of cell wall proteins expressed during the early steps of ectomycorrhiza development. New Phytol 133:78–85

Tagu D, Nasse B, Martin F (1996) Cloning and characterization of hydrophobins encoding cDNAs from the ectomycorrhizal basidiomycete *Pisolithus tinctorius*. Gene 168:93–97

Uetake Y, Peterson RL (1997a) Changes in microtubule arrays in symbiotic orchid protocorm during fungal colonization and senescence. New Phytol 135:701–709

Uetake Y, Peterson RL (1997b) Changes in actin filaments arrays in protocorm cells of the orchid species, *Spiranthes sinensis*, induced by the symbiotic fungus *Ceratobasidium cornigerum*. Can J Bot 75:1661–1669

van Buuren ML, Trieu AT, Blaylock LA, Harrison MJ (1999) Novel genes induced during an arbuscular mycorrhizal (AM) symbiosis formed between *Medicago truncatula* and *Glomus versiforme*. Mol Plant Micr Inter 12(3):171–181

Vera P, Lamb C, Doerner PW (1994) Cell cycle regulation of hydroxyproline-rich glycoprotein HRGPntt3 gene expression during the initiation of lateral root meristems. Plant J 6:17–727

Verma DPS, Fortin MG (1989) Nodule development and formation of the endosymbiotic compartment. In: Shell J, Vasil IK (eds) Cell culture and somatic cell genetics of plants, vol VII. The molecular biology of nuclear genes. Academic Press, San Diego, pp 329–353

Wegel E, Schauser L, Stougaard J, Parniske M (1998) Mycorrhiza mutants of *Lotus japonicus* define genetically independent steps during symbiotic infection. Mol Plant Micr Inter 11:933–936

Wessels JGH (1996) Fungal hydrophobins: proteins that function at an interface. Trends Plant Sci 1:9–15

Wymer C, Lloyd c (1996) Dynamic microtubules: implications for cell wall patterns. Trends Plant Sci 1(7):222–228

Zelmer CD, Currah RS (1995) Evidence for a fungal liaison between *Corallorhiza trifida* (orchidaceae) and *Pinus contorta* (Pinaceae). Can J Bot 73:862–866

5 Lipids of Mycorrhizae

M. SANCHOLLE[1], Y. DALPÉ[2], and A. GRANDMOUGIN-FERJANI[1]

CONTENTS

[1] Laboratoire Mycologie/Phytopathologie/Environnement (L.M.P.E.), Université du Littoral Côte d'Opale, B.P. 699-62228 Calais Cedex, France
[2] Eastern Cereal and Oilseed Research Centre (ECORC), Wm. Saunders Building, Central Experimental Farm, 960 Carling Avenue, Ottawa, Ontario K1A 0C6, Canada

I. Introduction

Research dealing with the composition, diversity and role of lipids in mycorrhizal associations has, during the past decades, attracted the interest of an increasing number of investigators from diverse disciplines. First considered as a biochemical tool for the quantitative evaluation of mycorrhizae in a given system (Seitz et al. 1979; Salmanowicz and Nylund 1988; Salmanowicz et al. 1990; Schmitz et al. 1991; Davis and Lamar 1992; Nylund and Wallender 1992; Bermingham et al. 1995; Olsson et al. 1995), the composition and the transformation of plant and fungal lipids during the establishment of the symbiosis has been gradually pursued (Olsson et al. 1998) and is now, to a certain extent, considered as a potential biochemical approach for the comprehension of evolutionary (Bentivenga and Morton 1994a; Weete and Gandhi 1997) and chemotaxonomic studies (Sancholle and Dalpé 1993; Bentivenga and Morton 1994b; Graham et al. 1995, Grandmougin-Ferjani et al. 1996, 1999).

The present chapter is the first review that assembles information available on the lipids of mycorrhizal fungi and on the possible implication of these substances in mycorrhizal associations. The fungi and plants involved in mycorrhizal associa-

The Mycota IX
Fungal Associations
Hock (Ed.)
© Springer-Verlag Berlin Heidelberg 2001

tions belong to such diverse classes and groups of organisms that the proposed review on lipids in mycorrhizal fungi and symbiotic associations is presented through research topics instead of in mycorrhizal categories. This organisation avoids repetition and can more easily provide readers with a better overview of the total research activities.

The different aspects covered by this synthesis deal with the lipid composition within the different categories of mycorrhizal fungi, from the ectomycorrhizae fungal organisms to the ericoid, arbuscular and orchid ones. In addition. Actinomycetes are considered. The known qualitative and quantitative transformations that occur during the establishment of mycorrhizae are subdivided as those found within the root systems, as well as within shoots, leaves and seed portions of plants. Carbon cycling and mineral impacts are presented mainly for arbuscular mycorrhizae, for which category most of the research has been done. Fungal lipid changes through germination, mycelial growth and fungal maturation are presented together with the impact of pesticides and other exogenous substances on the lipid metabolism of mycorrhizal fungi. The use of specific sterols and fatty acids for the estimation of mycorrhizal fungi biomass and root colonization is presented and their potential for future research is discussed (Table 1).

II. Lipid Profiles of Mycorrhizal Fungi

A. Ectomycorrhizae

The majority of fungal organisms known for their involvement in ectomycorrhizal symbiosis belong to the classes Ascomycotina and Basidiomycotina with a few representatives within Zygomycotina (Bonfante-Fasolo and Scannerini 1977) and Deuteromycotina (Trappe 1972).

1. Fruiting Bodies

Total lipid contents of dried basidiocarps of various ectomycorrhizal Hymenomycetes range between 7.8 and 21% (Zhuk et al. 1981; Dzamic et al. 1985; Kutaf'eva and Tsapalova 1989; Dembitskii and Pechenkina 1991). The phospholipid fractions of the fruiting bodies of boletes and chanterelles were estimated by Zhuk et al. (1981) as between 8.7 and 25.4%. Phosphatidylcholine

and phosphatidyletholamine were found to be the major constituents of *Laccaria amethystina*, *Lactarius flexuosus* and *Clitocybe inversa* (Dembitskii and Pechenkina 1991).

The total lipid contents of ectomycorrhizal ascocarps ranged from 1.8 to 5% of their dry weight for *Tuber* and *Terfezia* species (Al-Shabibi et al. 1982; Weete et al. 1985a; Bokhary and Parvez 1993), and 1.78% of dried *Morchella conica* var. *costata* fruiting bodies (Karaboz and Oner 1988). Linoleic acid makes up more than 50% of the neutral lipids and about 20% of the polar lipid fraction of *Terfezia claveryi* and *T. hafizi* ascocarps while oleic acid makes up 11–14% of the neutral lipids and 25–34% of the polar ones (Al-Shabibi et al. 1982).

For the majority of ectomycorrhizal fungi, as for saprophytic mushrooms, the major fatty acids belong to the C18 mono and di-unsaturated acids such as the linoleic (C18:2) and oleic (C18:1) acids followed by the palmitic acid (C16:0). Those major components when combined can form up to 90% of the total fatty acids of their fruiting bodies (Melhuish et al. 1972; Wathelet et al. 1972; Morelli et al. 1981; Martin et al. 1984a; Senatore 1988; Senatore et al. 1988; Solberg 1989; Endo et al. 1991); (Table 1). Ergosterol remains the predominant sterol, in most cases, making up more than 50% of the total sterols (De Simone et al. 1979; Cerri et al. 1981; Yokokawa and Mitsuhashi 1981; Senatore et al. 1988; Solberg 1989) (Table 2).

Specific lipid compositions have, however, sometimes been detected within mycorrhizal fungi either from Ascomycetes or Basidiomycetes. Within the Ascomycotina, brassicasterol accounted for 98% of the sterols of a commercial truffle species of the genus *Terfezia*. It has also been found abundantly (26 to 44% of total sterols) in the black truffle *Tuber melanosporum* and also in *T. brumale* (Weete et al. 1985a). Ascocarps of *Morchella esculenta* also contain large amounts of brassicasterol (Yokokawa 1994). Basidiocarps of *Lactarius chrysorheus* synthetize the 6-oxo-decanoic acid, (Hiroi 1978), and those of *Boletus edulis* contain a diacylglycero-homoserine (DGTS) (Vaskovsky et al. 1991). This latter constituent is an unusual polar lipid component for mushrooms and could be of importance for chemotaxonomic recognition of ectomycorrhizae. The polyisoprenoid squalene was detected in *Boletus edulis* and *Lactarius rufus* fruiting bodies (Solberg 1989), 6-keto-stearic acid has been isolated from *Lactarius rufus* and *L. chrysorrheus*

Table 1. Data of the fatty acid profiles of some fructification of ectomycorrhizal fungi

	C12:0	C14:0	C15:0	C16:0	C16:1	C16:2	C17:0	C18:0	C18:1	C18:2	C18:3	C18:X	C24:0	Reference
Amanita porphyria	2.2	1.0		11.0	tr			tr	58.7	30.2		tr		De Kok et al. (1982)
Amanita rubescens		tr[a]	tr	+++	tr	tr	tr	+	+	+++				Melhuish et al. (1972)
Amanita rubescens				10.3	0.4			tr	61.5	27.2		tr		De Kok et al. (1982)
Boletus calopus		0.4		15.6	0.05			8.4	22.1	52	0.8			Senatore (1988)
Boletus edulis	0.6	0.1		9.6	2.0			0.7	9.2	60.7	1.6			Abo and Kurkela (1978)
Boletus regius	0.2	tr		19.5	0.03			4.3	17.9	56.2	1.3			Senatore (1988)
Boletus sardous	0.04	0.03		19.1	0.03			5.2	15.4	57.5	1.5			Senatore et al. (1988)
Cantharellus cibarius (fresh)				10.1	0.5			3.0	13.2	46.7		25.0		De Kok et al. (1982)
Cantharellus cibarius (dry)				6.3				1.2	18.0	41.2		32.3		De Kok et al. (1982)
Cantharellus cibarius			0.8	24.7	0.6		1.6	18.7					46.2	Prostenik et al. (1978)
Cenococcum graniforme				32					37	31				Martin et al. (1984b)
Clitocybe nebularis				15.5				1.9	25.0	57.4				Morelli et al. (1981)
Cortinarius largus morensis	tr	0.05	tr	21.6	1.6			5.4	22.3	48.1	0.03			Senatore (1988)
Dentinum repandum	0.03			16.4	tr		2.6	6.2	19.9	50.6	3.6			Senatore (1988)
Dentinum repandum				15.4	0.3			2.2	31.8	19.9	0.4	12.0		De Kok et al. (1982)
Dentinum repandum				13.0				4.2	37.7	32.8	2.3			Morelli et al. (1981)
Hebeloma crustuliniforme				9					17	74				Martin et al. (1984b)
Hebeloma sarcophyllum		tr	tr	+++	tr	tr	tr	+	+	+++				Melhuish et al. (1972)
Hydnum rufescens		1.2		24.7	7.9			2.0	54.3	6.2				Prostenik et al. (1978)
Hydnum rufescens				15.4	2.4	0.1		0.3	51.7	27.0				De Kok et al. (1982)

Table 1. Continued

	C12:0	C14:0	C15:0	C16:0	C16:1	C16:2	C17:0	C18:0	C18:1	C18:2	C18:3	C18:X	C24:0	Reference
Hygrophoropsis aurantiaca				10.1	0.7	0.1		2.2	56.9	23.7				De Kok et al. (1982)
Hypholoma sublateritium		1.7		24.9	11.4		1.0	3.3	47.9	3.9	0.7			Prostenik et al. (1978)
Lactarius deliciosus				9.9				68.6						Prostenik et al. (1978)
Lactarius rufus	0.03	0.4	0.9	2.6	9.1		0.8	5.4	16.5	33.0	0.4			Aho and Kurkela (1978)
Lactarius vellereus		0.2		7.1	0.3			53.8	21.6	14.3	2.3			Prostenik et al. (1978)
Leccinum quercinum	0.03	1.0	0.02	17.8	0.09			11.9	13	54.1	1.3			Senatore (1988)
Morchella rotunda		7		17					21	54				Wathelet et al. (1972)
Pisolithus tinctorius		tr	tr	+++	tr	tr	tr	+	+	+++				Melhuish et al. (1972)
Russula cyanoxantha	0.05	tr		17.6	1.1			4.8	26.7	48.7	0.03			Senatore et al. (1988)
Russula xerampelina	0.06	tr	0.03	19.7	1.4			4.6	21.1	52.1	tr			Senatore (1988)
Suillus brevipes	1.2	0.5	0.8	10.5	2.3		1.0	0.9	29.9	50.7	0.4			Sumner (1973)
Suillus granulatus	0.02	tr	tr	18.2	0.04			6.7	15.2	58.1	0.09			Senatore et al. (1988)
Suillus granulatus	0.2	0.3	0.4	11.3	3.2		0.4	0.9	16.2	67.1	0.2			Sumner (1973)
Suillus luteus		tr	0.03	tr	16.2	0.06			4.7	19.6	56.3	2.06		Senatore et al. (1988)
Tricholoma acerbum	0.8	0.5	tr	16.2	1.1			4.9	5.8	65.6	6.9			Senatore et al. (1988)
Tricholoma batschii	tr	0.2		23.0	0.5			5.3	7.1	58.5	1.9			Senatore et al. (1988)
Tricholoma pardinum	0.4	0.4		18.2	2.7			10.1	4.0	61.4	1.3			Senatore et al. (1988)
Tricholoma portentosum				8.5	tr			tr	22.8	68.3				De Kok et al. (1982)
Tricholoma terreum		1.7		17.0	1.9			2.7	42.7	27.4	3.9			Prostenik et al. (1978)

[a] Truce amount.

Table 2. Sterol profiles of some fructifications of ectomycorrhizal fungi

	C27 Δ^5	C27 $\Delta^{5,24}$	C28 Δ^5	C28 $\Delta^{5,22}$	C28 $\Delta^{5,7}$	C28 Δ^7	C28 $\Delta^{7,22}$	C28 $\Delta^{5,7,22}$	C28 $\Delta^{5,24(24)}$[1]	C29 Δ^5	C29 $\Delta^{5,22}$	C29 $\Delta^{5,24(24)}$[1]	non id.	Reference
Amanita caesarea	11.0				17.5	4.4	4.4	82.5						2
Armillaria mellea		0.09			11.1	4.4	0.75	80.0						4
Boletus calopus	0.15		tr	0.09	0.9	1.2	3.9	70.1	tr			0.15		1
Boletus luridus					9.8	0.42	0.91	85.0						4
Boletus regius	0.12	0.1			0.56			75.3				0.36		1
Boletus sardous	0.1	0.15	tr	0.16	0.71	0.63	0.71	73.0	0.1			0.14		1
Clitocybe nebularis	0.08	tr[a]						40.3			7.0	42.0		6
Cortinarius largus morensis					0.77	0.9	0.65	71.3				0.47		1
Dentinum repandum	0.19	0.03	0.25	0.7	0.85	0.69	0.59	62.8	0.05		0.02	0.35		1
Hygrocybe punicea					21.2			78.8						2
Lactarius controversus					10.6	10.8	13.1	65.3						5
Lactarius sanguifluus	7.4				19.1	16.6	7.3	56.6						5
Lactarius torminosus	5.8				17.0	13.7	8.3	60.5						5
Leccinum quercinum	0.03		0.14	0.42	0.48	0.79	1.63	59.2		0.11	0.23	0.25		1
Leucopaxillus giganteus	2.9				7.3	6.8		85.9						2
Russula cyanoxantha	0.1		0.07	0.11	1.28	1.45	0.4	63.2		0.05		0.22		1
Russula foetens					29.1			70.9						2
Russula nigricans					16.2			83.8						2
Russula senecis								84.5						2
Russula xerampelina	0.15			0.08	1.12	1.3	0.53	65.5		0.08		0.19		1
Scleroderma aurantium								++						3
Suillus granulatus	0.08			0.8	0.95	0.76		70.2	0.08	0.13		0.18		1
Suillus luteus	0.02				0.83	1.21	1.00	66.2	0.11	0.1		0.11		1
Terfezia sp.				9.8			2.0							8
Tricholoma acerbum	2.3			5.5	13.4	2.7		60.8						7
Tricholoma batschii	1.9			9.1	10.2	2.8		59.3						7
Tricholoma pardinum	1.5			8.2	15.4	2.4		62.2						7
Tuber brumale				33.0				60.0						8
Tuber melanosporum				39.0				61.0						8

C27 D[5]	cholesterol
C27 D[5,24]	desmosterol
C28 D[5]	24-methylcholesterol
C28 D[5,22]	brassicasterol
C28 D[5,7]	22-dihydroergosterol
C28 D[7]	fungisterol
C28 D[7,22]	5-dihydroergosterol
C28 D[5,7,22]	ergosterol
C28 D[5,24(24)1]	24-methylenecholesterol
C29 D[5]	24-ethylcholesterol
C29 D[5,22]	24-ethylcholest 5,22 dienol
C29 D[5,24(24)1]	funcosterol

1 = Senatore et al. (1988)
2 = Yokokawa and Mitsuhashi (1981)
3 = Vrkoc et al. (1976)
4 = De Simone et al. (1979)
5 = Cerri et al. (1981)
6 = Morelli et al. (1981)
7 = Senatore (1988)
8 = Weete et al. (1985a)

[a] Trace amount.

(Turner and Aldridge 1983), and *cis*-polyisoprene has been detected in some *Lactarius* species (Tanaka et al. 1994). Pyrophosphatidic acid was detected in the fructification of several ectomycorrhizal fungi (Sugai et al. 1986). Solberg (1989) published a good review of the lipid constituents of several ectomycorrhizal Agaricales.

2. Mycelium

Within the context of the present study on the lipids involved in mycorrhizae, the lipid profiles of mycelia had particularly caught our attention, as hyphae constitute the basic link between plant and fungal associates. Unfortunately, very little documentation exists on the lipid composition of the mycelium of ectomycorrhizal fungi either when originating from pure cultures or when obtained from mycorrhizal associations.

In *Pisolithus tinctorius*, a well-known and well-studied ectomycorrhizal fungus, lipids from basidiospores and mycelia were suspected to be an important substrate for endogenous respiration (Taber and Taber 1982). Ergosterol was found by Ruzic et al. (1997) as the major sterol of *P. tinctorius*, and its concentration was influenced by low-frequency magnetic fields. Lipid droplets observed in pure *Cenococcum graniforme* and *Hebeloma crustuliniforme* mycelial cultures contained high amounts of triacylglycerol (Martin et al. 1984a). The nature of these reserves varied with fungal species and growth phase (Martin et al. 1984b). Triglyceride concentrations increased with maturity and remained constant until senescence. Lipid mycelium content is estimated at 5 to 10% of the dry weight of the mycelia and considered as a carbon storage strategy (Martin et al. 1984a,b). In a study of the characterization of the different forms of phosphorus in the mycelium of the ectomycorrhizal fungus *Hebeloma cylindrosporum*, Rolin et al. (1984) demonstrated by TLC studies that most of the inorganic phosphorus absorbed by the fungus is dedicated to the synthesis of both DNA and phospholipids.

B. Arbuscular Mycorrhizae

The arbuscular mycorrhizal fungi (AMF) are characterized by thick-walled, large-diameter spores where lipids are the predominant cytoplasmic constituent. Ultrastructural microscopy (Sward 1981a) and biochemical evaluations con-

firmed that lipids constitute almost 50% of the AMF spore dry weight (Beilby 1983; Amijee and Stribley 1987; Schubert et al. 1992). This rich lipid composition led Gaspar et al. (1994) to give them the designation oleaginous fungi. When observed under the microscope, fresh, ungerminated spores and attached hyphae appeared filled with multiple lipid droplets (Sward 1981a) which react positively to lipid staining (Casana and Bonfante-Fasolo 1982). With spore maturity, lipid droplets tended to coalesce into larger globules that exude from the crushed spores in a uniform matrix of oily material.

All the different AMF propagules, either hyphae, vesicles or arbuscules, and also the auxiliary cells of the Gigasporaceae representatives, were found to contain lipids (Nemec 1981; Jabaji-Hare et al. 1984, 1986; Selvaraj and Subramanian 1990). Moreover, AM vesicles in the root intercellular hyphal swellings that were believed to be only storage reserve propagules, were proven, under in vitro culture conditions, to have the potential of regenerating the organisms (Diop et al. 1994). Their lipid composition was evaluated at 58% of their dry weight by Jabaji-Hare et al. (1984). The histochemical characterization of AMF propagules by Nemec (1981) revealed that arbuscular lipids were reactive to phosphoglycerides and glycolipids tests but, when isolated by enzymatic maceration from *Ornithogallum umbellatum* roots, arbuscular hyphae were found unreactive in the presence of lipid reagents (Scannerini and Bonfante-Fasolo 1975; Casana and Bonfante-Fasolo 1982). Liu and Zhou (1987) also detected lipids within arbuscular material but did not specify their nature.

Spore walls of AMF comprised chitinous, polysaccharidic, proteinic (Bonfante-Fasolo and Schubert 1987; Bonfante-Fasolo and Grippiolo 1984) and lipidic material (Sward 1981a) for which the histochemical investigations made by Sward (1981a) suggested a neutral lipid composition. In fact, the neutral lipid fraction largely predominates over polar lipid in either spores, vesicles, or auxiliary cells total lipids (Jabaji-Hare et al. 1984; Jabaji-Hare 1988). Major saturated fatty acids were palmitic (C16:0), stearic acid (C18:0) and monounsaturated fatty acids, which account together for almost 50% of the total fatty acids and are represented by both *cis* and *trans* isomers of the C16:1 and C18:1 fatty acids. Of the monoenes, oleic acid (*cis* C18:1Δ5) dominated in several of the species studied particularly among Gigasporaceae representatives (Beilby 1980; Jabaji-Hare 1988; Sancholle and Dalpé 1993; Graham et al. 1995). Alpha and gamma linolenic acids were found only as trace elements. The C16:1Δ11 hexadecenoic acid, that makes up to 80% of the AMF total spore fatty acids (Nordby et al. 1981; Gaspar et al. 1994; Graham et al. 1995; Grandmougin-Ferjani et al. 1996; Larsen et al. 1998), is mainly confined in the neutral lipid fraction. It seems that this unusual fatty acid is specific to AMF, as it has never been found in any other fungal species.

The fatty acid profiles were demonstrated by Bentivenga and Morton (1994a) to be stable and heritable through several generations of pot culturing and long-term storage, and found not to be affected by the identity of the host plant. These verifications allow fatty acids to be considered as potential taxonomic and possibly systematic tools for the distinction of species. However, the detailed investigations now available on the lipid and fatty acid composition of Glomales (Sancholle and Dalpé 1993; Gaspar et al. 1994; Graham et al. 1995; Grandmougin-Ferjani et al. 1996) permit segregation of species only at the family level.

24-Ethylcholesterol was found to be by far the major sterol (80–85%) of the AMF spore (Beilby 1980; Beilby and Kidby 1980; Grandmougin-Ferjani et al. 1996, 1999) followed by cholesterol (11–15%). 24-methylcholesterol, Δ5-avenasterol and 24-ethylcholesta-5,22-dienol were also detected.

Contradictory results have been published on the effective presence of ergosterol into AMF propagules. Nordby et al. (1981) detected no ergosterol in their analysis of spores in several *Glomus*, species. In contrast, Frey et al. (1992, 1994) identified ergosterol in VAM-infected roots and in the extraradical mycelia. The minute amounts of ergosterol (ng g^{-1} of dry roots) extracted from AMF roots and hyphae remain suspicious due to the fact that experiments were conducted with material obtained from pot cultures susceptible to having been contaminated, even though at low rates, with some other fungal organisms. The higher concentration of ergosterol obtained with AMF colonized roots compared to non-colonized ones may also be proportional to the higher diversity of rhizosphere (fungal) organisms that occupy the neighbourhood of AMF colonized root systems (Brussaard et al. 1997).

Quite recently, α-amyrin, a common vascular plant triterpene, was detected in all the AMF spores species studied (Grandmougin et al. 1999).

The close similarity observed between AMF species sterol composition, as the one found for fatty acid profiles revealed metabolic relationships between these organisms which are currently related taxonomically on the basis of their obligately symbiotic way of life.

C. Actinorhizae

The term actinorhiza is now widely used to designate the non-leguminous nitrogen-fixing interaction between woody plants and Actinomycetes (filamentous bacteria) from the genus *Frankia*. It is well known that fatty acid composition of Actinomycetes differ considerably from other microbial organisms in that they synthetize, out of the usual unsaturated fatty acids, iso-branched, cyclopropane, methyl-branched fatty acids, and also highly complex lipids such as mycolic acid (Asselineau 1966; Kroppenstedt 1985; Kroppenstedt and Kutzner 1978; Lechevalier 1977). Lipid contents of the nitrogen-fixing *Frankia* species have been surveyed mainly for taxonomical purposes and the growth-stimulating effect of the endophyte. Polar lipids from pure cultures of *Frankia* species were reported to contain phosphatidylinositol, diphosphatidylglycerol and phosphatidyl inositol mannosides (Lechevalier et al. 1982; Lopez et al. 1983); iso-hexadecanoic (iso C16:0), palmitic, oleic acids (Lechevalier et al. 1983) and also 17:1Δ8 (Tunlid et al. 1989) represented the major fatty acids. Lipid droplets were observed in *Frankia* hyphae and also in the specialized nitrogen-fixing structures called symbiotic vesicles (Newcomb et al. 1978; Newcomb and Wood 1987). These symbiotic vesicles are encapsulated in a thick laminated envelope made of several lipid monolayers (Torrey and Callaham 1982; Lamont et al. 1988; Harriott et al. 1991) whose lipid composition has not yet been precisely determined.

Comparative analysis between the lipid composition of *Frankia* vegetative cells and nitrogen fixing vesicles showed that symbiotic vesicles contained two to three times less polar lipids than vegetative cells, but two to five times more neutral lipids (Tunlid et al. 1989). The higher proportion of monounsaturated fatty acids (25–75%) and cyclopropane fatty acids (20–23%) in the polar lipid fraction of vesicles compared to that of vegetative cells is considered by Tunlid et al. (1989) as a possible adaptative mechanism to preserve the physical integrity and the physiological function of the vesicle membranes. *Frankia* species, as other Actinomycetes and the great majority of bacteria, are recognized as not able to synthetize their own sterols (Ourisson and Rohmer 1982).

Based on the unique lipid composition of Actinomycetes, two biochemical tools are currently used as qualitative and quantitative indicators of soil Actinomycetes. The methyl branched tuberculostearic acid (10Me 18:0), recognized to occur exclusively in Actinomycetes (Lechevalier 1977; Kroppenstedt 1985) and the phospholipid fatty acids (PLFA) spectra have been used as tools for the evaluation of microbial population changes in polluted soils (Baath et al. 1992; Frostegaard et al. 1993). None of these indicators is, however, specific to *Frankia* actinorhizal species.

III. Plant Lipid Transformations Induced by Mycorrhizae

A. Comparative Analysis of Lipids Between Mycorrhizal and Non-Mycorrhizal Plants

Fungal lipids differ considerably from plant lipids. However, when analyzing roots of a colonized plant the fungal, partner lipids mix together with those of the root, at least partially. Then, the results of the analysis will reflect the composition of the more abundant of the organisms. Moreover, some physiological changes may be induced in the metabolism of the host plant resulting in a modified lipid composition. As a result, several questions need to be answered:

1. Does the fungal colonization induce any changes in the lipid composition of the host plant?
2. Are these quantitative or qualitative changes, resulting from the stimulation or the inhibition of the lipid synthesis?
3. Which parts of the plant are essentially affected by these changes?
4. Are these changes dependent upon the type of fungal colonization?

Several papers report analyses of different colonized plants, either ectomycorrhizae (pines, birch, alder), or endomycorrhizae with orchids, and arbuscular mycorrhizae (AM) (maize, safflower, leek). Soybean has been interestingly studied when dually colonized by *Bradyrhizobium* and AM fungi (Pacovsky and Fuller 1987, 1988),

as has been maize, when cocolonized by *Azospirillum* and *Glomus* sp. (Pacovsky 1989).

1. Compared Content of Colonized Versus Non-Colonized Roots

The mycelia of mycorrhizal fungal symbionts are located either inside (arbuscular mycorrhizae, orchid, ericoid), or outside, or in between cortical cells (ecto). Sometimes special structures, such as nodules, vesicles and arbuscules are differentiated by the colonizing organism (actinorhizae, arbuscular mycorrhizae). It is then likely that the major changes in the lipid contents will be located in the plant root.

a) Quantitative Changes

Both the increase and the decrease of total lipid or of total fatty acid and sterol will be considered here separately.

i. Total Lipid and Fatty Acids

Abundant oil droplets were detected in root cells of *Betula verrucosa* and *Pinus radiata*, respectively, associated with *Lactarius pubescens* and *Rhizopogon luteolus* (Bevege et al. 1975). Similar lipid droplets can be easily observed in the mycelium or in the spores of most AMF, being essentially located inside the oldest parts of the infection units, near the entry points, in both intercellular and external hyphae (Cooper and Lösel 1978). These authors reported that there are no visible lipids in young arbuscules, but they were present in the older branches of the arbuscules through aging. The presence of lipid droplets in the mycelium and spores of *Glomus fasciculatum* inside the roots of *Sesamum indicum* was highlighted by histochemistry (Selvaraj and Subramanian 1990). The accumulation of lipids in colonized roots has been found to be directly proportional to the number of AMF vesicles detected inside the roots (Gaspar et al. 1994; Pacovsky and Fuller 1988; Gnekov and Marschner 1989; Johansen et al. 1996). The lipid contents of the mycelium of *Glomus mosseae* may reach up to 43.8 mg. g^{-1} fresh weight (Cooper and Lösel 1978). As a consequence, roots of colonized plants which contain such lipid-rich mycelium must contain more lipids than uncolonized ones. Such an increase has been reported by several authors (Cooper and Lösel 1978; Nagy et al. 1980; Nordby et al. 1981; Pacovsky 1989; Bago et al. 1995;

Graham et al. 1996; Gaspar et al. 1997). The mycorrhizal and nonmycorrhizal roots contained, respectively, 3.4 and 2.7 mg g^{-1} fresh weight (onion), 4.1 and 3.3 (clover) or 6.6 and 3.7 (ryegrass) of lipid (Cooper and Lösel 1978). Similarly, the colonization of roots of *Malus pumila* by *Glomus macrocarpum* increased their lipid contents from 0.9 to 4.5 mg g^{-1} (Gnekov and Marschner 1989). Roots of alfalfa showed an increase in lipid content proportional to their level of colonization by *Glomus versiforme* (Gaspar et al. 1997). The total fatty acids content of the roots of *Citrus* is higher when they are mycorrhizal (Peng et al. 1993); these results agree with those of Nagy et al. (1980) obtained with several cultivars of the same plant. By measuring the caloric content of mycorrhizal roots of *Malus pumila*, Gnekov and Marschner (1989) showed that the lipid content (0.9–4.5% of the dry matter) should be attributed to the presence of the vesicles corresponding to 1.6–8.2% of the total root caloric contents. The lipid content of soybean seeds can be modified by the presence of AM fungi in the roots of the plant. Depending upon phosphorus availability, this content can be increased (Bethlenfalvay et al. 1994) or decreased (Bethlenfalvay et al. 1997). The importance of the changes can depend on the mycorrhizal fungus and be related to its aggressivity. The lipid content of *Citrus* roots was found lower with the less aggressive *Glomus etunicatum* than with *Glomus claroideum* and *Glomus intraradices* (Graham et al. 1996) or *Glomus mosseae* (Nordby et al. 1981). The increase in lipid content was reported as being much smaller with plants colonized by Gigasporaceae fungi than with plants colonized by *Glomus* species. These data may be explained by the fact that Gigasporaceae species do not differentiate vesicles which are particularly lipid rich propagules (Graham et al. 1995).

Actinorhizae may also induce changes in the roots of *Alnus* species nodulated by *Frankia* species. Comparative analyses of laboratory grown nodulated versus *Frankia*-free roots of the same age (3 or 6 months old) were analyzed by Maudinas et al. (1982). On the basis of dry weight, nodulated roots contain more total lipids than unnodulated ones (3-month-old) or are unchanged (6-month-old); but the fatty acid contents of these lipids is very different independent of the age, much lower in lipids from nodules (71.8–78.3 mg g^{-1} total lipids) than in lipids from free roots (162–193 mg g^{-1} total lipids). This very strange decrease in the total lipid fatty acid content probably results either from an increased synthesis of fatty acid poor lipids, like phospholipids, di- or monoacylglycerols or from an intense utilization of the root triacylglycerols. Unfortunately, no qualitative analyses have been performed, and therefore, we only can hypothesize about this strange phenomenon.

ii. Sterols

Changes in sterol content of mycorrhizal plants have been also investigated but the published results remained controversial according to the studied material. An increase in the sterol content of mycorrhizal roots of *Zea mays* was reported by Ho (1977). A similar increase was observed by Fontaine (unpublished results) with in vitro cultures of *Glomus intraradices* on transformed roots of *Daucus carota*. Qualitative changes in the sterol contents appear more significant than quantitative ones.

b) Qualitative Changes

These are the most important and most significant for both fatty acid and sterol contents.

i. Lipid Classes and Specific Fatty Acids

Unusual fatty acids have been reported in *Ophrys* mycorrhizal roots (Barroso et al. 1986). Most of the available data are derived from arbuscular mycorrhizae, probably due to the interesting and very specific fatty acid profiles observed in the *Glomales*. The most peculiar fatty acid found in AM fungi is certainly the C16:1Δ11 (Nordby et al. 1981; Jabaji-Hare 1988; Sancholle and Dalpé 1993; Bentivenga and Morton 1994a,b; Gaspar et al. 1994; Grandmougin-Ferjani et al. 1996). A ω-6 polyunsaturated fatty acid (C18:3 Δ6,9,12 = γ-linolenic acid), which is considered almost specific to the *Zygomycetes* (Weete 1980), has also been described (Lösel 1988; Lösel and Sancholle 1996). It seems likely that the presence of these fatty acids in the root tissue should be considered as a "signature" of the presence of the AMF. However, things are not always so evident. For example, roots of *Zea mays* colonized with *Glomus etunicatum* were found to contain many very specific fatty acids (C18:3Δ8,9,12; C20:3Δ8,11,14; C20:4Δ5,8,11,14 or C20:5Δ5,8,11,14,17), which are absent from uncolonized roots (Pacovsky 1989). Surprisingly, these fatty acids were not detected in the spores of *Glomus etunicatum* (Bentivenga and

Morton 1994b; Grandmougin-Ferjani et al. 1996). The C16:1Δ11 is essentially located in the triacylglycerols, which are considered as storage lipids (Nordby et al. 1981; Olsson et al. 1998). Since triacylglycerols are the major lipids in the roots of colonized plants (Cooper and Lösel 1978; Lösel 1980; Nagy et al. 1980; Gaspar et al. 1994), this fatty acid can become one of the most important lipids of colonized roots (Pacovsky 1989; Gaspar et al. 1994). There is a concomitant increase in C16 and C20 fatty acids in the roots of soybean colonized with *Glomus fasciculatum* (Pacovsky and Fuller 1988). The analysis of microsomal fractions of mycorrhizal roots of *Helianthus annuus* and *Allium cepa* showed the presence of C16:1Δ11, whose amount increases with the fungal root invasion, suggesting the possibility of using this molecule as a marker of colonization. An increase in root colonization is known to stimulate the synthesis of new membranes that contain considerable amounts of unsaturated fatty acids, mainly C18:2. The result is a higher membrane fluidity, and consequently a better membrane permeability induced by the symbiotic status (Bago et al. 1995). When the colonizing organism belongs to the genus *Gigaspora*, there is no increase in C16:1Δ11 but one in C18:1Δ9 which comprises 35–55% of the total profile whereas C16:1Δ11 represents less than 2% of it (versus 35–75% in *Glomus*). Three other fatty acids may be used to delimit taxonomic boundaries at the family level: 3-OH-C18:0 (0–2% in Glomaceae, absent from Gigasporaceae); C20:0 (10–16% in Glomaceae, <1% in Gigasporaceae) and cis C20:1Δ11 (2–15% in Gigasporaceae, absent from Glomaceae) (Bentivenga and Morton 1994b).

The observed changes in lipid classes can also provide valuable information on the biosynthetic events at the mycorhizal interface. Phosphatidyl choline (PC) is the major phospholipid in plants while phosphatidyl ethanolamine (PE) is the most abundant one in *Glomus mosseae* (Cooper and Lösel 1978) and *Glomus versiforme* (Gaspar et al. 1994). It is common knowledge that phospholipids are essentially located in membranes. In colonized roots, PC increases more rapidly than PE, suggesting a more active synthesis of host membranes versus endophyte membranes (Cooper and Lösel 1978; Gaspar et al. 1997). The synthesis of fungal membranes can be masked by the faster increase in host membranes (Lösel 1980). Cooper and Lösel (1978) reported an increase in all molecular types of phospholipids in mycorrhizal roots, not only PC and PE, but also diphosphatidylglycerol (DPG), phosphatidyl serine (PS), phosphatidyl glycerol (PG), phosphatidic acid (PA) and phosphatidyl inositol (PI). This increase in PA and PI in mycorrhizal roots is of interest since these molecules have not been detected in the isolated mycelium of *Glomus mosseae*. Additionally, in plant/pathogen interactions, PI is considered to play an important role in plant cell signaling; this may explain its increase in the colonized roots.

When AM roots are dually colonized, the presence of a second symbiont may influence the root fatty acid composition. When roots of soybean are double-inoculated, with *Glomus fasciculatum* + *Bradyrhizobium*, the fatty acids considered as characteristic of the presence of the mycosymbiont are less abundant than in the presence of *G. fasciculatum* alone. The respective percentage of fatty acids are 1.48–3.55% for C16:1Δ11 and 0.28–0.35% for C18:3Δ6,9,12 (Pacovsky and Fuller 1987). In contrast, the presence of *Azospirillum* together with *Glomus etunicatum* in the roots of *Zea mays* increases the amount of the fungal specific fatty acids: 5.7–3.18% for C16:1Δ11 and 0.42–0.35% for C18:3Δ6,9,12 (Pacovsky 1989). In both cases, the second symbiont is presumed to provide the plant with nitrogen, and therefore the comparison is significant. It would be of interest to know whether the presence of *Azospirillum* stimulates fungal growth and whether *Bradyrhizobium* inhibits it.

Although analysis of lipid classes from nodulated versus free roots of *Alnus glutinosa* has not been performed, qualitative data relative to the fatty acids are available (Maudinas et al. 1982). C14:0 is the major fatty acid in the nodules, while the most important fatty acid in the roots is C18:2. Whatever the roots are, either free or nodulated, C14:0 decreases with the age while C16:0 increases concomitantly.

ii. Sterols

These structural membrane constituents are considered partially responsible for membrane fluidity (Weete 1989) and therefore play a very important role in the physiology of the organism. Most fungi have ergosterol as their major sterol (Weete 1980) in contrast to higher plants. Therefore, just as specific fatty acids can be indicators, the presence of specific sterols could be considered a signature for the presence of the corresponding fungus. However, not all fungi contain ergosterol as primary sterol and their sterol composition is less clearcut than their fatty acid content (Weete 1989). Comparing infected and uninfected *Ophrys lutea* tubers, Barroso et al. (1987) demonstrated that free sterols are synthesized during the infection process. These authors noticed an accumulation of sitosterol and campesterol in infected tubers correlated to an interesting decrease in ergosterol and 24-ethylcholest-7,22-dienol. They also found in these infected tubers a significant concentration of cycloartenol, which is a 4.4-dimethylsterol, precursor of Δ5 sterols. As most of the orchid mycorrhizal fungi (e.g., *Rhizoctonia*) are anamorphs of ergosterol-producing *Basidiomycotina*, the observed decrease in this sterol cannot be related to the accumulation of the fungus inside the plant tissue but rather to a change in sterol metabolism due to the symbiotic association. Unfortunately, the absence of any data on the endophyte sterol composition did not permit the authors to determine the level of contribution of the fungal sterol to the sterol pool of the association.

When studying the effects of nitrogen fertilization on mycorrhizal infection in *Calluna vulgaris*, Caporn et al. (1995) found that the concentrations of ergosterol were significantly higher in fine root hairs than in thicker roots and also higher in the surface horizons than deeper in the soil core. This means that the presence of the mycosymbiont induces physiological changes in the root and that these changes can be influenced by the depth at which the root is located.

Ergosterol was studied in *Pinus sylvestris/Paxillus involutus* ectomycorhizal associations, as a toll for determining the biomass of extramatrical mycelium, and yielded good correlations. This correlation was not as clear with the actinorhizal association *Alnus incana/Frankia* sp. (Ekblad et al. 1995); this is not surprising since *Frankia* is a prokaryote (see Sect. VI).

For triacylglycerols and phospholipids, once again, more results are available with arbuscular mycorrhizae. AM colonized roots were found to contain higher amounts of sterol esters (Cooper and Lösel 1978). Mycorrhizal roots of *Zea mays* contained more cholesterol and campesterol than the nonmycorrhizal ones (Ho 1977). The presence of AMF inside the roots of several *Citrus* species can also be characterized by changes in their sterol contents. In uncolonized roots, sitosterol, campesterol and stigmasterol are the major sterols, accompanied by smaller amounts of cholesterol

and isofucosterol. When these roots are colonized by *Glomus mosseae*, a consequent decrease in sitosterol, cholesterol and isofucosterol can be observed while campesterol increases concomitantly. Stigmasterol stays roughly unchanged (Nagy et al. 1980); although the sterol content of the mycelium has not been analyzed, it is unlikely that the campesterol increase should be related to the campesterol contained inside the mycorrhizal mycelium. We must not forget that campesterol is also a major sterol in the uncolonized roots; therefore, the observed changes are possibly due to alterations in the sterol metabolism of the host plant. Campesterol accumulates in roots of sweet orange colonized by *Glomus mosseae* correlated with a cholesterol and sitosterol decrease (Nagy et al. 1980). Grandmougin-Ferjani et al. (1995) also observed a significant increase of 24-methylcholesterol in *G. mosseae* colonized *Allium porum* roots.

In a recent investigation on the major sterols of several species of *Glomus*, 24-ethylcholesterol was found to be the major sterol of most of the species (Grandmougin-Ferjani et al. 1997, 1999). Spores of the different species grown on leek roots in soil do not contain ergosterol. Recently, Fontaine et al. (1998), using the in vitro transformed root technique (Bécard and Fortin 1988) to grow *Glomus intraradices*, were able to analyze separately the extraradical mycelium, the mycorrhizal roots, and the non-mycorrhizal ones coming from similar systems. Working under dark conditions in order to avoid ergosterol degradation, their investigation demonstrated beyond doubt that *G. intraradices* does not contain ergosterol. This is contradicted by the results published by Frey et al. (1992, 1994), but these authors used soil-grown nonaxenic cultures of *Zea mays* and *Trifolium alexandrinum* or *Trifolium pratense*. Moreover, neither spores nor mycelium were analyzed separately to offer suitable confirmation of the results.

Grandmougin-Ferjani et al. (1995) reported that uncolonized leek roots contain classical plant sterols like sitosterol, stigmasterol, 24-methylcholesterol and a little cholesterol. The only sterol specific to colonized roots was 24-methylenecholesterol. Interestingly, 24-methylenecholesterol is present in neither uncolonized root nor the spores of *G. mosseae*. 24-methylenecholesterol was previously reported as associated with campesterol in mycorrhizal roots (Schmitz et al. 1991). Therefore, this sterol should be considered as synthesized by the mycorrhizal association. Such specific production by plant-microorganism symbioses are known. In infected legume nodules, *Rhizobia* species are known to be responsible for the synthesis of nitrogenase while the nodule plant cells synthesize leghemoglobin, but neither nitrogenase nor leghemoglobin is recognised as a structural molecule.

2. Compared Content of Leaves from Colonized Versus Non-Colonized Plants

Most scientists believe that changes induced by mycorrhiza in parts of the plant differing from those containing the fungus are unlikely. In fact, there are very few papers dealing with the changes that occur within the aerial parts of mycorrhizal plants. The first mention of changes in fatty acid composition of the leaves of a mycorrhizal plant concerned *Citrus* colonized by *Glomus mosseae*: leaves of colonized plants were found to contain more (C16 + C17) and less C18 fatty acids than those of uncolonized ones (Nordby et al. 1981). Leaves of soybean whose roots have been inoculated with *Glomus fasciculatum* contain more lipid per unit of dry weight than noninoculated ones; the increase is still higher when plants are dually inoculated with *Glomus fasciculatum + Bradyrhizobium* (Pacovsky and Fuller 1987). On a qualitative basis, leaves of mycorrhizal soybean contained higher levels of C18:2 and C18:1 than leaves of nonmycorrhizal plants (Pacovsky and Fuller 1987). An accumulation of C16:0 in the pods of these mycorrhizal plants was also reported in the same paper. These results were corroborated by those of Grandmougin-Ferjani et al. (1995), who found increase of C18:2 and C18:3 with a correlated decrease of C16:0 in leaves of leek plants colonized by *Glomus mosseae*. In leaves of *Glomus etunicatum* colonized maize, there is a general decrease in fatty acid contents while leaves of dually inoculated plants (*Glomus + Azospirillum*) contain more C18:3 and C16:0 than the controls (Pacovsky 1989); this suggests that there are substantial interactions between both symbionts.

Changes can also affect the sterol composition of leaves of mycorrhizal plants. Leaves of *Linum usitatissimum* colonized with *Glomus intraradices* accumulate sitosterol and stigmasterol, but the campesterol contents remained unchanged (Dugassa et al. 1996).

Since in mycorrhizal plants the presence of the mycosymbiont is restricted to the root tissue, the observed changes in the fatty acid and/or sterol contents of leaves cannot be attributed to the presence of substances coming from the mycelium. There is evidence that root colonization may deeply alter the metabolism of the host plant. These alterations, specific for every plant/mycosymbiont association, can be observed in all parts of the plant body. This suggests the possibility that changes in the contents of seeds can be considered as metabolic sinks for the plant.

3. Compared Lipid Content of Seeds Produced by Mycorrhizal Versus Non-Mycorrhizal Plants

While studying the seedling nutrient dynamics as influenced by mycorrhizal infection of the maternal generation, Lu and Koïde (1991) suggested that seeds produced by mycotrophic plants may be modified specifically in relation to the relative abundance of some of their main storage products: proteins and oils. The lipid contents of seeds of *Pisum sativum* colonized by *Glomus mosseae* is higher than for seeds from non-mycorrhizal plants (Bethlenfalvay et al. 1994). The same observation was made with mycorrhizal durum wheat, whose kernels contain more lipid than non-mycorrhizal ones (Al-Karaki et al. 1998). Opposite results were obtained with soybean seeds. When *Glycine max* roots are colonized by *Glomus mosseae*, *Glomus etunicatum* or *Gigaspora rosea*, they produce seeds with a lower lipid content than controls. As the seed protein contents stay unchanged, it resulted in higher protein/lipid ratios in the seeds of AM plants (Bethlenfalvay et al. 1997). All these results and observations show that the lipid metabolism of the host plants can be deeply altered by the presence of the mycosymbiont, probably through the modified supply of mineral nutrients like nitrogen or phosphorus.

B. Impact of Mineral Nutrients on Lipid Metabolism

Fungi are known as mineral scavengers. This is true for "free fungi" which can be used as metallic ion trappers (Nandan and Raisuddin 1992). Thus, when fungi are involved in mycorrhizal associations, they also can be very efficient in ion uptake, particularly of phosphorus (P), enhancing the uptake of P by the host plant, operating as extensions of the root system (Rolin et al. 1984). The amount of P uptake per unit of carbon allocated below ground is then expected to be higher in mycorrhizal than in non-mycorrhizal roots (Jakobsen 1991). Non-mycorrhizal *Pinus sylvestris* does not respond to P limitation by a proportional increase of root production. This might be a reflection of an obligate dependency on mycorrhiza for effective P uptake, since the production of extramatrical biomass of the mycosymbiont *Paxillus involutus* peaks when P is low and other nutrients are high (Ekblad et al. 1995). Very little is known about the transport of phosphorus from the fungus to the host (Rolin et al. 1984). As phosphorus is required for the synthesis of phospholipids, several aspects have to be clarified:

1. Does the increase of phosphorus uptake by mycorrhizal fungi alter their own lipid metabolism?
2. Is the lipid metabolism of the host plant modified?
3. Do other mineral nutrients exert any influence on these lipid metabolisms?

When trying to verify if and how phosphorus uptake can influence lipid metabolism, we should keep in mind that the mineral requirements of living organisms are not restricted to phosphorus and should include nitrogen, which is involved as well and acting in close relationship with phosphorus. The influence of both P and N has to be taken into account.

1. Repercussion of P Uptake on Lipid Metabolism

Working with the ectomycorrhizal fungus *Hebeloma cylindrosporium*, Rolin et al. (1984) showed that the P taken up by the fungus was stored as orthophosphate, a form easily mobilized for DNA and phospholipid synthesis. Soluble P can also be mobilized, but to a lesser extent. Unfortunately, these authors did not perform any analysis of the lipid contents, either in the roots or in the seeds.

a) Roots

The amount of P available is soil may act directly, by either limiting or exhausting fungal growth. It may also act indirectly by altering the lipid metabolism. The application of P to the soil reduces the total fatty acid content originating from the fungus in the roots of AM *Citrus* plants (Peng et al. 1993). Even if the mycorrhizal colonization of *Citrus* roots does not appear significantly reduced by high P supply, the relative content of the fungal C16:Δ11 fatty acid is significantly lower in plants grown with high P concentrations (Graham et al. 1997). Olsson et al. (1997) showed that P applications resulted in an increase in P concentration in both the shoots and the roots of non-mycorrhizal plants of *Cucumis sativa* while plants colonized by *Glomus caledonium* demonstrated a decrease in P concentration in the shoots but not in the roots. The percent of root colonization decreased as the applications of P increased. The mycorrhizal root length was not affected by the application of P up to 60 mg kg⁻¹ whereas it tended to decrease at 90 mg kg⁻¹. Using C16:Δ11 fatty acid as a signature, they showed that its concentration decreased in roots and soil with an increase in P application. By the study of the evolution of the neutral/phospholipid ratio which showed a dramatic decrease with the applications of 15 mg of P kg⁻¹, they showed that P supplementation may negatively influence the lipid metabolism, particularly the synthesis of storage lipids. Colonization of roots of *Sorghum vulgare* and *Citrus aurantium* is inhibited by high P fertilization. High P content induces an increase in the phospholipid content of the membranes, rendering them less permeable, then resulting in a reduced leakage of sugars and amino acids which are essential for the establishment of the mycorrhiza (Ratnayake et al. 1978). There is probably a need for the regulation of the P and N content since roots dually inoculated (*Azospirillum brasilense* + *Glomus etunicatum*) contain more specific fatty acids than when inoculated with only one or none of the symbiotic partners (Pacovsky 1989).

b) Seeds

The physiological effects of N and P supplementation were also studied on *Glycine max* colonized by either *Glomus fasciculatum*, *Bradyrhizobium japonicum* or by both symbionts. The response to root colonization is particularly spectacular in the seed lipid fraction which is twice higher with dually inoculated plants than with non-inoculated ones supplied with N or P fertilizers (Pacovsky and Fuller 1988). This experience demonstrated that the regulation of P uptake by mycorrhizae strongly influences the synthesis of storage lipids in seeds. The lipid contents of seeds produced by mycorrhizal *Pisum sativum* is also highly correlated to the seed P contents. The level of AM root colonization has been shown to affect both the protein and the lipid composition of pea seeds, acting the same way as changes of the soil P availability (Bethlenfalvay et al. 1994). A negative correlation between seed P and lipid concentration was clearly observed with *Glycine max* (Bethlenfalvay et al. 1997). Lipid concentrations are more affected by P supplementation of AM root colonization than protein ones and tend to decrease with higher levels of root colonization. This results in higher protein/lipid ratios in the seeds of highly colonized plants. On the contrary, *Glomus* inoculated soybeans contain more lipids per unit of dry weight in the leaves and roots than the P-fertilized ones. However, highest lipid contents are obtained with dually (*Bradyrhizobium* + *Glomus* sp.) inoculated plants (Pacovsky and Fuller 1988).

2. Effects of Nitrogen on Lipid Content

Ammonium supply to *Candida apicola* is known to increase yields and stimulate sophorose lipids when transforming the organism fatty acid profile (Hommel et al. 1994). With mycorrhizal fungi, the availability of nitrogen has been shown to alter the lipid metabolism of colonized plants. Changes can be induced, not only by the level of available nitrogen, but also by the type of the supplied nitrogen. Ergosterol seems to be particularly sensitive to nitrogen sources being amended. As with phosphorus, an increased supply of mineral nitrogen results in a decrease of ergosterol content in roots of *Picea abies* colonized by *Telephora terrestris* and *Cenococcum geophilum*. This is true when inorganic nitrogen is supplied (ammonium nitrate), but not with keratin, an organic form (Wallenda et al. 1996). With *Pisolithus tinctorius*, total fatty acids levels were found to decrease under extreme nitrogen input but their concentration increased with nitrogen increase (Melhuish and Janerette 1979). It thus seems that the type of nitrogen supplied to the plant can greatly influence both the level and the profile of its lipids. In *Calluna vulgaris* mycorrhizal plants, the ergosterol content was found maximal at an application of $80\,kg\,ha^{-1}\,year^{-1}\,NH_4NO_3$, but variable throughout the growing season (Caporn et al. 1995).

3. Influence of Other Minerals on Lipid Content of Mycorrhizal Plants

Some other minerals, such as K, Ca and Mg, may influence the lipid composition of mycorrhizal plants (Lynd and Ansman 1989). Strullu et al. (1983) showed that the vesicles of AMF contain K, Ca, Mg and Na as principal elements, in addition to lipids. Aluminum was found to affect fungal membrane fluidity of the mycorrhizal fungus *Amanita muscaria* through the arrangement and dynamics of lipid molecules in plasmalemma (Zel et al. 1993).

Even if they are not directly incorporated into the lipid molecules, mineral nutrients may influence the lipid content of the mycorrhizal fungi by interfering with each other.

C. Carbon Cycling

In most types of mycorrhiza, the host plant represents the major carbon source for the fungus which is heterotroph for such nutrient sources. The inability to culture VA mycorrhizal fungi saprophytically (Jakobsen 1991) has brought research to focus on the physiological studies of symbiotic fungi and on their interactions with plants. Fungal survival and growth depend on the plant supply for specific nutrients (Smith and Gianinazzi-Pearson 1998). Several observations and experiments lead to the conclusion that precursors for the metabolism of the mycosymbiont undoubtedly come from the host plant, even if the enzymes of the dark fixation of CO_2, like pyruvate carboxylase and phophoenol-pyruvate carboxykinase, have been detected in filamentous fungi (reported by Jakobsen 1991). Using labeled ^{13}C and ^{14}C, Martin et al. (1984a,b) showed the presence of these enzymes in ectomycorrhizal fungi, while Bécard and Piché (1988) obtained the same results with arbuscular mycorrhizal fungi. This direct CO_2 assimilation is certainly not the main metabolic pathway for mycorrhizal fungi development. Carbon cycling through the host plant seems to be a much better research avenue. Symbiotic fungi play an important role in nutrient cycling through the uptake, transformation and release of nutrients (Frey et al. 1994). Of course, the host photosynthetic plants provide the mycosymbionts with enough variety and quantity of photosynthates to cover their carbon requirements (Ho and Trappe 1973; Bevege et al. 1975; Cox et al. 1975; Lösel and Cooper 1979). As the mycorrhizal association improves plant access to soil water and minerals, the carbon fixation activities are consequently stimulated (Paul and Kucey 1981; Trent et al. 1989). Several authors showed that mycorrhizae may increase below-ground C allocation by 4–20% (Pang and Paul 1980; Snellgrove et al. 1982; Kucey and Paul 1983; Douds et al. 1988; Wang et al. 1989; Jakobsen and Rosendahl 1990). When studying the carbon economy of sour orange in relation to mycorrhizal colonization, Eissenstat et al. (1993) concluded that mycorrhizae increased the root biomass fraction, the root length/leaf area ratio and the percentage of ^{14}C recovered from below-ground components.

All these results showed clearly that the presence of mycorrhizae stimulates the transfer of photosynthates to the roots. Using soybean-*Rhizobium-Glomus* associations, Harris et al. (1985) showed that the rate of specific ^{14}C uptake was greater in symbiotic plants, but the starch contents of their leaves was reduced by half of that in control plants. One can suppose that most of the missing starch is used as carbon supply by the microbial symbionts. There is no doubt that the development of a functional mycorrhiza simultaneously depends on the availability of photosynthates and enhances their availability to the host plant (Hampp et al. 1994). Provided there is full compensation for the carbon drain by the mycobiont in the form of an increased photosynthetic rate, plant growth should not be affected by the carbon drain (Jakobsen 1991). Graham et al. (1997) showed that colonized *Citrus* plants contained more sucrose and less starch than non-mycorrhizal ones. Sucrose concentration is consistently lower in root tissues of P-fed mycorrhizal than non-mycorrhizal plants indicating a substantial utilization of sucrose by the fungus (Peng et al. 1993). These authors found significantly higher fatty acids concentrations in mycorrhizal than in non-mycorrhizal fibrous roots. There was a 37% difference in daily total root/soil respiration, two thirds of which was associated with the construction of lipid-rich roots, to the maintenance of the root fungal tissues and to the growth maintenance of extramatrical hyphae. This demonstrated that sucrose can be converted into lipid. An increased capacity for new sucrose synthesis was observed in mycorrhizal seedlings of spruce (Hampp et al. 1994). The review of Smith and Gianinazzi-Pearson (1988) presented a good overview of carbon distribution in the symbionts and arrived at the conclusion that the fungi can convert host metabolites into specific fungal compounds, such as lipids, that are found particularly abundant in AM fungal vesicles. The vesicles of *Glomales* possess high lipid contents and may act as reser-

voirs for many elements (Strullu et al. 1983); this is in correlation with Bevege et al. (1975), who considered that lipid fractions, in the hyphae, may have an important storage function. Analysis of root fatty acid methyl esters (FAME), not only provided a measure of colonization development but also served as an index of carbon allocated to intraradical fungal growth and lipid storage (Graham et al. 1995).

Since the transfer of carbohydrates from the host plant to the mycosymbiont is well documented, one can be surprised that the transfer of carbon from carbohydrates into lipids took so much time to be studied. In this sense, only a few studies, using ^{14}C labeled precursors, were performed during the 1980s. These studies demonstrated the incorporation of a substantial proportion of host photosynthate into fungal lipids, providing short or long-term carbon sinks and membrane components for an often extensive inter- and intracellular fungal thallus (Lösel 1980). Ho and Trappe (1973) recorded the accumulation of ^{14}C-labeled photosynthate in the lipid-filled chlamydospores of *Endogone* (*Glomus*) *mosseae*. Using autoradiography of mycorrhizal roots and external mycelium, Cox et al. (1975) showed that after translocation of ^{14}C-labeled photosynthate lipid-rich hyphae, vesicles and arbuscles contained a greater proportion of ^{14}C. These authors suggested that the synthesis of lipid by mycorrhizal fungi could provide an alternative sink for photosynthate. Bevege et al. (1975) investigated the distribution of ^{14}C-labeled photosynthate in infected and uninfected roots of mycorrhizal *Araucaria cinninghamii* and *Trifolium subterraneum* plants and concluded that a high proportion of the ^{14}C-lipid fraction was found in the extraradical fungal hyphae.

Lösel and Cooper (1979) followed the time course of the incorporation by leaves of labeled photosynthate in mycorrhizal versus non-mycorrhizal onion roots. Even in roots of plants rapidly extracted after photosynthesis, radioactivity appeared in lipid and non-lipid, in both soluble and insoluble fractions. During the subsequent dark period, the total radioactivity increased and was higher in the mycorrhizal than in non-mycorrhizal roots. Radioactivity accumulated to the greatest extent in ethanol-soluble carbohydrates. This fraction showed consistently higher labeling in mycorrhizal than in non-mycorrhizal plants. More labeled $^{14}CO_2$ was accumulated in the lipid fraction of leaves originating from mycorrhizal than from non-mycorrhizal plants. With [2-^{14}C]acetate, there was a significantly greater labeling of lipids of infected roots than in uninfected ones, but the incorporation of labeled sucrose into lipid was not significantly different. In the case of ectomycorrhizae (*Pinus sylvestris/Suillus luteus*), the proportions of total host photosynthate incorporated into the lipid of the roots was not significantly different, based on the mycorrhizal status of the plant, although mycorrhizal plants assimilated more ^{14}C through photosynthesis (Lösel 1980). Studying the carbohydrate storage by the use of ^{13}C NMR with the mycelia of three ectomycorrhizal fungi, Martin et al. (1984a) showed that the triacylglycerols represented only 20% of the accumulated carbon. Recently, using compartimentalized monoxenic cultures of *G. intraradices* grown on Ti-DNA-transformed carrot roots Pfeffer et al. (1998) showed that labeled trehalose was transferred into the extraradical mycelium. Using labeled glucose, they obtained mycelium-labeled lipids and concluded that lipids were synthesized in the roots before being transferred into the hyphae.

Actually, these results appear, for the moment, somewhat controversial. However, they clearly demonstrate transfers of photosynthates from the host plant to the fungal lipids. It should be of great interest to evaluate these transfers and the way in which they are processed, both quantitatively and qualitatively. The promising molecular biology methods should certainly be helpful for this kind of study, and the fate of the host-plant photosynthate into the mycorrhizal fungi lipids seems to be a very good research avenue.

IV. Fungal Lipid Changes Through the Growth Process

A. Spore Germination

The fate of lipids during spore germination is a major question the answer to which could provide valuable information on the eventual control of mycorrhiza establishment and colonization processes. The involvement of lipids in the germination process of ectomycorrhizal fungal spores is a topic on which information remains sparse. While spore lipids of the ectomycorrhizal fungus *Tuber melanosporum* are known to com-

prise only 1% of their dry weight (Fonvieille et al. 1990) no information is available on the way in which they are transformed during germination. Tree root exudates are otherwise known to stimulate the spore germination of ectomycorrhizal fungi (Melin 1962; Birraux and Fries 1981; Hua et al. 1995). Fries et al. (1985) demonstrated the lipidic nature of some of these germination inducers, but without determining precisely the identity of the active substances.

The germination process and control of AMF spore germination represent crucial points for the use of mycorrhizae in agriculture as these organisms are obligate symbionts within most of the vascular plants. Beilby (1983) studied lipid behavior during the very first germination stage, i.e., the lipid transformations when spores are subjected to liquid imbibition prior to germination per se. Results showed that, within a few hours, ^{14}C acetate source was recovered mainly in the triacylglycerides, as well as in the free fatty acids and later in the phospholipid and the sterol fractions. After 7 days of imbibition, the triacylglyceride contents decreased as the bounded fatty acids and the sterol esters fractions. These results demonstrated that, simultaneously, both de novo synthesis and breakdown of lipids occur during the first stages of germination (Beilby and Kidby 1980; Beilby 1983; Gaspar et al. 1994). Beilby and Kidby (1980) demonstrated that fatty acid, sterol, diacylglyceride, and phospholipid concentrations increased during the first 7 days following *Glomus caledonium* spore germination. Gaspar et al. (1994), in their study on *Glomus versiforme* spores, clearly demonstrated the consumption of triacylglycerides during the first stages of germination together with a de novo synthesis of phospholipids and free fatty acids. In fact, they evaluated at 75 and 50%, respectively, the decrease in total and neutral lipids in germinated spores. Free fatty acid contents increased with the production of more elongated carbon chain fatty acids and of phospholipids. These observations indicated the existence of hydrolytic enzymes dedicated to the degradation of TG and simultaneously the use of the released carbon for the synthesis of phospholipids. After spore germination, lipid bodies are observed ultrastructurally in the subapical zone of the germ tube (Sward 1981b), then dispersed in the cytosol of the whole germinating hyphae (Meier and Charvat 1992), and also observed in any of the differentiated fungal propagules, such as the auxiliary cells of

Gigaspora margarita (Jabaji-Hare 1988), the secondary spores of *Glomus mosseae*, and the intraradical vesicles (Hepper 1979).

B. Mycelial Development

The only information available in the literature on lipid transformations that occur during the development of ectomycorrhizae deals with ergosterol which is by far their major sterol (Table 3). Ekblad et al. (1995) estimated the ergosterol contents of *Pinus sylvestris* roots colonized with *Paxillus involutus* at 155 to $250 \mu g g^{-1}$ of root dry weight. If we assume that the ergosterol concentration of the mycelium is comparable to that found in fruiting bodies, the fungal biomass associated with mycorrhizal roots would easily reach between 18 to 40% of the fungal dry weight. However, no information is available on the behavior of those lipids during the mycorrhizal process except that ergosterol levels inside roots may reflect the fungal root biomass involved in the symbiosis.

With arbuscular mycorrhizae, it has been pointed out by several workers that the total lipid content of mycorrhizal roots is usually found to be proportional to the root colonization levels (Cooper and Lösel 1978; Graham et al. 1996). Moreover, a direct correlation between lipid content and the number of vesicles inside roots has been made by Gnekow and Marschner (1989), Johansen et al. (1996), and Pacovsky and Fuller (1988).

The early work of Ho and Trappe (1973) demonstrated that ^{14}C photosynthates were translocated from *Festuca* colonized plants into the lipid globules of *G. mosseae* spores. Subsequent investigations revealed that a high proportion of assimilated ^{14}C photosynthates was transferred to the hyphae, vesicles and arbuscule structures, even though applied on the plant leaves or in the soil (Bevege et al. 1975; Cox et al. 1975). Dehydrogenase activity was detected in young arbuscules as in intraradical hyphae whereas peroxidase and catalase activities observed in senescing arbuscules were attributed by Nemec (1981) to the oxidation of the fungal fatty acids.

Histochemical studies carried out on *Asparagus* mycorrhizae by Leu and Chang (1989), on *Citrus* by Nemec (1981), and on *Sesamum* by Selvaraj and Subramanian (1990) pointed out the positive reaction of vesicular lipids with neutral lipid reagents, mainly triglycerides. The lipid contents of vesicles had been previously estimated by Jabaji-Hare et al. (1984) at 58% of their dry mass, with glycolipids and sphingolipids as the major lipid fractions. As glycolipids are known to be major components of plant lipids, and as these results contradict most of the other AMF lipid investigations, the possibility that the extracted vesicle material may have included residues of plant tissues should not be excluded. With the development of in vitro propagation techniques of AM fungi (Bécard and Piché 1992; Diop et al. 1994), investigations on the lipid transformations occurring inside the different propagules and during the different growing stages of AM fungi may provide more precise and useful information on the physiology of the fungal-plant interaction.

Table 3. Ergosterol contents of various ectomycorrhizal associations

Plant	Mycorrhizae	Ergosterol ($\mu g\,mg^{-1}$ root dry weight)	Age of plants (weeks)	Reference
Betula populifolia	Smooth orange	1.95 ± 0.10	20	Antibus and Sinsabaugh (1993)
Betula populifolia	Smooth white	1.02 ± 0.05	20	Antibus and Sinsabaugh (1993)
Betula populifolia	Smooth	1.34 ± 0.09	20	Antibus and Sinsabaugh (1993)
Betula populifolia	Tomentose	2.08 ± 0.12	20	Antibus and Sinsabaugh (1993)
Pinus sylvestris	Smooth brown	1.16 ± 0.02	20	Antibus and Sinsabaugh (1993)
Pinus sylvestris	*Paxillus involutus*	$0.155 - 0.22$	13	Ekblad et al. (1995)
Pinus contorta var. *latifolia*	*Hebeloma crustuliniforme*	$0.62 - 0.95$	35	Johnson and McGill (1990)
Eucalyptus globulus	*Pisolithus tinctorius*	0.31 (f.w.)	4 days	Martin et al. (1990)
Pinus sylvestris	*Piloderma* sp.	0.68 ± 0.24	Seasonal mean	Wallander et al. (1997)
Pinus sylvestris	*Piloderma* sp.	1.36 ± 0.47	Seasonal mean	Wallander et al. (1997)

C. Storage Lipids

1. Ectomycorrhizae

The accumulation of lipids as storage reserves has been observed regularly in almost all types of mycorrhizal associations. Within ectomycorrhizal associations, several ultrastructural observations confirmed the presence of lipid deposits in mycorrhizal fungal propagules that are directly in contact with the plant partners. For example, the sclerotia of *Paxillus involutus*, as of *Pisolithus tinctorius*, when produced either in pure cultures or in association with *Pinus strobus*, were found to contain lipids in both their cortical and medullary zones. (Grenville et al. 1985a,b; Moore et al. 1991). Similar observations were made by Fox (1986) on the sclerotium-like bodies of *Hebeloma sacchariolens* when associated with birch trees (*Betula* sp.). In this latter case, histochemical tests revealed the neutral lipid nature of the observed bodies. Unfortunately, no information on the comparative lipid accumulation between sclerotia produced in pure culture and those that are mycorrhizae-associated has ever been published.

Fungal rhizomorphs, that are also considered storage structures, are implied in the long-term survival of fungi. Using electron spectroscopic imaging, important hyphal lipid inclusions, together with polysaccharide deposits, were detected in rhizomorphs associated with *Picea abies* ectomycorrhizae (Franz and Acker 1995). Stored lipid droplets have been repetitively detected in Hartig net cells of several plant mycorrhizal partnerships such as the "pseudoectomycorrhizal" association described by Bonfante-Fasolo and Scannerini (1977) between *Endogone flammicorona* and *Pinus strobus*, also in *Alnus rubra-Alpova diplophloeus* ectomycorrhizas (Massicotte et al. 1989) as in *Quercus robur* and *Fagus grandifolia* ectomycorrhizas (Edwards and Gessner 1984; Herr and Peterson 1996). Mantle hyphae of the tuberculate ectomycorrhiza of *Eucalyptus pilularis* contained lipid reserves and were described as being "cemented together with an interhyphal matrix of carbohydrates and lipids" (Dell et al. 1990).

Triglyceride accumulation within intact mycelia of pure cultured *Cenococcum graniforme*, *Hebeloma crustuliniforme* and *Laccaria proxima* was observed by Martin et al. (1984a) using ^{13}C NMR spectroscopy. These neutral lipids, that constituted around 80% of the observed carbon, should be involved in storage reserves as well as in carbon sources for structural fungal organization and energy. The authors confirmed the vacuolic droplet organization of fatty acids based on the fact that most of the ^{13}C NMR resonances of fatty acids arise from the lipids located in the refringent droplets (Martin et al. 1984b).

2. Arbuscular Mycorrhizae

Since early research on arbuscular mycorrhizae, lipidic globules have been considered as storage compounds, and Harley (1989) suggested that lipids in AM fungi play the same role as carbohydrates play for other fungi. Refringent lipid droplets can be observed in all the existing AM fungal propagules, either within extraradical hyphae and spores or inside intramatrical vesicles and arbuscules (Boullard 1957; Mosse and Bowen 1968; Harley and Smith 1983). Studies of the lipid profiles of arbuscular mycorrhizal roots pointed out the presence and the abundance of an unusual fatty acid, the C16:1Δ11 hexadecenoic acid, extracted exclusively from mycorrhizal roots and legitimately attributed to the root-associated fungal symbiont (Nordby et al. 1981; Gaspar and Pollero 1994; Gaspar et al. 1994; Graham et al. 1995; Grandmougin-Ferjani et al. 1996). This fatty acid has been regularly found an important storage reserve of the mycorrhizal roots and this for a variety of cultivated plants (Nordby et al. 1981; Pacovsky and Fuller 1988; Pacovsky 1989; Bago et al. 1995; Johansen et al. 1996). However, the close association of AM fungal walls with Gram-negative soil bacteria has to be taken into account, as part of the hexadecenoic acid isolated from AMF spores could be the result of a contamination by the fatty acids of endophytic bacteria (Frostegaard et al. 1993).

3. Actinorhizae and Ericoid Mycorrhizae

Lipid-like inclusions were observed in the intranodular hyphae of a soil Actinomycete associated with *Comptonia peregrina* (Newcomb et al. 1978). *Frankia* species, when associated with actinorhizal plants such as the *Alnus*, developed within the nodules some type of vesicles which possessed a multilayered envelope composed almost essentially of lipids (Huss-Danell 1997). Lipid droplets were also observed in hyphae of ericoid mycorrhizal fungi either within extramatrical mycelium or intraradical hyphae coils as well (Allen et al. 1989).

V. Impact of Exogenous Additives on Mycorrhizae

A. Impact of Pesticides on Lipid Metabolism of Mycorrhizae

Very little information is available on the effects of pesticides on endo- and ectomycorrhizae, with regard to lipid metabolism of the symbiotic fungus and/or the host plant as well. The impact of fungicides with target sites in fungal sterol biosynthesis has been particularly studied on pathogenic fungi. The biochemistry of various classes of inhibitors has been described in numerous articles (Burden et al. 1989; Weete 1989; Köller 1992). Sterol biosynthesis inhibitors (SBI) fungicides are known to inhibit fungal growth by blocking specific sites in the sterol pathway. These compounds interfere mainly with the synthesis of structural sterols and can also induce major alterations in the functioning of fungal membranes. The sensitivity or the resistance of some pathogens to fungicides depends greatly on their membrane sterol composition (Siegel 1981; Sancholle et al. 1984a).

Some Oomycetes differ from other fungi in being unable to synthesize sterols. Some species of *Pythium* and *Phytophthora* can grow on sterol-free medium, producing mycelium and asexual spores which contain no sterol. It is well known that vegetative growth and asexual reproduction are the preferred target sites for antifungal agents. Therefore, an important implication of the lack of sterol synthesis by Oomycetes is their resistance to SBI fungicides. Another important implication is the resistance to polyen antibiotics because polyens form complexes with sterol and disrupt membrane function by this mechanism (Köller 1992; Lösel and Sancholle 1996).

The effects of SBI fungicides on fungi containing either ergosterol or a functionally equivalent sterol like brassicasterol in Ascomycetes was reviewed by Weete (1989). The reaction of the Ascomycete *Taphrina deformans* to triazoles (fungicides inhibiting C14 demethylation in the sterol pathway) were intensively studied in relation with the fungal parasite fatty acid metabolism, ultrastructure and cellular permeability of this fungus (Sancholle et al. 1984a,b, 1988; Weete et al. 1985b, 1991).

The effect of SBI fungicides on the fungal growth, sporulation, and root colonization of arbuscular mycorrhizal fungi has been investigated up to now by several workers (Nemec 1985; Hetrick et al. 1988; Dodd and Jeffries 1989; von Alten et al. 1993; Kling and Jakobsen 1997). As for other research performed on the AM reaction to fungicides, published results are often controversial mainly due to the great variations of experimental conditions, namely the sensitivity of fungal and plant symbionts, the diversity in the mode of application of fungicides and the fungicide concentrations. Such diversity in experimental procedures resulted in strong inconsistencies in the data but nevertheless brought forward interesting matters for discussion.

In one of the first investigations on SBI done by Nemec (1985), propiconazole, a systemic fungicide from the triazole family, was shown to severely inhibit both soil spore populations of Glomales and plant root colonization when applied directly to *Citrus* plant shoots colonized by *G. intraradices* and *G. mosseae*. Von Alten et al. (1993) using the same fungicide at high concentration but several other host-fungi systems (barley roots/*G. etunicatum*, *G. mosseae*, and *G. intraradices*) confirmed these results. However, much less severe or no inhibition at all were respectively registered with other sterol-inhibiting fungicides such as the plifenate, triadimefon, and triforine, all involved in the C14 sterol demethylation (Nemec 1985). Dodd and Jeffries (1989) studied the effect of field application of two fungicides containing also triazoles derivatives: Sportak (prochloraz) and Tilt Turbo (propiconazole) on spore germination in three species of *Glomus mosseae*, *G. monosporum* and *G. geosporum*. These fungicides, when applied at recommended field rates and at appropriate crop growth stages, decreased colonization of wheat roots and reduced spore production. Dodd and Jeffries (1989) also showed that tridemorph, a fungicide belonging to the morpholine family and known as Δ14 reductase and Δ8-Δ7 isomerase inhibitor, increases the root colonization rate of the winter wheat.

In a recent experiment done by Kling and Jakobsen (1997) under greenhouse conditions using compartmented pot cultures, propiconazole, when applied on soil free of roots at recommended field rates, showed no inhibitory activity on the soil hyphal density, on succinate dehydrogenase levels, or on ^{32}P uptake by *Pisum sativum*. These results were similar with all three arbuscular fungal strains (*G. intraradices*, *G. claroideum* and *G. invermaium*) tested. It is difficult to compare both the Dodd and Jeffries (1989) and Kling and Jakobsen (1997) experiments, as different host/fungi systems were studied and different fungicides were applied. Tilt Turbo fungicide contains not only propiconazole as the active matter but also tridemorph, a fungicide from the morpholine family. Sportak fungicide contains no propiconazole but also no other types of triazole derivatives. These results, as discussed by Kling and Jakobsen (1997), Dodd and Jeffries (1989), and Schreiner and Bethlenfalvay (1997), revealed the importance of studying host-fungi responses using fungicides at recommended doses and conditions in order to evaluate efficiently the fungicidal impact on AM symbiosis under natural conditions.

The anti-Oomycete fungicide fosetyl-Al (aluminum tris-O-ethylphosphonate) is largely used in European countries. The mode of action of this fungicide is not well known so far. It is assumed to result from the combination of both a direct and an indirect effect. The direct mode of action alters the morphology and the intermediate metabolism of the fungus through deep changes in polyphosphate and triacylglycerols. The indirect mode of action stimulates the plant defense responses, thus inhibiting the pathogen development (Griffith et al. 1992). Fosetyl-Al applications to leek plants affected neither the mycorrhizal nor the nonmycorrhizal plant growth (Despaties et al. 1989). The symplastic transport of phosphoric acid the unique Fosetyl-Al fungicide residue may be promoted by mycorrhizal root colonization. Moreover, foliar treatment of mycorrhizal leek plants with fosetyl induced an increase in total lipids of roots compared to nonmycorrhizal plants treated with fosetyl (Jabaji-Hare and Kendrick 1985). It should be of interest to see if fosetyl has a direct influence on the lipid metabolism of the host plant and/or its associated VAM fungus.

Fungicides which have been described as stimulating endomycorrhizal development should be of interest for the stimulation of in vitro inoculum production (spores and mycelia). Indeed, these fungi, whose development is very

slow, are difficult to grow in vitro under axenic conditions.

Very little information on the effects of pesticides on the lipid metabolism of ectomycorrhizae is available. Triadimefon, an SBI fungicide from the triazole family, is known to modify ectomycorrhizal development. This fungicide, used for the control of the fusiform rust fungus in pine nurseries, significantly suppresses ectomycorrhizal development and reduces the viability of the inoculum even at doses lower than those recommended for rust control. (Marx et al. 1986; Cordell and Marx 1994)

B. Impact of Exogenous Lipids on Mycorrhizae

Lipids including fatty acids and sterols, when incorporated in the culture media, are able to stimulate the growth of several fungi (Dijkstra et al. 1972; Lehrian et al. 1976; Lindeberg and Lindeberg 1974; Lösel 1989; Wardle and Schisler 1969). This means that these fungi are able to use lipids as carbon source for fungal growth and that extracellular lipases are synthesized accordingly (Caldwell et al. 1991; Sancholle and Lösel 1995). It has been demonstrated that a fatty acid (palmitic acid), the predominant component of palm oil, was a good carbon source for several thermophilic fungi (Ogundero 1981). *Aspergillus flavus*, that produces aflatoxin, can use lipids such as palmitic acid and tripalmitoylglycerol as carbon source (Fanelli et al. 1980) and can also degrade hydrocarbons (Raper 1965). The addition of the lipidic fraction of elm wood or of oleic and linoleic acid to the growing media of *Ceratocystis* species induced both sexual and asexual fructifications (Dalpé and Neumann 1977). Sterols, such as cholesterol and some of its derivatives, when incorporated in the growing media of several species of *Pythium* and *Phytophthora*, enhanced vegetative growth and spore production. These fungi have been shown to possess an absolute requirement for exogenous sterols for their sexual reproduction (Elliott 1977). These fungi are the sole cases of auxotrophy toward sterols known within fungal species.

For mycorrhizal fungi, very little information has ever been published on their capacity to use lipids as a carbon source. Hutchinson (1990) described *Amanita* strains that, when grown axenically in the presence of fatty acids esterified with polyethylene sorbitol (Tweens), produced lipases and were capable of utilizing lipids as a sole source of carbon. Under in vivo conditions, the main carbohydrate source for *Amanita muscaria* is glucose (Chen and Hampp 1993). Short-chain fatty acids such as butyric acid were found to induce the basidiospore germination of several ectomycorrhizal fungal species (Ohta 1988). The ectomycorrhizal fungus *Boletus*, when cultivated in pure cultures in the presence of safflower oil, a rich source of mono-unsaturated oleic acid, showed an increase in mycelial growth (Schisler and Volkoff 1977). Otherwise, *Boletus variegatus* when grown in the presence of short chain fatty acids was inhibited by a leakage of its cell material and blockage of respiration (Pedersen 1970). The inhibitory effect of shorter-chain fatty acids has been attributed to their interaction with membrane components. Inhibitory effects of medium- and shorter-chain fatty acid on fungal growth were discussed by Lösel (1988).

Some research has also been directed to the carbon requirement of Actinomycetes. Root-extracted lipids appeared to be essential for the isolation of *Frankia* species from the nodules of their associate plant roots (Quispel et al. 1983). In some cases, this requirement for root-lipid extracts remains obligate throughout their cultivation process, as they never adapt to any artificial addition of carbon to their growing media. Special fatty acids bounded with diglyceride and phospholipid fractions of the root extracts were reported to be the active factors in the growth of *Frankia*. However, all the component fatty acids isolated from diglyceride and phospholipid root fractions failed to support *Frankia* growth in pure culture (Quispel et al. 1983). Blom (1981) reported that some *Frankia* species can use carbon sources in the form of C2 to C18 saturated fatty acids, but never the mono-unsaturated C18:1 oleic acid. Preferred fatty acid carbon sources for *Frankia* appeared to be those esterified with polyethylene sorbitol (Tween) instead of those esterified with glycerol (Blom 1983).

The presence of lipids in root exudates has been demonstrated by several workers (Hale et al. 1981; Thompson and Hale 1983; Fries et al. 1985). However, the biological significance of those lipid exudates on the propagation, survival, and selection of mycorrhizal fungi by the host plant has almost never been studied. Differential reactions of ectomycorrhizal fungi grown axenically in media where root exudates were added permitted Fries et al. (1985) to suggest the idea of a selective growth promotion of fungal symbionts by the differential composition of root exudates. This study is the only one that revealed the potential of lipid root exudates in stimulating ectomycorrhizal fungal growth. Fries et al. (1985) found that exudates from 6-month-old *Pinus sylvestris* roots contained lipids at the level of 2.1% of the fresh weight of the seedling roots. The fatty acid composition of pine root exudates appeared to be predominantly linoleic acid (42% of the total fatty acids) followed by oleic (23%), stearic (18%) and palmitic (14%) acids. Once added to the culture media of axenically grown ectomycorrhizal fungi, the lipid fraction of root exudates increased the mycelial growth of *Laccaria bicolor* and *L. amethystina* strains by 150 to 250% but had reduced growth-promoting effects (17–21%) on *Leccinum aurantiacum*. Unfortunately, the complete lipid exudate composition has never been studied, and there is no possibility of determining from these investigations the nature of the active lipidic component responsible for the fungal growth stimulation. More recently, Hua et al. (1995), in a study on the improvement of mycorrhization in pine nursery, demonstrated that pine root exudations can stimulate the germination and growth of mycorrhizal fungi. Unfortunately, the root exudate factors that may be responsible for this stimulation were not identified.

The composition of root exudation from endomycorrhizal plants has been reported by several workers to contain mainly reducing sugars, soluble amino acids (Ratnayake et al. 1978) organic acids, and phenols (Marschner 1996). However, none of the analyses performed on this topic mentioned or detected any lipid components in either mycorrhizal or nonmycorrhizal root plant exudates. Hyphal growth of AM fungi appeared to be slightly more stimulated by pea root exudates than by water (Samra et al. 1996). However, when exudates produced by roots grown under reduced P conditions and abundant P conditions were compared for growth-promotion activities, it was clearly shown that exudates produced by phosphorus-stressed roots promoted hyphal growth and induced significantly more hyphal branching than exudates originating from roots grown in the presence of medium with high concentrations of P (Nagahashi et al. 1996; Tawaraya et al. 1996). It is widely accepted that root phosphorus content is proportional to soil phosphorus supply.

In a study on the relationship between root exudation and plant phosphorus supply, Ratnayake et al. (1978) demonstrated that a decrease in root P content corresponded to an increase in the root membrane permeability and consequently a decrease in the phospholipid content of root tissues, together with an enhancement of sugar and amino acid root exudation. These observations unfortunately provide no pertinent information about the possible involvement of lipid components in root exudates of arbuscular mycorrhizal roots. However, these results clearly demonstrated that the P nutrition of host plants influences the composition of root exudates and thereby the mycorrhizal fungal growth. If, as for ectomycorrhizae, lipid components could be involved in the composition of root exudates, exploratory research on the lipidic nature of root exudates should bring substantial information on plant-root interactions, as on selection and competitivity of AM fungal partners.

VI. Lipids as Indicators of Root Colonization

A. Specific Sterols

1. In Ectomycorrhizae

Ergosterol appears to be the major sterol of most species of Ascomycotina, Basidiomycotina, and Deuteromycotina, with the exception of Uredinales and Taphrinales (Lösel 1988; Weete 1989). Some Ascomycotina species from the orders Taphrinales and Tuberales contain in place of ergosterol a functionally equivalent sterol, brassicasterol (Weete 1989). Ergosterol remains absent from the most primitive fungi, such as Chytridiomycetes, Hyphochytriomycetes, and Oomycetes. This sterol is present neither in prokaryotic cells nor in higher plants. Being a fungal-specific component of higher fungi, ergosterol has been retained as an indicator of fungal biomass and, as such, recognized as an efficient tool for the estimation of ectomycorrhizal root colonization. Moreover, as ergosterol is rapidly degraded in dead fungal tissues, its evaluation is currently used to quantify the viable fungal biomass either in saprophytes or in plant-decaying fungi (Newell et al. 1988; Seitz et al. 1977, 1979; Whipps et al. 1982). With modern technologies, the possibility of detecting ergosterol quantitatively, even at very low concentrations, gives ergosterol

another major advantage for the evaluation of the intramatrical biomass of mycorrhizal fungi (Frankland and Harisson 1985; Salmanowicz and Nylund 1988; Johnson and McGill 1990; Martin et al. 1990; Antibus and Sinsabaugh 1993) or as mycorrhizal fungal biomass in soils (Sung et al. 1995)

The amount of ergosterol contained in the mycelium of pure mushroom cultures, including some mycorrhizal species, ranged from 1.5 and $15\,\mu g\,mg^{-1}$ of fungal dry weight (Martin et al. 1990; Antibus and Sinsabaugh 1993; Sung et al. 1995). The mycorrhizal fungal biomass in soils evaluated through ergosterol assay reaches between 149 and $750\,\mu g\,l^{-1}$ (Sung et al. 1995). The range of ergosterol concentrations inside ectomycorrhizal-colonized root tissues varies between 0.62 and $2.08\,\mu g\,mg^{-1}$ of root dry weight (Table 3) and no variation was registered in ergosterol content when axenic cultures were grown on different substrates or under a range of temperatures (Sung et al. 1995).

Since sterols are essentially membrane components, the ergosterol content of a colonized root should be proportional to the amount of fungal biomass present in that root. When compared to other methods used for the evaluation of the fungal biomass inside colonized roots, the ergosterol assay appears highly reliable. It allows quantitative measurements that are not available through the microscopic observation of stained roots (Sylvia et al. 1993). However, ergosterol content does not necessarily represent the exact fungal biomass. In fact, ergosterol content can vary with the age of fungal cultures, different substrates, and growing conditions (Bermingham et al. 1995; Sung et al. 1995). When grown under poor conditions, a fungus usually tends to differentiate highly vacuolated hyphae. This cellular behavior clearly stimulates membrane synthesis, especially at the hyphal tips, increasing consequently the percent of ergosterol content (Cooke and Rayner 1984; Wallander et al. 1997). However, ergosterol is, in fact, found much more abundantly at the margin of fungal colonies where hyphae are younger and fully viable (Wallander et al. 1997).

Antibus and Sinsabaugh (1993) reported a decrease of ergosterol contents from 2 to up to 50% in freeze-dried fungal mycelia. On the contrary, while analyzing ergosterol from very young roots and fungal cultures, Sung et al. (1995) recovered equivalent ergosterol contents before and after freeze drying. As stated by the authors, the loss of ergosterol reported by Antibus and

Sinsabaugh (1993) may be due as much to an inadequate freeze-drying process as to the high sensitivity of ergosterol to ultraviolet light (Newell et al. 1988).

The ratios of ergosterol/sistosterol have been proposed as a more appropriate measurement for the evaluation of the fungal biomass inside plant tissues (Antibus and Sinsabaugh 1993). Ergosterol content corresponds to the fungal biomass and sitosterol to the host plant. The ergosterol/sitosterol ratio may differ substantially according the plant-fungi partners. For example, it reached, respectively 0.45 and 0.54, in *Betula populifolia* and *Pinus sylvestris* colonized roots (Antibus and Sinsabaugh 1993).

Chitin, a cell wall component found in ectomycorrhizal fungi, has also been used to determine the level of root fungal colonization (Hepper 1977; Whipps et al. 1982; Plassard et al. 1983; Vignon et al. 1986). In comparison with ergosterol usage, the chitin assay is quantitatively less sensitive (Johnson and McGill 1990) but more sensitive to chemical interferences (Matcham et al. 1985) and may strongly interfere with root aldehydes (Plassard et al. 1983; Matcham et al. 1985). Chitin assays measure both viable and dead fungal tissues and appear to be much more affected by seasonal changes than the ergosterol assays (Johnson and McGill 1990; Wallander et al. 1997).

Better than the ergosterol content of mycorrhizal fungi, which is known to increase with increased phosphorus fertilization (Johnson and McGill 1990), the ergosterol/chitine ratio, which represents the ratio between living and dead fungal tissues, can be used to estimate fungal biomass (Newell et al. 1988). However, according to Wallander et al. (1997), who used this ratio, ergosterol remains the best indicator for the estimation of active root fungal biomass.

Other useful methods for evaluating the living fungal biomass in ectomycorrhizal roots are the staining of active hyphae with fluorescein diacetate (Söderstrom 1977) and the measurement of ^{14}C movement into the fungal partner (Jakobsen and Rosendahl 1990). Unfortunately, this latter method does not distinguish carbon bondings originating from living and non-living fungal cells.

2. In Endomycorrhizae

Root staining methods, followed by microscopic examination, and traditionally used to evaluate the AM root colonization levels, are tedious and time-consuming activities (Sylvia et al. 1993). Several workers have tried to find simple biochemical methods to quantify AM root colonization. Until now, such reliable methods remain to be developed.

In the early 1990s, ergosterol was proposed by Frey et al. (1992) as a potential indicator of AM fungal biomass inside colonized root tissues, as it has been done for more than a decade with ectomycorrhizal fungi. Ergosterol appeared as a minor peak eluting very close to other minor components of the sterol fraction. Minute amounts of ergosterol (0.052–$0.072\,\mu g\,mg^{-1}$ root dry weight) were detected in mycorrhizal roots of *Zea mays* and *Trifolium alexandrinum*, respectively, and appeared to be seven to nine times more abundant than in non-mycorrhizal roots. Antibus and Sinsabaugh (1993) also detected traces of ergosterol in *Plantago major* and *Acer rubrum* mycorrhizal roots (0.01–$0.13\,\mu g\,mg^{-1}$ root dry weight). In both cases, the respective ergosterol contents appeared to be 13 to 40 times lower than ergosterol contents found in ectomycorrhizal roots (Table 3).

The known sterol compositions of isolated AM mycelium and spores are not in agreement with the previous published results. The primary major sterols of AM fungi were found to be the 24-ethylcholesterol and the 24-methyl cholesterol (Beilby 1980; Beilby and Kidby 1980; Grandmougin-Ferjani et al. 1995, 1996, 1999). Beilby (1980) detected no ergosterol in spores of *Acaulospora laevis*. Similar results were obtained by Grandmougin-Ferjani et al. (1999) from analysis of the spores of 16 different Glomales species. Nordby et al. (1981) reported only traces of ergosterol in spores of *Glomus mosseae* and estimated their contents at less than 0.1% of the total sterol. Similarly, Frey et al. (1994) evaluated the ergosterol content of *G. intraradices* extraradical mycelia at $0.063\,\mu g\,mg^{-1}$, a value comparable to the ergosterol contents measured in *T. alexandrinum* and *Z. mays* colonized roots (Frey et al. 1992). No ergosterol was detected by Grandmougin-Ferjani et al. (1995, 1996) in any of the roots colonized with *G. intraradices*, either in those grown monoxenically on Ri T-DNA transformed roots (Bécard and Fortin 1988) or in those obtained from pot cultures (Grandmougin-Ferjani et al. 1995, 1996). These results suggest that the very low ergosterol contents detected by Frey et al. (1992) and Antibus and Sinsabaugh (1993) may have originated from some mycoparasites of AMF spores (Jeffries and Young 1994) or from some coinhabiting saprophytic fungal organisms. For these reasons, ergosterol should not be considered a suitable indicator for quantifying AM fungal biomass, in either soil or colonized roots.

Schmitz et al. (1991) proposed campesterol (24-methylcholesterol) and 24-methylenecholesterol as more appropriate indicators for AM fungal biomass. The 24-methylcholesterol concentrations were found significantly higher in AM colonized roots than in other root extracted free sterols (Ho 1977; Nagy et al. 1980; Nordby et al. 1981). Grandmougin-Ferjani et al. (1995) detected the presence of 24-methylenecholesterol inside roots of *Allium porum* colonized by *Glomus mosseae* seemingly a result out of the symbiotic association.

3. In Actinorhizae

Alnus incana, when associated with *Frankia* species, showed a growth decrease compared to the nonmycorrhizal plants (Ekblad et al. 1995). Ergosterol contents, used to evaluate the root fungal biomass, reached 0.015 to 0.15 $\mu g\,mg^{-1}$ root dry weight in colonized roots. The correlation found between fungal biomass and ergosterol concentration with the actinorrhizal association *Alnus incana/Frankia* sp. was found to be weak when compared to the ectomycorrhizal association of *Pinus sylvestris* with *Paxillus involutus* that were studied simultaneously. In fact, these results should be carefully analyzed because *Frankia* species, being prokaryotes, are assumed to contain no sterols (Ourisson and Rohmer 1982).

B. Specific Fatty Acids

Several fatty acids have been reported as specific to arbuscular mycorrhizal fungi. The C16:1Δ11 fatty acid has, until now, never been detected in any fungi other than the AMF. As such, it may be considered a reliable tool for the determination of fungal biomass in colonized roots. Attempts to quantify root fungal biomass were described by Graham et al. (1995) using *Citrus* roots. The amount of 16:1Δ11 found in *Citrus* roots was effectively correlated with the level of root colonization but not with the fungal dispersion inside roots. Graham et al. (1995) suggested that the 16:1Δ11 fatty acid concentration could be used to estimate the amount of carbon allocated to fungal growth and to lipid storage within roots. The 16:1Δ11 fatty acid may also be a valuable tool for the estimation of AMF biomass in soil. Olsson et al. (1995) reported an interesting correlation between the length of *Glomus* hyphae in soil and the combined levels of the C16:1Δ11 and the polyunsaturated C20:5 fatty acids. However, knowing that C16:1Δ11 fatty acid is commonly isolated from the soil Gram negative bacteria (Frostegaard et al. 1993; Olsson et al. 1995), its use as a lipid indicator may face severe limitations when used under natural conditions.

VII. General Conclusions and Future Research Avenues

The knowledge of mycorrhizal fungal lipids has improved tremendously during the past 20 years. Research was first oriented toward the development of biochemical tools for the evaluation of the fungal biomass inside plant root tissues or in soil. Sterols were the first components tentatively used. Now, ergosterol content is considered a reliable measurement, well established in both fundamental and applied research on ectomycorrhizae. With arbuscular mycorrhizal fungi, some unusual fatty acids such as the C16:1Δ11, C20:1, and C20:5, which are sometimes specific to Glomales species, may have the potential to become efficient markers of root an soil mycorrhizae. Unfortunately, until now, tentative tests were not found sufficiently reliable. Being essentially located in the storage lipids, their concentration with regard to dry fungal and plant root weight is subjected to large variations. Moreover, their presence in soil bacteria may confirm their use as soil fungal biomass indicators. Therefore, sterols remain the most promising avenue, provided that the specific sterol composition of the studied fungus is well known.

Knowing the obligate symbiotic status of many mycorrhizal fungi, exciting questions remain unanswered:

– Are mycorrhizal fungi able to synthesize their own lipids?
– Do they import them from their plant symbionts?
– Do plants provide mycorrhizal fungi with sterol precursors and let the fungal symbiont elabo-

rate their own lipid constituents and storage material?

– Is the extraradical mycelium capable of taking up xenobiotic compounds and using them as precursors for the biosynthesis of their own fatty acids and sterols?

In fact, the biosynthetic pathways of either fatty acids and sterols remain poorly known and, as such, should attract the interest of pluridisciplinary research teams. Scarce and controversial results highlight the need of more research in this area. The possibility of growing in vitro, even though at slow rates, pure cultures of ectomycorrhizal, ericoid, or orchid mycorrhizal fungi provide sufficient material to make this type of research possible.

As early as 1979, the relationship between plant photosynthesis and mycorrhizae was studied by Lösel and Cooper. Their impressive pioneer work now needs to be rediscovered and taken into account within studies of the carbon transfer from the atmosphere into the fungal lipids through the plant photosynthates. Novel technologies such as molecular biology and precise high tech measurement apparatus should considerably simplify and accelerate such research works. The lipid metabolism within the tripartite symbiotic association (legume/*Rhizobium*/ *Glomus*) should also provide valuable information, as would studies of the influence of mineral nutrition on lipid metabolism.

Probably due to their impact on the growth and health of the majority of cultivated plants, research efforts toward mycorrhizal lipid metabolism until now have been mainly oriented toward the study of arbuscular mycorrhizae. However, ectomycorrhizae, as well as ericoid and orchid mycorrhizae, should not be neglected, because they also provide excellent models for the development of research on the metabolism of lipids within mycorrhizal associations and as such must be encouraged.

Acknowledgments. The authors are grateful to Drs. J. Cayouette and S. Redhead from the Eastern Cereal Oilseed Research Centre, Agriculture and Agrifood Canada for their revision of the manuscript, to Drs. D.M. Lösel and J.L. Hilbert for providing them with many references, and to Mr. Joël Fontaine for his help. Special thanks are also addressed to Mrs. Corinne Ait-Said for her expert secretarial assistance.

References

Aho L, Kurkela R (1978) Free fatty acids of some wood mushrooms. Nahrung 7:603–607

Al-Karaki GN, Al-Raddad A, Clark RB (1998) Water stress and mycorrhizal isolate effect on growth and nutrient acquisition of wheat. J Plant Nutr 21:891–902

Allen WK, Allaway WG, Cox GC, Valder PG (1989) Ultrastructure of mycorrhizas of *Dracophyllum secundum* R. Br. (Ericales: Epacridaceae). Aust J Plant Physiol 16:147–153

Al-Shabibi MMA, Toma SJ, Haddad BA (1982) Studies of Iraqi truffles. I. Proximate analysis and characterization of lipids. Can Inst Food Sci Technol J 15:200–202

Amijee F, Stribley DP (1987) Soluble carbohydrates of vesicular-arbuscular mycorrhizal fungi. Mycologist 21:20–21

Antibus RK, Sinsabaugh RL (1993) The extraction and quantification of ergosterol from ectomycorrhizal fungi and roots. Mycorrhiza 3:137–144

Asselineau J (1966) Bacterial lipids. Hermann, Paris

Baath E, Frostegaard A, Fritze H (1992) Soil bacterial biomass, activity, phospholipid fatty acid pattern, and pH tolerance in an area polluted with alkaline dust deposition. Appl Environ Microbiol 58:4026–4031

Bago B, Donaire JP, Azcon-Aguilar C (1995) Biochemical characterization of membranes in arbuscular mycorrhiza: fatty acid analysis. In: 4th Eur Symp on Mycorrhizae, Granada, Spain (11–14 July 1994) pp 211–214

Barroso J, Neves HC, Pais MSS (1986) Isolation and chemical characterisation of odd fatty acids present in *Ophrys lutea* roots during invasion by the endophyte. In: Gianinazzi-Pearson V, Gianinazzi S (eds) Physiological and genetical aspects of mycorrhizae. INRA Dijon, pp 437–440

Barroso J, Neves HC, Pais MSS (1987) Production of free sterols by infected tubers of *Ophrys lutea* Con.: identification by gas chromatography-mass spectrometry. New Phytol 106:147–152

Bécard G, Fortin JA (1988) Early events of vesicular-arbuscular mycorrhiza formation on Ri T-DNA transformed roots. New Phytol 108:211–218

Bécard G, Piché Y (1988) New aspects on the acquisition of biotrophic status by a vesicular-arbuscular mycorrhizal fungus, *Gigaspora margarita*. New Phytol 112:77–83

Bécard G, Piché Y (1992) Establishment of vesicular-arbuscular mycorrhiza in root organ culture: review and proposed methodology. Methods in microbiology, vol 24. pp 90–108

Beilby JP (1980) Fatty acid and sterol composition of ungerminated spores of the vesicular arbuscular mycorrhizal fungus *Acaulospora laevis*. Lipids 15:949–952

Beilby JP (1983) Effects of inhibitors on early protein, RNA, and lipid synthesis in germinating vesicular-arbuscular mycorrhizal fungal spores of *Glomus caledonium*. Can J Microbiol 29:96–601

Beilby JP, Kidby DK (1980) Biochemistry of ungerminated and germinated spores of the vesicular-arbuscular mycorrhizal fungus, *Glomus caledonium*: changes in neutral and polar lipids. J Lipid Res 21:739–750

Bentivenga SP, Morton JB (1994a) Stability and heritability of fatty acid methyl ester profiles of glomalean endomycorrhizal fungi. Mycol Res 98:1419–1426

Bentivenga SP, Morton JB (1994b) Congruence of fatty acid methyl ester profiles and morphological characters of arbuscular mycorrhizal fungi in Gigasporaceae. Proc Natl Acad Sci USA 93:659–662

Bermingham S, Maltby L, Cooke RC (1995) A critical assessment of the validity of ergosterol as an indicator of fungal biomass. Mycol Res 99:479–484

Bethlenfalvay GJ, Mihara KL, Schreiner RP (1994) Mycorrhizae alter protein and lipid contents and yield of pea seeds. Crop Sci 34:998–1003

Bethlenfalvay GJ, Schreiner RP, Mihara KL (1997) Mycorrhizal fungi effect on nutrient composition and yield of soybean seeds. J Plant Nutr 20:581–591

Bevege DI, Bowen GD, Skinner MF (1975) Comparative carbohydrate physiology of ecto and endomycorrhizas. In: Sanders FE, Mosse B, Tinker PB (eds) Endomycorrhizas. Academic Press, London, pp 149–174

Birraux D, Fries N (1981) Germination of *Thelephora terrestris* basidiospores. Can J Bot 59:2062–2064

Blom J (1981) Utilization of fatty acids and NH_4 by *Frankia* AvcI1. FEMS Microbiol Lett 10:143–145

Blom J (1983) Carbon and nitrogen source requirements of *Frankia* strains. FEMS Microbiol Lett 13:51–55

Bokhary HA, Parvez S (1993) Chemical composition of desert truffles *Terfezia claveryi*. J Food Compos Anal 3:285–293

Bonfante-Fasolo P, Scannerini S (1977) Cytological observations on the mycorrhiza *Endogone flammicorona – Pinus strobus*. Allionia 22:23–54

Bonfante-Fasolo P, Grippiolo R (1984) Cytochemical and biochemical observations on the cell wall of the spore of *Glomus epigaeum*. Protoplasma 123:140–151

Bonfante-Fasolo P, Schubert A (1987) Spore wall architecture of *Glomus* spp. Can J Bot 65:539–546

Boullard B (1957) La mycotrophie chez les ptéridophytes. Sa fréquence, ses caractères, sa signification. Botaniste 45:5–185

Brussaard L, Behan-Pelletier VM, Bignell DE, Brown VK, Didden W, Folgarait P, Fragoso C, Wall D, Gupta VSR, Hattori T, Hawksworth DL, Klopatek C, Lavelle P, Malloch DW, Rusek J, Söderstrom B, Tiedje JM, Ross AV (1997) Biodiversity and ecosystem functioning in soil. Ambio 26:563–570

Burden RS, Cooke DT, Carter GA (1989) Inhibitors of sterol biosynthesis and growth in plants and fungi. Phytochemistry 28:1791–1804

Caldwell BA, Castellano MA, Griffiths RP (1991) Fatty acid esterase production by ectomycorrhizal fungi. Mycologia 83:233–236

Caporn SJM, Song W, Read DJ, Lee JA (1995) The effect of repeated nitrogen fertilization on mycorrhizal infection in heather *Calluna vulgaris* (L.) Hull. New Phytol 129:605–609

Casana M, Bonfante-Fasolo P (1982) Intercellular and arbuscular hyphae of *Glomus fasciculatum* (Thaxter) Gerd. and Trappe isolated after enzymic maceration. Allionia 25:17–25

Cerri R, De Simone F, Senatore F (1981) Sterols of three *Lactarius* species. Biochem Syst Ecol 19:247–248

Chen Xy, Hampp R (1993) Sugar uptake by protoplasts of the ectomycorrhizal fungus, *Amanita muscaria* (L. ex. Fr.) Hokler. New Phytol 125:601–608

Cooke RC, Rayner ADM (1984) Ecology of saprophytic fungi. Longman, London

Cooper KM, Lösel DM (1978) Lipid physiology of vesicular-arbuscular mycorrhiza. I. Composition of lipids in roots of onion, clover and ryegrass infected with *Glomus mosseae*. New Phytol 80:143–151

Cordell CE, Marx DH (1994) Effects of nursery cultural practices on management of specific ectomycorrhizae on bare root seedlings. In: Pfleger FL, Lindermann RG (eds) Mycorrhizae and plant health St. Paul (Minnesota). APS Press, pp 133–151

Cox G, Sanders FE, Tinker PB, Wild JA (1975) Ultrastructural evidence relating to host-endophyte transfer in a vesicular-arbusular mycorrhiza. In: Sanders FE, Mosse B, Tinker PB (eds) Endomycorrhizas. Academic Press, London, pp 297–311

Dalpé Y, Neumann PJ (1977) L'induction chez *Ceratocystis* de fructifications de types *Graphium* et *Leptographium* par des acides gras insaturés. Can J Bot 55:2159–2167

Davis MW, Lamar RT (1992) Evaluation of methods to extract ergosterol for quantification of soil fungal biomass. Soil Biol Biochem 24:189–198

De Kok LJ, Kuiper PJC, Bruins AP (1982) A polyunsaturated octadecanoic acid derivative, a major fatty acid in sporophores of *Cantharellus cibarius* Fr. In: Wintermans JFGM, Kuiper PJC (eds) Biochemistry and metabolism of plant lipids. Elsevier, Amsterdam, pp 47–50

Dell B, Malajczuk N, Thompson GT (1990) Ectomycorrhiza formation in *Eucalyptus*: VA tuberculate ectomycorrhiza of *Eucalyptus pilularis*. New Phytol 114:633–640

Dembitskii VM, Pechenkina EE (1991) Phospholipid and fatty acid compositions of higher fungi. Chem Nat Compd 27:155–156

De Simone F, Senatore F, Sica D, Zollo F (1979) Sterols from Badisiomycetes. Phytochemistry 18:1572–1573

Despaties S, Furlan V, Fortin JA (1989) Effects of successive applications of fosetyl-Al on growth of *Allium cepa* L. associated with endomycorrhizal fungi. Plant Soil 113:175–180

Dijkstra FI, Scheffers WA, Wiken TO (1972) Submerged growth of the cultivated mushroom *Agaricus bisporus*. Antonie van Leeuwenhoek Microbiol Serol 38:329–340

Diop TA, Plenchette C, Strullu DG (1994) Dual axenic culture of sheared-root inocula of vesicular-arbuscular. Mycorrhiza 5:17–22

Dodd JC, Jeffries P (1989) Effects of fungicides on three vesicular-arbuscular mycorrhizal fungi associated with winter wheat (*Triticum aestivum* L.). Biol Fertil Soils 7:120–128

Douds DD Jr, Johnson CR, Koch KE (1988) Carbon cost of the fungal symbiont relative to net leaf P accumulation in a split-root VA mycorrhizal symbiosis. Plant Physiol 86:491–496

Dugassa GD, Von Alten H, Schönbeck F (1996) Effects of arbuscular mycorrhiza (AM) on health of *Linum usitatissimum* L. infected by fungal pathogens. Plant Soil 185:173–182

Dzamic M, Miljkovic B, Zoric D (1985) Dry matter carbohydrates and lipids in some edible mushrooms in Toplica country, Yugoslavia. Agrohemija 0:143–158

Edwards HH, Gessner RV (1984) Light microscopy and transmission electron microscopy of English oak (*Quercus robur*) ectomycorrhizal short roots. Can J Bot 62:1327–1335

Eissenstat DM, Graham JH, Syvertsen JP, Drouillard DL (1993) Carbon economy of sour orange in relation to mycorrhizal colonization and phosphorus status. Ann Bot (Lond) 71:1–10

Ekblad ALF, Wallander H, Carlsson R, Huss-Danell K (1995) Fungal biomass in roots and extramatrical mycelium in relation to macronutrients and plant biomass of ectomycorrhizal *Pinus sylvestris* and *Alnus incana*. New Phytol 131:443–451

Elliott CG (1977) Sterols in fungi. Their function in fungal growth and reproduction. Adv Microb Physiol 15:121–173

Endo S, Zhiping G, Takagi T (1991) Lipid components of seven species of Basidiomycotina and three species of Ascomycotina. J Jpn Oil Chem Soc 40:574–577

Fanelli C, Fabbri AA, Passi S (1980) Growth requirements and lipid metabolism of *Aspergillus flavus*. Trans Br Mycol Soc 75:371–375

Fontaine J, Grandmougin-Ferjani A, Hartmann MA, Sancholle M (1998) Is the vesicular arbuscular mycorrhizal fungus *Glomus intraradices* able to synthesize its own lipids? In: Sanchez J, Cerra-Olmedo E, Martinez-Force E (eds) Advances on Plant Lipid Research. Universidad de Sevilla. Secretario de Publicaciones Sevilla, pp 560–563

Fonvieille JL, Touzé-Soulet JM, Kulifaj M, Montant C, Dargent R (1990) The composition of ascospores of *Tuber melanosporum* and of their isolated walls. CR Acad Sci III 30:557–563

Fox FM (1986) Ultrastructure and infectivity of sclerotium-like bodies of the ectomycorrhizal fungus *Hebeloma sacchariolens*, on birch (*Betula spp.*). Trans Br Mycol Soc 87:359–369

Frankland JC, Harisson AF (1985) Mycorrhizal infection of *Betula pendula* and *Acer pseudoplatanus*: relationship with seedling growth and soil factors. New Phytol 101:108–112

Franz F, Acker G (1995) Rhizomorphs of *Picea abies* ectomycorrhizae: ultrastructural aspects and elemental analysis (EELS and ESI) on hyphal inclusions. Nova Hedwigia 60(1–2):253–267

Frey B, Buse HR, Schüepp H (1992) Identification of ergosterol in vesicular-arbuscular mycorrhizae. Biol Fertil Soils 13:229–234

Frey B, Vilarino A, Schüepp H, Arines J (1994) Chitin and ergosterol content of extraradical and intraradical mycelium of the vesicular-arbuscular mycorrhizal fungus *Glomus intraradices*. Soil Biol Biochem 26:711–717

Fries N, Bardt M, Serck-Hanssen K (1985) Growth of ectomycorrhizal fungi stimulated by lipids from a pine root exudate. Plant Soil 86:287–290

Frostegaard A, Tunlid A, Baath E (1993) Phospholipids fatty acid composition biomass and activity of microbial communities from two soil types experimentally exposed to different heavy metals. Appl Environ Microbiol 59:3605–3617

Gaspar ML, Pollero RJ (1994) *Glomus antarcticum*: the lipids and fatty acid composition. Mycotaxon 51:129–136

Gaspar ML, Pollero RJ, Cabello MN (1994) Triacylglycerol consumption during spore germination of vesicular-arbuscular mycorrhizal fungi. J Am Oil Chem Soc 71:449–452

Gaspar L, Pollero R, Cabello M (1997) Variations in the lipid composition of alfalfa roots during colonization with the arbuscular mycorrhizal fungus *Glomus versiforme*. Mycology 89:37–82

Gnekow M, Marschner H (1989) Role of VA mycorrhiza in growth and mineral nutrition of apple (*Malus pumila* var. *domestica*) rootstock cuttings. Plant Soil 119:285–293

Graham JH, Hodge NC, Morton JB (1995) Fatty acid methyl ester profiles for characterization of glomalean fungi and their endomycorrhizae. Appl Environ Microbiol 61:58–64

Graham JH, Drouillard DL, Hodge NC (1996) Carbon economy of sour orange in response to different *Glomus* spp. Tree Physiol 16:1023–1029

Graham JH, Duncan LW, Eissenstat DM (1997) Carbohydrate allocation patterns in citrus genotypes as affected by phosphorus nutrition, mycorrhizal colonization and mycorrhizal dependency. New Phytol 135:335–343

Grandmougin-Ferjani A, Dalpé Y, Veignie E, Hartmann MA, Rafin C, Sancholle M (1995) Infection by arbuscular mycorrhizal fungus *Glomus mosseae* of leek plants (*Allium porrum* L.) effects of lipids. In: Kader JC, Mazliak P (eds) Plant lipid metabolism. Kluwer, Dordrecht, pp 444–446

Grandmougin-Ferjani A, Dalpé Y, Hartmann MA, Laruelle F, Couturier D, Sancholle M (1996) Taxonomic aspects of the sterol and 11-hexadecenoic acid (C16:1Δ11) distribution in arbuscular mycorrhizal spores. In: Williams JP, Khan MU, Lem NW (eds) Physiology, biochemistry and molecular biology of plant lipids. Kluwer, Dordrecht, pp 195–197

Grandmougin-Ferjani A, Dalpé Y, Hartmann MA, Laruelle F, Sancholle M (1999) Sterol distribution in arbuscular mycorrhizal fungi. Phytochemistry 50:1027–1031

Grenville DJ, Peterson RL, Piché Y (1985a) The development, structure, and histochemistry of sclerotia of ectomycorrhizal fungi: I. *Pisolithus tinctorius*. Can J Bot 63:1402–1411

Grenville DJ, Peterson RL, Piché Y (1985b) The development, structure, and histochemistry of sclerotia of ectomycorrhizal fungi: II. *Paxillus involutus*. Can J Bot 63:1412–1417

Griffith JM, Davis AJ, Grant BR (1992) Target site of fungicides to control Oomycetes. In: Köller W (ed) Sites of fungicides action. CRC Press, Boca Raton, Fl, pp 69–100

Hale MG, Moore LD, Orcutt DM (1981) Effects of gibberellinA3 and 2,4-D on plant and root exudate lipids and susceptibility of *Pythium muriotylum*. Soil Biol Biochem 13:395–399

Hampp R, Schaeffer C, Wallenda T, Stulten C, Johann P, Einig W (1994) Changes in carbon partitioning of allocation due to ectomycorrhiza formation. Biochemical evidence. Can J Bot 73:448–556

Harley JL (1989) The significance of mycorrhiza. Mycol Res 92:129–139

Harley JL, Smith SE (1983) Mycorrhizal symbiosis. Academic Press, London

Harriott PT, Khairallah L, Benson DR (1991) Isolation and structure of the lipid envelopes from the nitrogen-fixing vesicles of *Frankia* sp. strain CpI1. J Bacteriol 173:2061–2067

Harris D, Pacovsky RS, Paul EA (1985) Carbon economy of soybean-*Rhizobium-Glomus* associations. New Phytol 101:427–440

Hepper CM (1977) A colorimetric method for estimating vesicular-arbuscular mycorrhizal infection in roots. Soil Biol Biochem 9:15–18

Hepper CM (1979) Germination and growth of *Glomus caledonium* spores: the effects of inhibitors and nutrients. Soil Biol Biochem 11:269–277

Herr DG, Peterson RL (1996) Morphology, anatomy and histochemistry of *Fagus grandifolia* Ehrh. (North

American beech) ectomycorrhizas. Bot Acta 109: 64–71

Hetrick BAD, Wilson GT, Kitt DG, Schwab AP (1988) Effects of soil microorganisms on mycorrhizal contribution to growth of big bluestem grass in non sterile soil. Soil Biol Biochem 20:501–507

Hiroi M (1978) Identification of 6-oxooctadecanoic acid in mushroom, *Lactarius chrysorheus* Fr lipid. J Agric Chem Soc Jpn 52:351–353

Ho I (1977) Phytosterols in root systems of mycorrhizal and non-mycorrhizal *Zea mays* L. Lloydia 40:476–478

Ho I, Trappe JM (1973) Translocation of ^{14}C from *Festuca* plants to their endomycorrhizal fungi. Nature (Lond) 244:30

Hommel RK, Stegner S, Weber L, Kleber HP (1994) The effect of ammonium ions on glycolipid production by *Candida (Torulopsis) apicola*. Cell Technol (Leipzig) 42:192–197

Hua X, Liu G, Zhang X, Yu L, Zeng P, Huang D (1995) Study on mycorrhization of pine in nursery and field by cutting off primary root apex of young seedlings. For Res 8:535–543

Huss-Danell K (1997) Transley Review No. 93. Actinorhizal symbioses and their N$_2$ fixation. New Phytol 136: 375–405

Hutchinson LJ (1990) Studies on the systematics of ectomycorrhizal fungi in axenic culture: II. the enzymatic degradation of selected carbon and nitrogen compounds. Can J Bot 68:1522–1530

Jabaji-Hare SH (1988) Lipid and fatty acid profiles of some vesicular-arbuscular mycorrhizal fungi contribution to taxonomy. Mycologia 80:622–629

Jabaji-Hare SH, Kendrick WB (1985) Effect of fosetyl-Al on root exudation and on composition of extracts of mycorrhizal and non-mycorrhizal leek roots. Can J Plant Pathol 7:18–126

Jabaji-Hare SH, Deschene A, Kendrick B (1984) Lipid content and composition of vesicles of a vesicular-arbuscular mycorrhizal fungus. Mycologia 76:1024–1030

Jabaji-Hare SH, Piché Y, Fortin JA (1986) Isolation and structural characterization of soil borne auxiliary cells of *Gigaspora margarita*, a vesicular-arbuscular mycorrhizal fungus. New Phytol 103:77–784

Jakobsen I (1991) Carbon metabolism in mycorrhiza. In: Varma AK (ed) Methods in microbiology, vol 23. Academic Press, London, pp 149–180

Jakobsen I, Rosendahl L (1990) Carbon flow into soil and external hyphae from roots of mycorrhizal cucumber plants. New Phytol 115:77–83

Jeffries P, Young TWK (1994) Ecological aspects of mycoparasitism. In: Jeffries P, Young TWK (eds) Interfungal parasitic relationships. CAB International, Wallingford, pp 147–180

Johansen A, Finlay RD, Olsson PA (1996) Nitrogen metabolism of external hyphae of the arbuscular mycorrhizal fungus *Glomus intraradices*. New Phytol 133:705–712

Johnson BN, McGill WB (1990) Comparison of ergosterol and chitin as quantitative estimate of mycorrhizal infection and *Pinus contorta* seedling response to inoculation. Can J For Res 20:1125–1131

Karaboz I, Oner M (1988) The chemical composition and use as single cell protein of *Morchella conica* var. *costata* Vent. mycelium grown in submerged culture. Doga, Turk Bioyol Dergisi 12(3):190–196

Kling M, Jakobsen I (1997) Direct application of carbendazim and propiconazole at field rates to the external

mycelium of three arbuscular mycorrhizal fungi species: effect on ^{32}P transport and succinate dehydrogenase activity. Mycorrhiza 7:33–37

Köller W (1992) Antifungal agents with target sites in sterol functions and biosynthesis. In: Köller W (ed) Target sites of fungicides action. CRC Press, Boca Raton, Fl, pp 119–206

Kroppenstedt RM (1985) Fatty acids and menaquinone analysis of Actinomycetes and related organisms. In: Goodfellow M, Minnikin DE (eds) Chemical methods in bacterial systematics. Academic Press, London, pp 73–199

Kroppenstedt RM, Kutzner HJ (1978) Biochemical taxonomy of some problem Actinomycetes. In: Mordarski M, Kurylowicz W, Jeljaszewicz J (eds) Proc Int Symp on *Nocardia* and *Streptomyces*. Gustav Fischer, Stuttgart, pp 125–133

Kucey RMN, Paul EA (1983) Vesicular arbuscular mycorrhizal spore populations in various Saskatchewan Canada soils and the effect of inoculation with *Glomus mosseae* on faba bean *Vicia faba* growth in greenhouse and field trials. Can J Soil Sci 63:87–96

Kutaľeva NP, Tsapalova IE (1989) The biochemical composition of little known edible fungi from Siberia: *Lyophyllum decastes* (Fr.) Sing. and *Tricholoma caligatum* (VIV.) Rick Rastit Resur 25:278–283

Lamont HC, Silvester WB, Torrey JG (1988) Nile red fluorescence demonstrates lipid in the envelope of vesicles from N$_2$-fixing cultures of *Frankia*. Can J Microbiol 34:656–666

Larsen J, Olsson PA, Jakobsen I (1998) The use of fatty acid signatures to study mycelial interactions between the arbuscular mycorrhizal fungus *Glomus intraradices* and the saprotrophic fungus *Fusarium culmorum* in root-free soil. Mycol Res 102(12): 1491–1496

Lechevalier MP (1977) Lipids in bacterial taxonomy – a taxonomist's view. Crit Rev Microbiol 5:109–210

Lechevalier MP, Horrière R, Lechevalier HA (1982) The biology of *Frankia* and related organisms. Dev Ind Microbiol 23:51–60

Lechevalier MP, Baker D, Horrière F (1983) Physiology, chemistry, serology, and infectivity of two *Frankia* isolates from *Alnus incana* subsp. *rugosa*. Can J Bot 61:2826–2833

Lehrian DW, Shisler LC, Patton S (1976) The effects of linoleate and acetate on the growth and lipid composition of mycelium of *Agaricus bisporus*. Mycologia 68:453–462

Leu SW, Chang DCN (1989) Physiological studies on *Asparagus* mycorrhizae III. Histochemical studies on asparagus mycorrhizae. Mem Coll Agric Natl Taiwan Univ 29:118–123

Lindeberg G, Lindeberg M (1974) Effect of short chain fatty acids on the growth of some mycorrhizal and saprophytic hymenomycetes. Arch Microbiol 101:109–114

Liu CY, Zhou X (1987) Studies on the endomycorrhiza of *Galeola faberi* Rolfe. J Wuhan Bot Res 5:101–110

Lopez MF, Whaling CS, Torrey JG (1983) The polar lipids and free sugars of *Frankia* in culture. Can J Bot 61:2834–2842

Lösel DM (1980) The effect of biotrophic fungal infection on the lipid metabolism of green plants. In: Mazliak P, Benveniste P, Costes C, Douce R (eds) Biogenesis and function of plant lipids. Elsevier/North Holland, Amsterdam, pp 263–268

Lösel DM (1988) Fungal lipids. In: Ratledge C, Wilkinson SG (eds) Microbial lipids, vol 1. Academic Press, London, pp 699–806

Lösel DM (1989) Functions of lipids: Specialized roles in fungi and algae. In: Ratledge C, Wilkinson SG (eds) Microbial lipids, vol 2. Academic Press, London, pp 367–437

Lösel DM, Cooper KM (1979) Incorporation of ^{14}C-labelled substrates by uninfected and VA mycorrhizal roots of onion. New Phytol 83:415–426

Lösel DM, Sancholle M (1996) Fungal lipids. In: Prasad R, Ghannoum MA (eds) Lipids of pathogenic fungi. CRC Press, Boca Raton, FL, pp 27–62

Lu X, Koïde RT (1991) *Avena fortma* L. seedling nutrient dynamics as influenced by mycorrhizal infection of natural generation. Plant Cell Environ 14:931–939

Lynd JQ, Ansman TR (1989) Effects of phosphorus, calcium with four potassium levels on nodule histology, nitrogenase activity and improve Spanco peanut yields. J Plant Nutr 12:65–84

Marschner H (1996) Mineral nutrient acquisition in non-mycorrhizal and mycorrhizal plants. Phyton 36:61–68

Martin F, Canet D, Marchal JP (1984a) In vivo natural abundance of carbon 13 NMR studies of the carbohydrate storage in ectomycorrhizal fungi. Physiol Veg 22:733–744

Martin F, Canet D, Marchal JP, Brondeau J (1984b) In vivo natural abundance ^{13}C nuclear magnetic resonance studies of living ectomycorrhizal fungi. Plant Physiol 75:151–153

Martin F, Delaruelle C, Hilbert JL (1990) An improved ergosterol assay to estimate fungal biomass in ectomycorrhizas. Mycol Res 94:1059–1064

Marx DH, Cordell CE, France RC (1986) Effects of triadimefon on growth and ectomycorrhizal development of loblolly slash pines in nursery. Phytopathology 76:824–831

Massicotte HB, Ackerley CA, Peterson RL (1989) Ontogeny of *Alnus rubra, Alpova diplophloeus* ectomycorrhizae: II Transmission electron microscopy. Can J Bot 67:201–210

Matcham SE, Jordan BR, Wood DA (1985) Estimation of fungal biomass in a solid substrate by three independent methods. Appl Microbiol Biotechnol 21:108–112

Maudinas B, Chemardin M, Gadal P (1982) Fatty acid composition of root nodules of *Alnus* species. Phytochemistry 21:1271–1273

Meier R, Charvat I (1992) Germination of *Glomus mosseae* spores: procedure and ultrastructural analysis. Int J Plant Sci 15:541–549

Melhuish JH, Janerette CA (1979) The effects of the carbon-nitrogen ratios on carbohydrate, protein, lipid and fatty acid production in *Pisolithus tinctorius*. Proc 4th North American Conf on Mycorrhizae, 67 pp Fort Collins, Colorado USA (24–28 June 1979)

Melhuish JH Jr, Bean GA, Hacskaylo E (1972) Fatty acids and sterols of some mycorrhizal fungi. Phytopathology 62:77–778

Melin E (1962) Physiological aspects of mycorrhizae of forest trees. In: Kozlowski TT (ed) Tree growth. Ronald Press, New York, pp 247–263

Moore AEP, Ashford AE, Peterson RL (1991) Reserve substances in *Paxillus involutus* sclerotia: determination by histochemistry and X-ray microanalysis. Protoplasma 163:67–81

Morelli I, Pistelli L, Catalano S (1981) Constituents of *Clitocybe nebularis* and of *Hydnum repandum*. Fitoterapia 52(1):45–47

Mosse B, Bowen GD (1968) A key to the recognition of some *Endogone* spore types. Trans Br Mycol Soc 51:469–483

Nagahashi G, Douds DD Jr, Abney GD (1996) Phosphorus amendment inhibits hyphal branching of the VAM fungus *Gigaspora margarita* directly and indirectly through its effect on root exudation. Mycorrhiza 6:403–408

Nagy S, Nordby HE, Nemec S (1980) Composition of lipids in roots of six *Citrus* cultivars infected with the vesicular-arbuscular mycorrhizal fungus, *Glomus mosseae*. New Phytol 85:377–384

Nandan R, Raisuddin S (1992) Fungal degradation of industrial wastes and wastewater. In: Arora DK, Elander RP, Mukerji KJ (eds) Handbook of applied mycology, vol 4, Fungal biotechnology. Marcel Dekker, New York, pp 931–961

Nemec S (1981) Histochemical characteristics of *Glomus etunicatum* infection of *Citrus limon* fibrous roots. Can J Bot 59:609–617

Nemec S (1985) Influence of selected pesticides on *Glomus* species and their infection in *Citrus*. Plant Soil 84:133–137

Newcomb W, Wood SM (1987) Morphogenesis and fine structure of *Frankia* (Actinomycetales): the microsymbiont of nitrogen-fixing actinorhizal root nodules. Int Rev Cytol 109:1–88

Newcomb W, Peterson RL, Callaham D, Torrey JG (1978) Structure and host actinomycete interactions in developing root nodules of *Comptonia peregrina*. Can J Bot 56:502–531

Newell SY, Arsuffi TL, Fallon RD (1988) Fundamental procedures for determining ergosterol content of decaying plant material by liquid chromatography. Appl Environ Microbiol 54:1876–1879

Nordby HE, Nemec S, Nagy S (1981) Fatty acids and sterols asociated with *Citrus* root mycorrhizae. J Agric Food Chem 29:396–401

Nylund JE, Wallender H (1992) Ergosterol analysis as a mean of quantifying mycorrhizal biomas. In: Norris JR, Redd D, Varmd AK (eds) Methods in microbiology, vol 24, pp 77–88

Ogundero VW (1981) Degradation of nigerian palm product by thermophilic fungi. Trans Br Mycol Soc 77:267–271

Ohta A (1988) Effects of butyric acid and related compounds on basidiospore germination of some mycorrhizal fungi. Trans Mycol Soc Jpn 29:375–382

Olsson PA, Baath E, Jacobsen I, Söderstrom B (1995) The use of phospholipid and neutral lipid fatty acids to estimate biomass of arbuscular mycorrhizal fungi in soil. Mycol Res 99:623–639

Olsson PA, Bääth E, Jakobsen I (1997) Phophorus effects on the mycelium and storage structures of an arbuscular mycorrhizal fungus as studied in the soil and roots by analysis of fatty acid signatures. Appl Environ Microbiol 63:3531–3538

Olsson PA, Francis R, Read DJ, Söderström B (1998) Growth of arbuscular mycorrhizal mycelium in calcareous dune sand and its interaction with other soil microorganisms as estimated by measurement of specific fatty acids. Plant Soil 201:9–16

Ourisson G, Rohmer M (1982) Prokaryotic polyterpenes: phylogenetic precursors of sterols. In: Bronner

F, Kleinzeller A (eds) Current topics in membranes and transport, vol 17. Academic Press, London, pp 153–182

Pacovsky RS (1989) Metabolic difference in Zea-Glomus-Azospirillum symbioses. Soil Biol Biochem 21:953–960

Pacovsky RS, Fuller G (1987) Lipids of soybean inoculated with microsymbionts. In: Stumpf PK, Mudd JB, Nes WD (eds) The metabolism structure and function of plant lipids. Plenum Press, New York, pp 349–351

Pacovsky RS, Fuller G (1988) Mineral and lipid composition of Glycine, Glomus, Bradyrhizobium symbioses. Physiol Plant 72:733–746

Pang PC, Paul EA (1980) Effects of vesicular-arbuscular mycorrhiza on C and N distribution in nodulated faba beans. Can J Soil Sci 60:241–250

Paul EA, Kucey RMN (1981) Carbon flow in microbial associations. Science 213:473–474

Pedersen TA (1970) Effects of fatty acids and methyl octanoate on resting mycelium of Boletus variegatus. Physiol Plant 23:654–666

Peng S, Eissenstat DM, Graham JH, Williams K, Hodge NC (1993) Growth depression in mycorrhizal Citrus at high phosphorus supply. Analysis of carbon costs. Plant Physiol 101:1063–1071

Pfeffer PE, Douds DD, Becart G, Brouillette J, Bago B, Shach AR, Hill Y (1998) The uptake, metabolism and transport of different carbon substrates in VA mycorrhizal carrot roots. 2nd Int Conf on Mycorrhiza Uppsala, Sweden, 136 pp (5–10 July 1998)

Plassard C, Coll A, Mousain D (1983) Dosage de la chitine fongique: application à l'estimation de la masse mycélienne présente dans les racines mycorhizées du pin maritime cultivé in vitro ou en pépinière. C R Acad Sci III 297:233–236

Prostenik M, Burcar I, Castek A, Cosovic C, Golem J, Jandric Z, Kljaic K, Ondrusek V (1978) Lipids of higher fungi. III The fatty acids and 2-hydroxy-fatty acids in some species of basidiomycetes. Chem Phys Lipids 22:97–103

Ouispel A, Burggraaf H, Borsj H, Tak T (1983) The role of lipids in the growth of Frankia isolates. Can J Bot 61:2801–2806

Raper KB (1965) The genus Aspergillus. Williams and Wilkins, Baltimore

Ratnayake M, Leonard RT, Menge JA (1978) Root exudation in relation to supply of phosphorus and its possible relevance to mycorrhizal formation. New Phytol 81:543–552

Rolin D, Le Tacon F, Larher F (1984) Characterization of the different forms of phosphorus in the mycelium of the ectomycorrhizal fungus Hebeloma cylindrosporum cultivated in pure culture. New Phytol 98:335–344

Ruzic R, Gogala N, Jerman I (1997) Sinusoidal magnetic fields: Effects on the growth and ergosterol content in mycorrhizal fungi. Electro Magnetobiol 16(2):129–142

Salmanowicz B, Nylund JE (1988) High performance liquid chromatography determination of ergosterol as a measure of ectomycorrhizal infection in Scots pine. Eur J For Pathol 18:291–298

Salmanowicz B, Nylund JE, Wallander H (1990) High performance liquid chromatography assay of ergosterol: a technique to estimate fungal biomass in roots with ectomycorrhiza. Agric Ecosyst Environ 28:437–440

Samra A, Dumas-Gaudot E, Gianinazzi-Pearson V, Gianinazzi S (1996) Soluble proteins and polypeptide profiles of spores of arbuscular mycorrhizal fungi.

Interspecific variability and effects of host (myc+) and non-host (myc–) Pisum sativum root exudates. Agronomie (Paris) 16:709–719

Sancholle M, Dalpé Y (1993) Taxonomic relevance of fatty acids of arbuscular mycorrhizal fungi and related species. Mycotaxon 49:187–193

Sancholle M, Lösel DM (1995) Lipids in fungal biotechnology. In: Kück (ed) The Mycota II Genetics and biotechnology. Springer, Berlin Heidelberg New York, pp 339–367

Sancholle M, Weete JD, Touzé-Soulet JM (1984a) Composition of a plasma membrane enriched fraction from Taphrina deformans. Effects of propiconazole. In: Siegenthaler PA, Eichenberger W (eds) Structure, function and metabolism of plant lipids. Elsevier, Amsterdam, pp 347–352

Sancholle M, Weete JD, Montant C (1984b) Effects of triazoles on fungi. I. Growth and celllular permeability. Pest Biochem Physiol 21:31–44

Sancholle M, Dargent R, Weete JD, Rushing AE, Miller KS, Montant C (1988) Effects of triazoles on fungi. IV. Ultrastructure of Taphrina deformans. Mycologia 80:162–175

Scannerini S, Bonfante-Fasolo PB (1975) Preliminary data on the ultrastructure of intracelllular vesicles in endomycorrhiza of Ornithogalum umbellatum L. Atti Accad Sci Torino 109:619–621

Schisler LC, Volkoff O (1977) The effects of safflower oil on mycelial growth of Boletacea in submerged liquid cultures. Mycologia 69:118–125

Schmitz O, Danneberg G, Hundeshagen B, Klingner A, Bothe H (1991) Quantification of vesicular-arbuscular mycorrhiza by biochemical parameters. J Plant Physiol 139:106–114

Schreiner RP, Bethlenfalvay GJ (1997) Mycorrhizae, biocides, and biocontrol 3. Effects of three different fungicides on developmental stages of three AM fungi. Biol Fertil Soils 24:18–26

Schubert A, Wys P, Wiemken A (1992) Occurrence of trehalose in vesicular-arbuscular mycorrhizal fungi and in in mycorrhizal roots. J Plant Physiol 140:41–45

Seitz LM, Mohr HE, Burroughs R, Sauer DB (1977) Ergosterol as an indicator of fungal invasion in grains. Cereal Chem 54:1207–1217

Seitz LM, Sauer DB, Burroughs R, Mohr HE, Hubbard JD (1979) Ergosterol as a measure of fungal growth. Phytopathology 69:1202–1203

Selvaraj T, Subramanian G (1990) Phenols and lipids in mycorrhizal and non-mycorrhizal roots of Sesamum indicum. Curr Sci 59:471–473

Senatore F (1988) Chemical constituents of some species of Agaricaceae. Biochem Syst Ecol 16:601–604

Senatore F, Dini A, Marino A, Schettino O (1988) Chemical constituents of some Basidiomycetes. J Sci Food Agric 45(4):337–345

Siegel RR (1981) Sterol-inhibiting fungicides: effects on sterol biosynthesis and sites of action. Plant Dis 65:986–989

Smith SE, Gianinazzi-Pearson V (1988) Physiological interactions between symbionts in vesicular-arbuscular mycorrhizal plants. Annu Rev Plant Phys Plant Mol Biol 39:221–244

Snellgrove RC, Splittstoesser WE, Stribley DP, Tinker PB (1982) The distibution of carbon and the demand of the fungal symbiont in leek plants with vesicular-arbuscular mycorrhizae. New Phytol 92:75–87

Söderstrom B (1977) Vital staining of fungi in pure culture and in soil with fluorescein diacetate. Soil Biol Biochem 9:59–63

Solberg Y (1989) A literature review of the lipid constituents of higher fungi new investigations of Agaricales species. Int J Mycol Lichenol 4:137–154

Strullu DG, Chamel A, Eloy JF, Gourret JP (1983) Ultrastructure and analysis, by laser probe mass spectrography, of the mineral composition of the vesicles of *Trifolium pratense* endomycorrhizas. New Phytol 94:81–88

Sugai A, Itoh T, Kanako H, Kinjoj N, Muramatsu T (1986) Pyrophosphatidic acid in mushrooms. Lipids 21:666–668

Sumner JL (1973) The fatty acid composition of Basidiomycetes. N Z J Bot 11:435–442

Sung SJS, White LM, Marx DH, Otrosina WJ (1995) Seasonal ectomycorrhizal fungal biomass development on loblolly pine (*Pinus taeda* L.) seedling. Mycorrhiza 5:439–447

Sward RJ (1981a) The structure of the spores of *Gigaspora margarita*. I. The dormant spore. New Phytol 87:761–768

Sward RJ (1981b) The structure of the spores of *Gigaspora margarita*. III. Germ-tube emergence and growth. New Phytol 88:667–673

Sylvia DM, Wilson DO, Graham JH, Maddo JJ, Millner P, Morton JB, Skipper HD, Wright SF, Jarstfer AG (1993) Evaluation of vesicular-arbuscular mycorrhizal fungi in diverse plants and soils. Soil Biol Biochem 25:705–713

Taber WA, Taber RA (1982) Nutrition and respiration of basidiospores and mycelium of *Pisolithus tinctorius*. Phytopathology 72:316–322

Tanaka Y, Kawahara S, Eng AH, Takei A, Ohya N (1994) Structure of *cis*-polyisoprene from *Lactarius* mushrooms. Acta Biochim Pol 41:303–309

Tawaraya K, Watanabe S, Yoshida E, Wagatsuma T (1996) Effect of onion (*Allium cepa*) root exudates on the hyphal growth of *Gigaspora margarita*. Mycorrhiza 6:57–59

Thompson LK, Hale MG (1983) Effects of kinetin in the rooting medium on root exudation of free fatty acids and sterols from roots of *Arachis hypogea* L. "Argentine" under axenic conditions. Soil Biol Biochem 15:125–126

Torrey JG, Callaham D (1982) Structural features of the vesicle of *Frankia* sp. CpI1 in culture. Can J Microbiol 28:749–757

Trappe JM (1972) Fungus associates of ectotrophic mycorrhizae. Bot Rev 28:508–606

Trent JD, Svejcar TJ, Christiansen S (1989) Effects of fumigation on growth, photosynthesis water relations and mycorrhizal development of winter wheat in the field. Can J Plant Sci 69:535–540

Tunlid A, Schultz NA, Benson DR, Steele DB, White DC (1989) Differences in fatty acid composition between vegetative cells and N2-fixing vesicles of *Frankia* sp. strain Cp11. Proc Natl Acad Sci USA 86:3399–3403

Turner WD, Aldrige B (1983) Fungal metabolites. Academic Press, London

Vaskovsky VE, Khotimchenko SV, Benson AA (1991) Identification of diacylglycero-4'-o-n n n-trimethylhomoserine in mushrooms. Lipids 326:254–256

Vignon C, Plassard C, Mousain D, Salsac L (1986) Assay of fungal chitin and estimation of mycorrhizal infection. Physiol Vég 24:201–207

von Alten H, Lindermann A, Schonnbeck F (1993) Stimulation of vesicular-arbuscular mycorrhiza by fungicides or rhizosphere bacteria. Mycorrhiza 2:167–173

Vrkoc J, Budesinsky M, Dolejs L (1976) Constituents of the basidiomycete *Scleroderma aurantium*. Phytochemistry 15:1782–1784

Wallander H, Massicotte HB, Nylund JE (1997) Seasonal variation in protein, ergosterol and chitn in five morphotypes of *Pinus sylvestris* L. ectomycorrhizae in a mature Swedish forest. Soil Biol Biochem 29:45–53

Wallenda T, Schaeffer C, Einig W, Wingler A, Hampp U, Seith B, George E, Marschner H (1996) Effects of varied soil nitrogen supply on Norway spruce (*Picea abies* [L.] Karst.) Plant Soil 186:361–369

Wang GM, Coleman DC, Freckman DW, Dyer MI, McNaughton SJ, Acra MA, Goeschl JD (1989) Carbon partioning patterns of mycorrhizal versus nonmycorrhizal plants: real-time dynamic measurements using CO_2. New Phytol 112:489–493

Wardle KS, Schisler LC (1969) The effects of various lipids on growth of mycelium of *Agaricus bisporus*. Mycologia 61:305–314

Wathelet JP, Severin M, Impens R (1972) Etude des lipides de *Morchella rotunda* Pers. Analyse des acids gras. Bull Rech Agron Gembloux 7:350–357

Weete JD (1980) Lipid biochemistry of fungi and other organisms. Plenum Press, New York

Weete JD (1989) Structure and function of sterols in fungi. Adv Lipid Res 23:115–167

Weete JD, Gandhi SR (1986) Biochemistry and molecular biology of fungal sterols. In: Bramble R, Marzluf GA (eds) The Mycota III. Springer, Berlin Heidelberg New York, pp 421–438

Weete JD, Gandhi SR (1997) Sterols of the phylum Zygomycota: phylogenetic implications. Lipids 32: 1309–1316

Weete JD, Kulifaj M, Montant C, Nes WR, Sancholle M (1985a) Distribution of sterols in fungi II Brassicasterol in *Tuber* and *Terfezia* species. Can J Microbiol 31:1127–1130

Weete JD, Sancholle M, Touzé-Soulet JM, Bradley J, Dargent R (1985b) Effects of triazoles on fungi. III. Composition of a plasma membrane-enriched fraction of *Taphrina deformans*. Biochim Biophys Acta 812:633–642

Weete JD, Sancholle M, Patterson KA, Miller KS, Huang MQ, Campbell F, Van den Reek M (1991) Fatty acid metabolism in *Taphrina deformans* treated with sterol biosynthesis inhibitors. Lipids 26:669–674

Whipps JM, Haselwandter K, McGee EEM, Lewis DH (1982) Use of biochemical markers to determine growth, development and biomass of fungi in infected tissues, with particular reference to antagonistic and mutualistic biotrophs. Trans Br Mycol Soc 79:385–400

Yokokawa H (1994) Sterol compositions of the fruit-bodies of higher fungi. 5th Int Mycological Congr Abstr, Vancouver BC, Canada, 250 pp (14–21 July 1994)

Yokokawa H, Mitsuhashi T (1981) The sterol composition of mushrooms. Phytochemistry 206:1349–1351

Zel J, Svetek J, Crne H, Schara M (1993) Effects of aluminium on membrane fluidity of the mycorrhizal fungus *Amanita muscaria*. Physiol Plant 89:172–176

Zhuk YT, Tsapalova IE, Stepanova EN (1981) Lipids of some edible fungi growing in Siberia. Rastit Resur 17:109–114 (in Russian)

6 Arbuscular Mycorrhiza – a Key Component of Sustainable Plant-Soil Ecosystems

P. Jeffries[1] and J.M. Barea[2]

CONTENTS

I. Introduction

Sustainable plant-soil ecosystems occur when utilization of mineral resources by plants is balanced by efficient biogeochemical cycling, such that nutrients are not rapidly exhausted and plant communities can exist in a stable form for prolonged periods. Sustainable ecosystems are the target of ecologically sound management strategies in agriculture as well as being the normal situation in natural, undisturbed ecosystems. The concept of sustainability in agriculture aims to conserve the productive capacity of the soil, minimizing energy and resource use, and optimizing the rate of turnover and recycling of matter and nutrients. Sustainability demands the efficient utilization of nutrients by plants and this process is facilitated through mycorrhizal associations. The critical role that mycorrhizal symbioses play in plant nutrition (Smith and Read 1997) is now widely accepted. Almost all plants form mycorrhizal associations and a variety of types are formed depending on the plant taxa involved. However, it is the arbuscular mycorrhiza (AM) relationship that is most common, and over 80% of plant species are capable of forming these structures when associated with arbuscular mycorrhizal fungi (AMF). The AM symbiosis has an ancient origin and the rhizomes of the first primitive terrestrial plants contained fungal structures almost identical to their modern counterparts (e.g., Remy et al. 1994). As a consequence of this coevolution, the benefits of the relationship became an integral component of plant ecology in both natural and agricultural ecosystems. In this chapter we consider the concepts implicit in sustainable ecosystems and describe how the AM symbiosis is crucial in maintaining sustainability. We also discuss how the biodiversity of AMF populations can influence plant community dynamics in both natural and agricultural environments, and give examples where manipulation of the AM symbiosis can help restore sustainability to disturbed environments. Adverse conditions of differing origin affect the stability of both natural and agricultural systems, and thus plants must be able to cope with these stresses. In sustainable agriculture the AM symbiosis plays a key role in helping the plant not only to survive but to be productive under adversity (Mosse 1986).

II. Sustainable Plant-Soil Ecosystems

The concept of sustainability is often used in relation to environmental changes that have been induced as a result of the activities of mankind. Indeed, sustainability may be defined as "the successful management of resources to satisfy changing human needs while maintaining or enhancing the quality of the environment and conserving resources" (Bohlool et al. 1992). In this context, sustainability also takes into account the protection of natural resources, such as soil water, non-renewable energy resources and environmental

[1] Research School of Biosciences, University of Kent, Canterbury, Kent CT2 7NJ, UK
[2] Departamento de Microbiología del Suelo y Sistemas Simbióticos, Estación Experimental del Zaidín, CSIC, Prof. Albareda 1, 18008 Granada, Spain

The Mycota IX
Fungal Associations
Hock (Ed.)
© Springer-Verlag Berlin Heidelberg 2001

quality (Ladha 1992). A sustainable ecosystem approach involves the rational use of natural resources rather than their exploitation. It is based on the use of renewable inputs and in the optimization of resource utilization to reach a balanced environmental relationship (Lal 1989). To achieve sustainability it is necessary to prevent both environmental pollution and depletion of agricultural and forestry resources, whilst preserving the structure and diversity of natural plant communities. A sustainable approach also aims to provide vegetation cover, thus reducing the susceptibility of soils to erosion, and also to lower the energy-based inputs (Lal 1989; Bethlenfalvay and Linderman 1992; Peoples and Craswell 1992). These concepts apply to the preservation of natural plant communities but the term sustainable has been increasingly applied to agriculture as attempts are made to provide long-term sustained yields through the use of ecologically sound management technologies such as crop diversification, organic soil management, and biological pest and disease control (Altieri 1994).

Sustainable agriculture can be envisaged as a key component of current and future trends in plant productivity for both developing and developed countries. In the 1990 Food, Agriculture, Conservation, and Trade Act (GAO/RCED 1992), the US Congress discussed sustainable agriculture as "an integrated system of plant and animal production practices having a site-specific application that will, over the long term: (a) satisfy human food and fibre needs; (b) enhance environmental quality and the natural resource base upon which the agricultural economy depends; (c) make the most efficient use of non-renewable resources and on-farm resources and integrate, where appropriate, natural biological cycles and controls; (d) sustain the economic viability of farm operations, and (e) enhance the quality of life for farmers and society as a whole".

Sustainable agriculture can thus be considered as the maintenance of soil fertility and structure over a long period of time such that the economic yields from crop plants can be achieved through the minimum inputs of fertilizer necessary to reach such yields. It must achieve economic and sustained production and yet also preserve the resource base. However, true sustainability in which outputs are balanced by inputs may only be achieved in stable natural ecosystems. The term sustainable agriculture is a contradiction, as agriculture inevitably involves the artificial manipulation of plant ecosystems for production of food. Inputs are thus no longer balanced to outputs, and natural relationships are disturbed, diversity is decreased, and the sustainability is altered. In agricultural development, priority is often given to increasing yields and this can result in a dependence on the high inputs of artificial readily soluble fertilizers that are necessary to sustain them. It is difficult to develop any form of agriculture that could be truly sustainable. Instead, a modification of existing strategies is necessary in sustainable agriculture such that fertilizer inputs are reduced but not eliminated, and that maximum use is made of the soil microbiota in efficient nutrient capture and in cycling nutrients to the plant root system. In this context, mycorrhizal relationships are particularly important and will be discussed in the next section.

III. Importance of Arbuscular Mycorrhiza in Soil Fertility

This topic has been reviewed in depth on a number of occasions (e.g., see Bethlenfalvay and Lindermann 1992; Jeffries and Barea 1994) but a number of recent publications have highlighted the extent that AMF play in natural plant communities (e.g., Clapp et al. 1995; Merryweather and Fitter 1998; van der Heijden 1998a,b) and thus contribute to the maintenance of soil fertility. These studies have reinforced the view that the soil microbiota is a key component of sustainable systems when used as a natural resource tool (Bethlenfalvay and Linderman 1992; Kennedy and Smith 1995). The diversity of the microbiota is a major issue (Kennedy and Smith 1995) and molecular tools now offer the means to study diversity (Pace 1997). Microorganisms conduct activities which are crucial to the establishment, development, nutrition and health of plants (Azcón-Aguilar and Barea 1992; Linderman 1992; Barea et al. 1993, 1997). Microbial populations in soil actively develop around plant roots, within the rhizosphere, where they are stimulated by root exudates, plant residues and other organic substrates supplied by the plant. The beneficial activities of the rhizosphere microbiota include the increased availability of plant nutrients, improvement of nutrient uptake, and protection against root pathogens (Kloepper et al. 1991; Azcón and Barea 1992; Azcón-Aguilar and Barea 1992, 1996;

Kloepper 1992; Linderman 1992; Glick 1995; Barea et al. 1997). Mycorrhizal associations play a major role in these processes. For example, colonization by AMF improves the nutritional status of the plant and enables it to cope more effectively with cultural or environmental stress (Jeffries 1987; Barea 1991; Bethlenfalvay 1992; Johnson and Pfleger 1992; Sylvia and Williams 1992; Barea et al. 1983, 1997; Abbott et al. 1995). In addition, the soil mycelium of AMF is known to develop bridges connecting the root with the surrounding soil particles to improve both nutrient acquisition by the plant and soil structure (Bethlenfalvay 1992; Miller and Jastrow 1992a,b). However, in the case of an intercrop of *Hevea brasiliensis* with *Pueraria phaseoloides*, there was not a significant transfer of nutrients between plants (Ikram et al. 1994). Fitter et al. (1998) suggest, at least in the case of C, that transfer of this element is significant between fungal colonies but does not have an impact on plant carbon budgets. Whatever the case, the hyphal network within the soil is a vital component of the soil ecosystem and is the functional organ for the uptake and translocation of nutrients to and from mycorrhizal stuctures. Many reviews (e.g., Smith and Read 1997) describe the well-established role of the extraradical mycelium in the uptake of water and mineral nutrients. The network is essential for the continued cycling of nutrients within the plant community, and, once it is lost, nutrient sequestration or leaching will occur at a faster rate than in its presence. The hyphal network is thus vital for the maintenance of sustainable plant yields, both in a natural ecosystem and in low-input agricultural situations where nutrients are limiting. In a study of a tropical grazing ecosystem, the amount of AM mycelium in the soil was found to be negatively correlated with soil fertility, but was positively correlated with the plant nutrient content (McNaughton and Oesterheld 1990). These results suggested that AM associations act to stabilize ecosystem nutrient fluxes across edaphic gradients, and that they functionally compensate for wide variation in soil fertility since the nutritional status of the plants was not proportionally affected by differences in soil nutritional status. Mycorrhizal plants at sites of low fertility maintained mineral contents similar to, or only slightly lower than, those found at sites of high fertility with low mycorrhizal abundance.

The mycorrhizal mycelium of an undisturbed soil is extensive and can provide the conduit for the uptake of slowly exchangeable soil ions and sparingly soluble mineral ions that form the basis of sustainable systems. Different genera of AMF have unique patterns of mycelial development (Boddington and Dodd 1998) and also different mechanisms for control of P transfer to the host. It is clear that P, C, N and other mineral nutrients can be transported from remote sources in the soil by hyphae of AMF (e.g., Cooper and Tinker 1978; Ames et al. 1983; Francis and Read 1984). The fungi take up P from the same pool of soluble ions as do roots, and thus act as an extension to the root system. There is evidence that phosphatase activity is higher in the rhizosphere soil around arbuscular mycorrhiza than around that of non-mycorrhizal roots (Dodd et al. 1987; Azcón and Barea 1997) but there is no clear evidence that this is a fungus-mediated phenomenon that allows alternative P sources to be accessed. Many studies have also demonstrated that interplant bridges formed by AMF can provide channels for direct nutrient transfer between the AM of different plants. This nutrient transfer may be sufficient to sustain significant enhancement of both growth and nutrient composition of receiver plants, in some cases within 6 weeks of commencement of experimentation (Francis et al. 1986). In other cases, this route of nutrient transfer may be insignificant (Ikram et al. 1994).

In order to function effectively, the mycelium must be allowed to spread into the soil and remain intact. Destruction of the hyphae can occur through soil disturbance. For example, tillage or similar agricultural practices which disrupt the mycelial network have serious effects on its capacity to translocate nutrients over any significant distance. Grazing by small invertebrates can also destroy parts of the mycelial network. As much of the soil biomass may consist of mycorrhizal hyphae, they provide a major food source for the mycophagic soil fauna (Fitter 1985).

Mycorrhizal hyphae are involved in the formation of stable soil aggregates, a process crucial for soil conservation (Bethlenfalvay and Schüepp 1994). This effect is critical in the development of sustainable ecosystems especially in eroding soils. The soil mycelium first develops as a skeletal structure which holds soil particles by simple entanglement. Later, the roots and hyphae provide conditions necessary to form microaggregates by means of physicochemical mechanisms which involve the incorporation of organic debris (Tisdall 1991). Binding agents, mostly of microbial origin, participate in the process of cementation and stabilization of microaggregates. Wright and Upadhyaya (1998) demonstrated that hyphae of AMF produce copious amounts of a recalcitrant glycoprotein termed glomalin, the presence of which was closely correlated with soil aggregate stability. Microaggregates themselves may then be held together in macroaggregates by an enmeshing ability of AM hyphae, and may also be further stabilized by biotically mediated cementation (Tisdall and Oades 1980; Oades 1984; Elliot and Coleman 1988; Gupta and Germida 1988; Jastrow and Miller 1991). The soil aggregates, containing organic residues, behave as a nutrient reserve (Miller and Jastrow 1992a). The activity of the mycelium of AMF in these processes is of great relevance in sustainable systems, and is paralleled by the activity of actinomycetes which act in the same way to bind smaller aggregates. Mycorrhizosphere interactions are crucial for aggregate formation and stability (Andrade et al. 1998c; Bethlenfalvay et al. 1997).

IV. Biodiversity of Arbuscular Mycorrhizal Fungi and Relevance to Sustainability

As discussed earlier, AM symbioses enhance the relative abilities of plants to compete for limiting nutrients. Integration of individual plants into the collective nutrient-gathering capacity of the community assures survival, and the role of the soil mycelium is again crucial. In natural ecosystems plants with AM characteristically grow as heavily infected individuals occurring within communities of mixed species (Read et al. 1976). In a stable ecosystem these individuals are linked by the soil mycelium of the mycorrhizal network. The roots of new seedlings which enter this mycotrophic ecosystem become rapidly colonized and thus integrated into the community, providing an enormous increase in the absorptive surface at a relatively low energy cost. This phenomenon can give the plant access to resources otherwise tapped exclusively by the established plants (Read 1991). Alternatively, the non-mycotrophic plants are at a disadvantage in this respect and do not easily invade undisturbed mycorrhizal communities.

Our understanding of the functional ecology of mycorrhiza has been limited until recently, but progress has been reported which partly explains

how AM population diversity can determine the aboveground plant community. Initial studies were made by Grime et al. (1987), who used laboratory microcosms to show that the presence of mycorrhiza leads to an increase in the diversity of plant species. This was either through suppression of non-mycotrophic dominants or by enhancing species abundance through an increase in the biomass of some of the mycotrophic subordinate species via hyphal interconnections (Hetrick et al. 1989). It has also been suggested that mycorrhizal presence may reduce diversity if the symbiosis is of relatively greater benefit to the dominant species within the community. To test mycorrhizal contributions to an early successional plant community in the field, Gange et al. (1990) used a fungicide to reduce AM formation during the first year of colonization of bare soil by plants. Fewer plant species established in communities treated with fungicide, supporting the idea that AM formation promotes plant diversity. Thus, mycotrophic relationships may influence patterns of species diversity and relative abundance, once a mycorrhizal status of a plant community has been restored in a previously disturbed ecosystem.

Definitive evidence has recently been obtained that demonstrates that mycorrhizal diversity determines plant community structure, plant diversity and ecosystem variability (van der Heijden et al. 1998a). Laboratory microcosms and field plots were used to investigate how the diversity of AMF in the soil influenced plant communities in two characteristically species-rich natural grasslands. Plant productivity and the diversity of plants were correlated with AMF diversity, and different AMF promoted the growth of different plant species. In the field experiments, from 0 to 14 different isolates of AMF were added per treatment. As the number of species added was increased, the collective biomass of plant shoots and roots increased, along with the diversity of plants in each treatment. These experiments give a clear indication of differential functional compatability between AMF with different host plants. This could be accomplished via increased P nutrition (better capture or transfer of P from fungus to plant) or through other beneficial effects of mycorrhization. The loss of AMF biodiversity that occurs in intensive agricultural systems could thus decrease both plant diversity and productivity when the soil is returned to non-agricultural uses or if the system is changed towards more sustainable low-input practices. These results were mirrored in further experiments using three plant species from a calcareous grassland inoculated with mixtures of four AMF (van der Heijden et al. 1998a). Again it was concluded that AMF species that co-occur as natural AMF communities have the potential to determine plant community structure as a result of their differential effects on plant growth. The results also indicated that a single measurement of mycorrhizal dependency of a particular plant may be too simplistic if based only on the response to a single AMF (van der Heijden et al. 1998a).

The influence of AM on the diversity of the plant communities has been discussed above, but the converse should also be considered, i.e., fungal biodiversity is influenced by the above-ground plant community. The maintenance of fungal biodiversity is important since, as discussed above, differential benefits can be conferred to the plant hosts by different AMF, depending on host plant or environmental conditions (Sieverding and Howeler 1985; Heijden van der et al. 1998a,b). It is also clear that different isolates of the same species differ in their effectiveness in conferring benefits to similar host plants grown under standard conditions (Bethlenfalvay et al. 1989). In a natural ecosystem these mixed populations coexist, with certain fungi becoming dominant in particular patches and subsequently being replaced as environmental conditions change. Once this equilibrium is disturbed, for example by crop monoculture or by the use of fungicides, the population dynamics are disrupted and a bias can develop towards a few or even one dominant fungus (Johnson et al. 1992). In some environments tillage and fertilizer use have led to fewer species of AMF being found in the soil (Schenck and Kinloch 1980; Daniels Hetrick and Bloom 1983), while, more unusually, in others agricultural use may lead to greater diversity (Abbott and Robson 1977). When non-mycotrophic crops such as sugarbeet (*Beta vulgaris*) are grown in monoculture, there is evidence for a general decline in mycorrhizal propagules. Similar results were obtained by Gavito and Miller (1998), who noted a delay in mycorrhiza development in maize after cropping with canola, a non-host plant species. The longer the soil is cropped with a non-host species, the more pronounced will be the loss of mycorrhizal infectivity and diversity within the soil.

Where sustainable agriculture is desired, it is necessary to maintain a diversity of mycorrhizal fungi as mixed crops will benefit more from a mixed population of symbionts. The importance of diversity in a minimum-input agricultural system is illustrated by the data of Dodd et al. (1990b). These workers examined changes in spore numbers of individual AMF within the native populations of a savanna ecosystem in Colombia that resulted from different management practices. Twelve spore types were identified in the original soils and it was noted that different spore populations developed rapidly under different crop regimes. For example, spores of *Acaulospora myriocarpa* and *Glomus occultum* increased in subplots of *Sorghum*, while those of *Acaulospora mellea*, *Acaulospora morrowae*, and *Entrophospora colombiana* dominated subplots where cowpea (*Vigna unguiculata*) was grown following a crop of kudzu (*Pueraria phaseoloides*) (Dodd et al. 1990b). Spore numbers reflected the relative abundance of individual species within the populations, but this did not necessarily relate to their infectivity or effectiveness in stimulating plant growth (Clapp et al. 1995; Merryweather and Fitter 1998).

In non-agricultural situations, plant diversity has been related to AM diversity by Bethlenfalvay et al. (1984), who concluded that where the majority of plants are AM hosts, diversity in the indigenous population of AMF is directly related to plant diversity. It follows that maintenance of a sustainable mixed plant population depends on the maintenance of a diverse AMF population and vice versa. In some natural ecosystems there is a close positive correlation between plant cover and spore numbers (Anderson et al. 1984). The abundance of AM spores and the diversity of AM fungal species are known to increase as sand dunes increase in age and stability (Koske 1975; Koske and Halvorson 1981; Read 1989). In natural soils dominated by perennial shrubs, spore numbers of AMF are generally smaller than in adjacent soils used for agriculture (Abbott and Robson 1977). In contrast, in virgin native grasslands, spore numbers were much greater than those under adjacent wheat crops (Kucey and Paul 1983). The factors which influence the occurrence of AM have been discussed by Abbott and Robson (1991). The relationships between the level of mycorrhizal colonization and soil chemical and physical properties are extremely variable and high levels of infection have been observed over a wide range of soil pH (Read et al. 1976) and soil P levels (Hayman et al. 1976; Jeffries et al. 1988). It is clear that there is a high diversity of

AMF in most natural soils, and that several fungi can be associated with each plant. For example, molecular probes were developed by Clapp et al. (1995) to demonstrate that roots of a common woodland plant were simultaneously colonized by at least three genera of AMF. These results were supported by careful morphological observation of AMF structures in the roots (Merryweather and Fitter 1998). The presence of a *Glomus* spp. within the roots could not have been predicted from spore data. Nevertheless, spore surveys can still provide useful information about biodiversity of AMF populations provided thorough surveys are conducted.

Direct examination of spores extracted from field soils is problematical as sporulation of dominant species may be absent (Clapp et al. 1995) or spore aging may make identification impossible. Spore counting can be criticized because the abundance and species distribution of AMF in the roots may be poorly related to the sporulating capacity of the soil mycelia. Fungi frequently sporulate in response to nutrient limitation or other stress and may not necessarily do so when mycelial growth is at its maximum. A more satisfactory alternative is the use of trap cultures where mycorrhizal-dependent plants are planted in the soil to stimulate colonization and sporulation of indigenous fungi over a prolonged time period. Stutz and Morton (1996) have emphasised the need to continue this process through several successive cycles. Using soils from arid desert habitats they showed that seven to nine species could be recovered from each site, comparable to the numbers found from many other plant communities. Seventy five percent of species found after three culture cycles were not detected in the first trap cultures, suggesting that a high proportion of the indigenous fungi were non-sporulating in the field soil at the time of collection (Stutz and Morton 1996).

It is clear that there is no single mycorrhizal effect on plant communities (Allen 1991). As a consequence, it is necessary to have information about the individual fungi that are functional within an ecosystem in order to be able to predict their behaviour. For this purpose, it is necessary to identify and discriminate individual AMF within the roots. This used to be a daunting prospect, but the molecular tools are now available for a challenging dissection of mycorrhizal population dynamics (Jeffries and Dodd 2000). The development of PCR-based techniques using molecular

primers offers great promise for the development of specific probes to differentiate AMF at a species or isolate level (cf. Franken and Requena, Chap. 2, this Vol.). Such approaches have already been used to fingerprint DNA from spores of a range of isolates and the technique has now be taken further to identify mycelia within roots (e.g., Simon et al. 1992, 1993; Clapp et al. 1995; Di Bonito et al. 1995; Abbas et al. 1996; Sanders et al. 1996; Simon 1996; Zeze et al. 1996; Edwards et al. 1997; Harney et al. 1997; Redecker et al. 1997; van Tuinen et al. 1998). However, this may not simplify such studies, as examination of DNA profiles from individual spores of *Glomus* spp. has already suggested that the diversity in natural AMF communities and the genetic diversity within individual spores might be much greater than originally thought (Sanders et al. 1995).

Besides the use of nucleic acid probes, comparison of isozyme profiles can also be used to differentiate AMF at a number of levels, including individual isolates (Hepper et al. 1988; Rosendahl and Sen 1992). In addition, they can also be used to identify infection by different AMF within plant roots (Hepper et al. 1988) and have been adapted for detection of *Glomus* spp. at different stages of seedling development (Tisserant et al. 1998). Isozyme detection has an advantage in that the presence of an enzyme indicates that the fungus is metabolically active and not merely present yet non-functional. The presence of a mycorrhiza-specific alkaline phosphatase was used to track the fate of AM of *Glomus aggregatum* established on the indigenous shrub *Anthyllis cytisoides* when inoculated seedlings were transplanted to the field. Immunological approaches, using polyclonal or monoclonal antibodies, also have applications (Aldwell et al. 1985; Wright et al. 1987; Hahn et al., Chap. 3, this Vol.), but careful screening procedures must be employed in order to determine cross-reactivity across a range of isolates. For example, the presence of *Scutellospora heterogama* could be distinguished in plant roots using polyclonal antisera (Thingstrup et al. 1995) from other Glomalean genera and from another species of *Scutellospora*, but this approach would need extensive screening before being applied to field-collected material.

In previous work, the absence of molecular technologies meant that most studies of populations of AMF relied on counts of spores in the soil. Despite the criticisms expressed above, it remains the only simple way to obtain quantitative data regarding species within the mycorrhizal community. An alternative approach is to stain and examine the plant roots as an indirect estimate of the relative mycorrhizal status of the soil. This approach is more relevant to plant growth and yields, but it gives little information about fungal diversity. It is also possible to conduct bioassays or to establish plant trap cultures with soils taken from the field followed by an assessment of colonization and spore formation after a few months of growth under controlled conditions. It should not be assumed, however, that these conditions allow colonization of the roots or sporulation to occur in direct proportion to that which would occur in the field, due to the destruction of established soil mycelia, which in itself influences the subsequent behaviour of the fungi. To assess the natural mycorrhizal potential in a degraded semiarid ecosystem, Requena et al. (1996) compared a soil dilution method with a less disruptive approach based on the use of mesh bags with differently sized holes. The results highlighted the importance of the mycelial network in maintaining soil infectivity. Finally, some workers have used morphological differences between fungal species within the root, enabling them to examine the relative spread and aggressiveness of AM fungal species. This technique is very useful but it requires specialized skills and is limited to certain fungi. All these considerations must be taken into account when assessing information about spore populations and mycorrhizal infectivity. Both traditional and modern techniques have their value in the study of AMF ecology. As Allen (1996) stated, mycorrhizal research into next century should be addressed to develop a better understanding of population genetics of the fungi and the relationships of this with the phylogeny and ecological functioning. It will be important to link mycorrhizal dynamics into the changing global conditions. For example, elevated CO_2 levels have been shown to cause an increase in arbuscular and vesicular colonization of roots but not in the proportion of functional structures as assessed by enzyme activities (Dhillion et al. 1996).

Consideration of AMF ecology is important in sustainable strategies of agriculture or restoration of natural ecosystems. Once the importance of the AM symbiosis is recognized, a decision must be made as to whether the native population of AMF suffices as the starting material from which to develop a sustainable system. If not, it will be necessary to augment the native species with inoculum of exotic or indigenous isolates. If possible, it is best to use indigenous isolates that are already adapted to the prevailing conditions at the field site. In a study involving reciprocal transplanting of AMF and hosts across the Great Basin, USA (Weinbaum et al. 1996), AMF had higher survival on the site from which they were collected and with the host plant population with which they were originally associated. In contrast, although transplanted AMF survived and spread for at least three growing seasons, the populations declined significantly at the exotic site and with exotic hosts.

In cases where the native population is adequate but of low infectivity, it may be possible to rapidly increase inoculum levels using appropriate management practices (see Barea and Jeffries 1995). Unfortunately, there is no easy way to assess the suitability of the indigenous population. Principal component analysis of a number of factors was used by Hamel et al. (1997) to show that indigenous populations of AMF and soil aggregate stability were the major determinants of the response of leek (*Allium porrum* L.) to inoculation with *Glomus intraradices* or *G. versiforme*. Alternatively, young, fresh roots of native plants growing in the soil may be collected and stained to determine AM infection, but this may not always be possible. Differences in the proportion of roots colonized by particular species of AMF can undergo considerable seasonal fluctuations, particularly in temperate climates (Jakobsen and Nielsen 1983; Dodd and Jeffries 1986), and an absence of infection at a particular time may not mean that the soil does not contain propagules of AMF. One of the best indicators of infectiveness occurs when highly mycotrophic trap plants are grown in the test soil. If they do not become infected, or become infected only slowly, the soil will require inoculation. Because of these difficulties in approach, some successful examples of assessing the need for inoculation will be described.

The occurrence of AM propagules was studied in a degraded semiarid savanna in Botswana to assess the need for inoculation with introduced AMF to aid the recovery of this ecosystem (Veenendaal et al. 1992). Seedlings of native grasses were readily colonized by AMF from the topsoil in the degraded savanna. Thus, there was sufficient native inoculum present in the soil to assure plant establishment, and artificial supplementation was not necessary. A comparison between the effectiveness of an introduced and a native population of AMF was made during an intensive study of crops planted in a savanna ecosystem in the acid-infertile soils of Colombia (Dodd et al. 1990a). Inoculation was shown to increase the levels of AMF propagules and plant development in the early growth stages of all crops tested during the first season of experimentation, demonstrating that large amounts of inoculum could improve plant growth in these soils even though native mycorrhizal fungi were present. During the second season, the levels of AM infection were also increased in precropped plots compared to the native savanna control, and this was associated with significant yield responses. The authors concluded that increasing the inoculum potential of AMF by inoculation or by precropping could greatly increase the rate of establishment of mycotrophic host plants. These experiments emphasize the need for an assessment of the native population of AMF. Introduction of exotic inoculum may only be necessary if a native population is either absent or has a low inoculum potential, or if the native fungi are ineffective for the crop species that are being planted. In the case of coffee planted in a low-fertility soil in Brazil (Siqueira et al. 1998), precolonization of outplanted seedlings with selected AMF enhanced early crop development and crop productivity. Diminishing mycorrhizal effects over a 6-year study period were related to colonization of non-inoculated seedlings by indigenous AMF and to a reduced external P requirement of the mature crop. Mycorrhizal effects were most marked at the first harvest and were estimated to be equal to $111 \, kg \, ha^{-1}$ P. Because mycorrhizal plants were more productive at an early crop stage, and showed no reduced yield in the developed crop when compared to those without pre-colonization, mycorrhizal technology was considered a considerable saving on fertilizer costs for coffee agriculture in Brazil (Siqueira et al. 1998). This enabled a more sustainable approach to be implemented with the addition of low to moderate rates of fertilizer at planting.

When the top layer of a soil profile is degraded or eroded, most AM propagules are lost. For example, Cuenca et al. (1998a) reported that the loss of the topsoil organic layer from previously undisturbed communities caused a sharp decrease in the numbers and diversity of AMF spores recovered from the underlying substrate. Soil disturbance is also responsible for the destruction of the AM hyphal network, thereby affecting inoculum potential (Evans and Miller 1990). Thus, consideration of the AM symbiosis is usually recommended in the establishment of plants in degraded soils (Skujins and Allen 1986; Stahl et al. 1988; Jasper et al. 1989a; Cuenca et al. 1998b). The effect of the AM symbiosis could be due to its role in enhancing nutrient availability (Evans and Miller 1988; Fairchild and Miller 1988, 1990) and in improving soil aggregation (Miller and Jastrow 1992a; Schreiner and Bethlenfalvay 1995).

As most of the spores and mycelium of AMF are found in the upper few centimetres of soil, subsoils excavated during mining operations usually lack propagules of AMF. Although mycorrhizal potential was often ignored during the establishment of vegetation on mine spoils (Danielson 1985), the importance of AM in this respect is now well recognized (Gould and Hendrix 1998; Jasper 1992; Saxerud and Funke 1991; Zak and Parkinson 1983). The role of AM in the restoration of surface-mined lands has been reviewed (Miller and Jastrow 1992a). Diversity studies concerning mycorrhizal activity following surface mine reclamation have been carried out showing changes in AM fungal population during the first 5 years of the reclamation process. At the seedling stage of reclamation, low diversities and numbers of spores were recorded. Over time, both parameters increased, with different populations gaining ascendency with time. Most changes occurred over the first few years (Gould and Hendrix 1998) and species richness had stabilized at about ten species by 5 years. In general, species with larger spores were the slowest to appear, possibly reflecting slower dispersal rates or maybe different strategies of colonization (e.g., via the soil mycelium). Similar problems were reported by Cuenca et al. (1998a) in tropical soils disturbed as a result of road-building. Spore numbers and diversity were very low in disturbed areas with only one species found at some sites. Revegetation with *Brachiaria decumbens* increased spore populations and species richness; however, after 5 years they were still lower than those of natural ecosystems. Large spore types (*Gigaspora, Scutellospora*) were again absent from the revegetated soils. Low infectivity of mine spoil is a common problem as the removal of growing plants, soil disturbance and topsoil storage all typically accompany mining activity, and result in a loss of infective AM propagules. In many cases the infectiveness of the resulting soils has dropped below a critical level and appropriate mycorrhizal management or inoculation is necessary to restore plant cover. In the absence of soil management, populations of AMF recover only after considerable time. Naturally revegetated strip mine sites were examined by Kiernan et al. (1983), who found many different spore types, of which 13 were identified to species. Plants sampled had spores from one to eight different mycorrhizal species associated with them. Since these mine spoil sites had been abandoned for over 30 years, a trend towards a natural level of species diversity could have been expected. In a semiarid Mediterranean site, it was estimated that spore populations in abandoned fields reached values similar to virgin soils only after 45 years (Roldan et al. 1997).

V. Factors Affecting the Functioning of the Arbuscular Mycorrhizal Symbiosis

The populations of indigenous AMF in agricultural soils can be manipulated by a variety of management strategies including crop rotation, precropping or intercropping with mycorrhizally responsive plants, the selective use of biocides, through the choice of appropriate inorganic P and N sources or through the use of organic fertilizers such as sewage sludge. These strategies were reviewed by Barea and Jeffries (1995) and will not be revisited except to update specific information. Soil management can be preferable to bioaugmentation with inoculant fungi. Johnson (1998) considered that manipulation of edaphic conditions to favour mycotrophy may be more cost-effective for unreclaimed taconite mine tailings than large-scale inoculation. In other situations, precropping combined with reduced tillage would be an efficient way to improve mycorrhizal functioning early in the growing season (Kling and Jakobsen 1998). Integrated management using reduced or selective use of fungicides may be employed in sustainable systems. Care must be taken, as biocide use can reduce plant growth in soils of low fertility as a result of inhibition of AMF. It has been noted (Schreiner and Bethlenfalvay 1997) that a mixture of three AMF was more tolerant as a community to fungicide applications than they were as individuals. As a consequence, the presence of the mixture of fungi modified the response to fungicide stress, resulting

in high levels of pea plant performance and soil aggregation.

The application of organic fertilizers is a common practice adopted in sustainable agriculture, and organic matter tends to increase the amount of mycelium of AMF in the soil (Joner and Jakobsen 1995). The addition of liquid dairy manure has been shown to increase the abundance of soil mycelium of AMF (Kabir et al. 1998) and farmyard manure can increase the sporulation of AMF (Verma and Arya 1998). Other experiments suggest that dairy manure treatment does not increase mycorrhizal colonization of corn or wheat in the field, nor does it increase sporulation, despite a positive effect in pot experiments (Tarkalson et al. 1998). The use of sewage sludge, however, has inherent problems due to the levels of toxic metals introduced into the soil. This can delay the development of AM such that plant growth is reduced (Koomen et al. 1990), although results show that the influence of AMF on plant metal uptake depends on the plant growth condition, on the fungal partner, and on the metal and cannot be generalized (Weissenhorn et al. 1995). In order to investigate this problem, Val et al. (1999) selected ecotypes of AMF from soils that had received long-term applications of metal-contaminated sewage sludge. Differential tolerances to heavy metals were observed across the isolates obtained, and the growth of the external mycelium was particularly affected. Selection of appropriate isolates will be necessary in order to guarantee the effectiveness of AM symbioses.

Disturbance of soil can reduce soil fertility due to destruction of the soil mycelium of AMF. Shoot P and N concentrations were shown to be much lower in maize plants grown in disturbed soil (Evans and Miller 1990) indicating that the effects were a result of the destruction of a preexisting mycelial network. Ploughing of the soil reduces the abundance of soil mycelium around maize plants whilst substantially higher hyphal densities occur in non-tilled soil compared to conventionally tilled soil (Kabir et al. 1998). Hyphal densities around corn plants increased steeply towards silking when P demands were high, and concentrations of P, Zn and Cu were higher in plants in the non-tilled treaments. In sustainable agriculture where minimal tillage is employed, the effects on mycorrhizal communities are thus minimized, hence promoting the benefits of the symbiosis. Reduced tillage can lead to soil compaction, yet Yano et al. (1998) found that AM formation by

Gigaspora margarita in pigeon pea (*Cajanus cajan* L.) grown in undisturbed soil could promote root elongation despite the fact that the soil was seriously compacted.

Another factor which can affect mycorrhizal functioning is the degree of grazing pressure from herbivorous animals. Reports of effects on mycorrhizal colonization can be contradictory (Allen 1991), possibly due to differential effects dependent on the plant species involved. Allsopp (1998) demonstrated that three grasses used in defoliation experiments responded differently with respect to changes in mycorrhizal dynamics. In *Digitaria* and *Lolium*, colonization by AMF declined but the soil hyphal densities were unaffected. Phosphorus accumulation and leaf regrowth were unaffected. In contrast, *Themeda*, which is more intolerant to grazing, was not able to sustain its hyphal network once defoliated.

Sustainable systems seek to maximise the beneficial effects of the natural soil microbiota. However, the interactions between individual microbes in the mycorrhizosphere are still poorly understood. In some cases microbes can be detrimental to AM or to AMF but there are many situations where beneficial relationships arise. There is also a possibility that different AMF can have synergistic effects on one another. A nested PCR approach was used by van Tuinen et al. (1998) to show that individual root fragments of laboratory-grown onion or leek often contained all four AMF used in the experiments. Root colonization from *Glomus mosseae* and *G. intraradices* was similar from individual and mixed inoculum, whilst the frequency of *Scutellospora castanea* and *Gigaspora rosea* increased in the presence of the two *Glomus* spp.

Specific interactions are known to develop in the AM hyphosphere (Andrade et al. 1997, 1998b). Certain bacteria are stimulated to increase in number and persist in the region around the hyphae of AMF. These beneficial interactions are the subject of current research aimed at the improvement of soil structure (Andrade et al. 1998c), the reestablishment of nutrient cycles after soil degradation, or the optimization of a balanced microbial composition in the rhizosphere which will control root pathogens. Recent reviews (Azcón-Aguilar and Barea 1992, 1996; Linderman 1994; Barea 1997; Barea et al. 1997) have discussed the relationships between AMF and specific microbial groups. The main points from these reviews, focused on sustainability, are summarized here.

Interactions of AMF and N_2-fixing bacteria are critical to benefit biological N input to the soil-plant systems. The appropriate *Rhizobium/* AM fungus combination must be selected (Azcón et al. 1991; Herrera et al. 1993). Actinorhizas formed with the N_2-fixing actinomycete *Frankia* are also colonized by AMF, forming a tripartite symbiosis of great ecological value (Cervantes and Rodríguez-Barrueco 1992). Nitrogen-fixing, free-living microorganisms, such as *Azotobacter*, appear to enhance mycorrhizal colonization and increase plant growth. The application of ^{15}N methods to measure N_2 fixation in sustainable systems (Danso et al. 1992) or to evaluate the AM role in these systems (Barea et al. 1992b, 1996; Toro et al. 1998) has been recently discussed. The isotope ^{15}N has been also used to demonstrate the N uptake from soils by the AM mycelial network (Barea et al. 1991; Johansen et al. 1992; Ibijbijen et al. 1996). Similarly, greenhouse and field experiments, using ^{15}N, have been carried out to ascertain the role of AM in N transfer (Johansen and Jensen 1996).

Many soil microorganisms are able to solubilize phosphate ions from sparingly soluble inorganic and organic P compounds. It has been suggested that the phosphate solubilized by these microbes could be taken up by mycorrhizal hyphae to develop a synergistic microbial interaction (Barea et al. 1983). Methodologies based on the use of the isotope ^{32}P (Zapata and Axmann 1995; Di et al. 1997) have been used to demonstrate such interactive microbial effects with regard to the use of rock phosphate by plants (Toro et al. 1997, 1998). Other plant growth-promoting rhizobacteria (PGPR) may also interact with AMF in a synergistc manner (Azcón-Aguilar et al. 1986; Meyer and Linderman 1986; Requena et al. 1997; Toro et al. 1997; Kim et al. 1998).

Although several studies have been made on the activity of AMF as biocontrol agents, there are few studies on the interactions of other microbial biocontrol agents and AMF aimed at developing mutualistic strategies against root disease. For example, dual inoculation of *Glomus* and *Trichoderma* improved the growth of plants infested with a pathogenic isolate of *Pythium* (Calvet et al. 1993). *Pseudomonas* strains used as inoculants for biocontrol of soil-borne fungal pathogens, including GM strains which overproduce antifungal substances, did not adversely affect mycorrhiza performance (Barea et al. 1998), and even benefit nodulation (Andrade et al. 1998a).

VI. Examples Where the Arbuscular Mycorrhizal Symbiosis Has Been Promoted

Having established that AMF have an important role in the maintenance of soil structure and fertility in sustainable practices, it is essential to consider their behaviour in strategies designed to establish sustainable practices in agriculture or restoration of natural ecosystems. In this chapter, we have chosen to use examples from the latter situation to demonstrate how mycorrhizal fungi can be used to benefit the process.

One of the most serious world problems affecting marginal agricultural lands is desertification. It is a complex and dynamic process which is claiming several hundred million hectares annually (Francis and Thornes 1990; Morgan et al. 1990). Human activities can cause or accelerate desertification and the loss of most plant species and their corresponding symbionts. Revegetation of desertified ecosystems is problematical but management practices are being developed (Herrera et al. 1993). The introduction of native species (reclamation) or exotic species (rehabilitation) must be accompanied by the introduction of appropriate symbionts, including mycorrhizal fungi, in order to aid the restoration of soil fertility. The use of fertilizer inputs to accelerate this process is not a feasible strategy, as fertilizer application encourages the growth of ruderal weeds which are not adapted to the local edaphic conditions. These weeds will outcompete and suppress the pioneer plants such that the resulting plant community will not survive in the absence of continued addition of fertilizer. Instead, the growth of highly mycotrophic pioneer species should be encouraged under conditions of low nutrient availability.

A number of plant species have been proposed for revegetation purposes. Among these, woody species, particularly legumes, have been widely used. Woody legumes are particularly useful for revegetation of water-deficient ecosystems which have a low availability of N, P and other nutrients (Olivares et al. 1988; Batzli et al. 1992; Danso et al. 1992). Although *Bradyrhizobium* or *Rhizobium* spp. have been isolated from root nodules of different woody legumes, little is known about the selection of microsymbionts in order to realize the full potential of a given woody legume/rhizobia combination, thus maximizing

biological N_2 fixation (Olivares et al. 1988; Danso et al. 1992). The scarcity of available P and the imbalance of trace elements in a desertified ecosystem can limit legume establishment and N_2 fixation (Barea et al. 1990a,b). Thus, AM have been found to benefit legume performance (Barea et al. 1992a; Haselwandter and Bowen 1996). Moreover, woody legumes are dependent on AM for active growth in stressed situations (Roskoski et al. 1986; Carpenter and Allen 1988; Habte et al. 1988; Sylvia 1990; Osonubi et al. 1991). The AM symbiosis is known to enhance the ability of plants to become established and cope with environmental stresses (e.g., nutrient deficiency, drought, trace element imbalance, soil disburbance) typical of desertified areas (Sylvia and Williams 1992). However, as soil disturbance often tends to reduce mycorrhizal propagules (Skujins and Allen 1986; Jasper et al. 1989a,b, 1991), it is critical to reintroduce them to improve the recovery rate of disturbed ecosystems.

Inoculation with AMF (Le Tacon and Harley 1990) is now being incorporated into agroforestry and silvicultural practices in tropical ecosystems (Haselwandter and Bowen 1996). Tree legume species belonging to the genera *Acacia*, *Gliricidia*, *Leucaena*, *Paraserianthes*, *Prosopis* and *Robinia*, commonly used in revegetation programs, have been hosts for dual *Rhizobium* and AMF inoculation (Manjunath et al. 1984; De La Cruz et al. 1988; Gardezi and Ferrera-Cerrato 1989; Jasper et al. 1989c; Michelsen and Rosendahl 1990; Colonna et al. 1991; Osonubi et al. 1991; Gueye et al. 1992; Osunde 1992). However, the information from field experiments is scarce. Some nursery studies (De La Cruz et al. 1990) and field trials in microplots (Roskoski et al. 1986) are representative examples that support the potential of dual microsymbiont inoculation in reforestation to improve legume performance. There is analytical evidence that with regard to soil organic matter, soil nutrients and the AMF communities, invading tropical grasslands may be as sustainable as the native forests they replace (Johnson and Wedin 1997). However, these conclusions need support from experimental studies in the field.

In agroforestry systems, the use of trees with strong AM associations will help the perennial sustainability of a mycorrhizal network in the soil regardless of the rotation of annuals grown between them. The trees will retain the inoculum potential of the soil and act as a reservoir for infection of interplanted crops. In alley systems the mycelium will be able to infiltrate back into the interalley regions as successive crops are grown and removed. The beneficial effects of AMF have been demonstrated, for example, for *Albizzia lebbek*, *Gliricidia sepium* and *Leucaena leucocephala* (Osonubi et al. 1991).

Some revegetation programmes depend on the use of shrubs. For example, when eroded areas have not previously supported trees and their edaphoclimatic conditions cannot sustain a forest ecosystem (Francis and Thornes 1990; Morgan et al. 1990). Similarly, where restoration or reclamation strategies follow the natural successional pattern of the ecosystem, it may be advisable to establish a shrub stage to improve the herbaceous vegetation, and to act as an intermediate stage to tree establishment (Skujins and Allen 1986). Several experiments have been carried out where shrub species were inoculated with AMF to improve their establishment and development. For example, Carpenter and Allen (1988) found that mycorrhizal inoculation (topsoil inoculum) of *Hedysarum boreale* increased growth, survival and the production of flowers and seeds by this shrub legume in the field. It was therefore expected that this would contribute to the establishment of a sustainable ecosystem.

Herrera et al. (1993) reported the results of a 4-year field revegetation trial carried out in a semi-arid region of Spain. A number of woody species, common in revegetation programs in Mediterranean regions (Barea et al. 1990a,b), were used. These include two native shrubs (*Anthyllis cytisoides* and *Spartium junceum*) and four exotic tree legumes (*Acacia caven*, *Medicago arborea*, *Prosopis chilensis* and *Robinia pseudoacacia*). Plant species and microsymbionts were screened for appropriate combinations, and a simple procedure to produce plantlets with an optimized mycorrhizal and nodulated status was developed. Results showed that: (1) only the native shrub legumes were able to establish under the local environmental conditions; and (2) inoculation with rhizobia and AMF improved plant survival and biomass development. Since these two shrubs are found in the natural plant community (López-Bermúdez and Albaladejo 1990), they can be used to revegetate desertified areas. A reclamation strategy has been proposed, using *Anthyllis cytisoides*, a particularly drought-tolerant species that is highly mycotrophic. This technique, involving the artificial acceleration of natural revegetation, could be accomplished by replanting randomly spaced groups of shrubs according to the natural pattern and structure of the undisturbed ecosystem (Francis and Thornes 1990; Morgan et al. 1990; Skujins and Allen 1986). Management of appropriate microsymbionts can help legumes to promote the stabilization of a self-sustaining ecosystem. The mycorrhizal shrubs, acting as a "fertile islands" (Skujins and Allen 1986), could serve as sources of inoculum for the surrounding area and to improve N nutrition for the non-N-fixing vegetation in semiarid ecosystems. *Anthyllis cytisoides* is very dependent on AM to reach optimal development in natural conditions (López-Sánchez et al. 1992) and the deployment of inoculated plants in a field site in a desertified semiarid ecosystem has been studied in Spain (Requena et al. 1996; Tisserant et al. 1998). In this context the use of multifunctional (PGPR, *Rhizobium* and AMF) inocula for *A. cytisoides* has been successfully assayed with regard to revegetation purposes (Requena et al. 1997). Selective and specific functional compatibility relationships in the plant response were observed to the different combinations of microbial inoculants. *Anthyllis cytisoides* may have more widespread applications and might also be used in areas contaminated with heavy metals, as inoculation with certain AMF have been shown to reduce the toxicity effects of Pb and Zn (Díaz et al. 1996). *Rosmarinus officialis* is an alternative Mediterranean shrub which may also aid reclamation processes in combination with appropriate mycorrhizal inoculum (Estaún et al. 1997).

VII. Conclusions

Ecosystem functioning is governed largely by soil microbial dynamics (Kennedy and Smith 1995) but these may fluctuate in non-sustainable systems. There is evidence that ecosystem function may be significantly impaired by a loss of soil biodiversity (Giller et al. 1997), yet the diversity of AMF is strikingly low in arable sites compared with natural woodland (Helgason et al. 1998). Mycorrhizal fungi are arguably the most important component of the soil microbiota in developing sustainable agricultural practices, because they enhance plant growth and nutrient uptake while at the same time stabilizing soil aggregates, making the soil less susceptible to erosion (Schreiner and Bethlenfalvay 1995). Thus it could be predicted that the loss of mycorrhizal diversity within the soil could have important effects on sustainability and plant community dynamics. In this chapter we have highlighted some of the evidence that has confirmed this prediction, and have discussed the ways in which AMF function can be affected by external factors. Mycorrhizal technology should be a component in sustainable strategies in the future. Protocols can be adapted, even for modern techniques such as micropropagation (Williams et al. 1992; Budi et al. 1998), whereby application of AMF can reduce fertilizer inputs yet promote healthy plant growth. Arbuscular mycorrhizal fungi can be considered as an essential natural resource for ensuring growth and health of plants and fully deserve their title as biological fertilizers. Several recent studies (e.g. Johnson 1998; Kahiluoto and Vestberg 1998) have suggested that utilization and management of indigenous AMF has more potential in sustainable practices and we expect to see an increase in the exploitation of this natural symbiosis in the future.

Acknowledgments. This chapter is the result of collaboration in the framework of the EU COST Actions on arbuscular mycorrhiza (8.21 and 8.38) and the ENVIRONMENT – REDMED project. J.M.B. is particularly grateful to the EU project INCO – MYRISME, and the CICYT projects in the context of the I & D en Medio Ambiente.

References

Abbas JD, Hetrick BAD, Jurgenson JE (1996) Isolate specific detection of mycorrhizal fungi using genome specific primer pairs. Mycologia 88:939–946

Abbott LK, Robson AD (1977) The distribution and abundance of vesicular-arbuscular endophytes in some Western Australian soils. Aust J Bot 25:515–522

Abbott LK, Robson AD (1991) Factors influencing the occurrence of vesicular-arbuscular mycorrhizas. Agric Ecosyst Environ 35:121–150

Abbott LK, Robson AD, Scheltema MA (1995) Managing soils to enhance mycorrhizal benefits in Mediterranean agriculture. Crit Rev Biotechnol 15:213–228

Aldwell FED, Hall JR, Smith JMB (1985) Enzyme-linked immunosorbent assay as an aid to taxonomy of the Endogonaceae. Trans Br Mycol Soc 84:399–412

Allen MF (1991) The ecology of Mycorrhizae. Cambridge University Press, Cambridge

Allen MF (1996) The ecology of arbuscular mycorrhizas: a look back into the 20th century and a peek into the 21st. Mycol Res 100:769–782

Allsopp N (1998) Effect of defoliation on the arbuscular mycorrhizas of three perennial pasture and rangeland grasses. Plant Soil 202:117–204

Altieri MA (1994) Sustainable agriculture. Encyc Agric Sci 4:239–247

Ames RN, Reid CPP, Porter LK, Canbardella C (1983) Hyphal uptake and transport of nitrogen from two ^{15}N labelled sources by *Glomus mosseae*, a vesicular-arbuscular mycorrhizal fungus. New Phytol 95: 381–396

Anderson RC, Liberta AE, Dickman LA (1984) Interaction of vascular plants and vesicular-arbuscular mycorrhizal fungi across a soil moisture gradient. Oecologia 64:111–117

Andrade G, Mihara KL, Linderman RG, Bethlenfalvay GJ (1997) Bacteria from rhizosphere and hyphosphere soils of different arbuscular-mycorrhizal fungi. Plant Soil 192:71–79

Andrade G, De Leij, FAAM, Lynch JM (1998a) Plant mediated interactions between *Pseudomonas fluorescens*, *Rhizobium leguminosarum* and arbuscular mycorrhizae on pea. Lett Appl Microbiol 26:311–316

Andrade G, Linderman RG, Bethlenfalvay GJ (1998b) Bacterial associations with the mycorrhizosphere and hyphosphere of the arbuscular mycorrhizal fungus *Glomus mosseae*. Plant Soil 202:79–87

Andrade G, Mihara KL, Linderman RG, Bethlenfalvay GJ (1998c) Soil aggregation status and rhizobacteria in the mycorrhizosphere. Plant Soil 202:89–96

Azcón R, Barea JM (1992) The effect of vesicular-arbuscular mycorrhizae in decreasing Ca acquisition by alfalfa plants in calcareous soils. Biol Fertil Soils 13:155–159

Azcón R, Barea JM (1997) Mycorrhizal dependency of a representative plant species in mediterranean shrublands (*Lavandula spica* L.) as a key factor to its use for revegetation strategies in desertification-threatened areas. Appl Soil Ecol 7:83–92

Azcón R, Rubio R, Barea JM (1991) Selective interactions between different species of mycorrhizal fungi and *Rhizobium meliloti* strains, and their effects on growth, N₂-fixation (^{15}N) and nutrition of *Medicago sativa* L. New Phytol 117:399–404

Azcón-Aguilar C, Barea JM (1992) Interactions between mycorrhizal fungi and other rhizosphere microorganisms. In: Allen MF (ed) Mycorrhizal functioning: an integrative plant-fungal process. Chapman and Hall, New York, pp 163–198

Azcón-Aguilar C, Barea JM (1996) Arbuscular mycorrhizas and biological control of soil-borne plant pathogens. An overview of the mechanisms involved. Mycorrhiza 8:457–464

Azcón-Aguilar C, Gianinazzi-Pearson V, Fardeau JC, Gianinazzi S (1986) Effect of vesicular-arbuscular mycorrhizal fungi and phosphate solubilizing bacteria on growth and nutrition of soybean in a neutral calcareous soil amended with ^{32}P ^{45}Ca tricalcium phosphate. Plant Soil 96:17–29

Barea JM (1991) Vesicular-arbuscular mycorrhizae as modifiers of soil fertility. Adv Soil Sci 15:1–40

Barea JM (1997) Mycorrhiza/bacteria interactions on plant growth promotion. In: Ogoshi A, Kobayashi L, Homma Y, Kodama F, Kondon N, Akino S (eds) Plant growth-promoting rhizobacteria, present status and future prospects. OECD, Paris, pp 150–158.

Barea JM, Jeffries P (1995) Arbuscular mycorrhizas in sustainable plant-soil systems. In: Hock B, Varma A (eds) Mycorrhizae: function, molecular biology and biotechnology. Springer, Berlin Heidelberg New York, pp 521–560

Barea JM, Azcón R, Azcón-Aguilar C (1983) Interactions between phosphate solubilizing bacteria and VA mycorrhiza to improve the utilization of rock phosphate by plants in non-acidic soils. 3rd Int Congr Phosphorus compounds, 4-6 Oct 1983, Brussels, pp 127–144

Barea JM, Salamanca CP, Herrera MA (1990a) The role of VA mycorrhiza at improving N₂-fixation by woody legumes in arid zones. In: Werner D, Müller P (eds) Fast-growing trees and nitrogen-fixing trees. Gustav Fischer, Stuttgart, pp 303–311

Barea JM, Salamanca CP, Herrera MA, Roldán-Fajardo BE (1990b) Las simbiosis microbioplanta en el establecimiento de una cubierta vegetal sobre suelos degradados. In: Albaladejo J, Stocking MA, Díaz E (eds) Soil degradation and rehabilitation in Mediterranean environmental conditions. CSIC, Murcia, pp 139–158

Barea JM, Azcón-Aguilar C, Azcón R (1991) The role of vesicular-arbuscular mycorrhizae in improving plant N acquisition from soil as assessed with ¹⁵N. Int Symp on the use of stable isotopes in plant nutrition, soil fertility and environmental studies. IAEA, Vienna, pp 209–216

Barea JM, Azcón R, Azcón-Aguilar C (1992a) Vesicular-arbuscular mycorrhizal fungi in nitrogen-fixing systems. In: Norris JR, Read DJ, Varma AK (eds) Methods in microbiology, vol 24. Academic Press, London, pp 391–416

Barea JM, Azcón R, Azcón-Aguilar C (1992b) The use of ¹⁵N to assess the role of VA mycorrhiza in plant N nutrition and its application to evaluate the role of mycorrhiza in restoring mediterranean ecosystems. In: Read DJ, Lewis DH, Fitter AH, Alexander IJ (eds) Mycorrhizas in ecosystems. CAB International, Cambridge, pp 190–197

Barea JM, Azcón R, Azcón-Aguilar C (1993) Mycorrhiza and crops. Adv Plant Pathol 9:167–189

Barea JM, Tobar RM, Azcón-Aguilar C (1996) Effect of a genetically modified Rhizobium meliloti inoculant on the development of arbuscular mycorrhizas, root morphology, nutrient uptake and biomass accumulation in Medicago sativa L. New Phytol 134:361–369

Barea JM, Azcón-Aguilar C, Azcón R (1997) Interactions between mycorrhizal fungi and rhizosphere microorganisms within the context of sustainable soil-plant systems. In: Gange AC, Brown VK (eds) Multitrophic interactions in terrestrial systems. Blackwell, Cambridge, pp 65–77

Barea JM, Andrade G, Bianciotto V, Dowling D, Lohrke S, Bonfante P, O'Gara F, Azcón-Aguilar C (1998) Impact on arbuscular mycorrhiza formation of Pseudomonas strains used as inoculants for the biocontrol of soil-borne plant fungal pathogens. Appl Environ Microbiol 64:2304–2307

Batzli JM, Graves WR, van Berkum P (1992) Diversity among rhizobia effective with Robinia pseudoacacia L. Appl Environ Microbiol 58:2137–2143

Bethlenfalvay GJ (1992) Mycorrhizae and crop productivity. In: Bethlenfalvay GJ, Linderman RG (eds) Mycorrhizae in sustainable agriculture. ASA Spec Publ, Madison, WI, pp 1–28

Bethlenfalvay GJ, Linderman RG (1992) Mycorrhizae in sustainable agriculture. ASA Spec Publ, Madison, WI, 124 pp

Bethlenfalvay GJ, Schüepp H (1994) Arbuscular mycorrhizas and agrosystem stability. In: Gianinazzi S, Schüepp H (eds) Impact of arbuscular mycorrhizas on sustainable agriculture and natural ecosystems. Birkhäuser, Basel, pp 117–131

Bethlenfalvay GJ, Dakessian S, Pacovsky RS (1984) Mycorrhizae in a southern California desert: ecological implications. Can J Bot 62:519–524

Bethlenfalvay GJ, Brown MS, Franson RL, Mihara KL (1989) The Glycine-Glomus-Bradyrhizobium symbiosis IX. Nutritional, morphological and physiological responses of nodulated soybean to geographic isolates of the mycorrhizal fungus Glomus mosseae. Physiol Plant 76:226–232

Bethlenfalvay GJ, Andrade G, Azcón-Aguilar C (1997) Plant and soil responses to mycorrhizal fungi and rhizobacteria in nodulated or nitrate-fertilized peas (Pisum sativum L.). Biol Fertil Soils 24:164–168

Boddington CL, Dodd JC (1998) A comparison of the development and metabolic activity of mycorrhizas formed by arbuscular mycorrhizal fungi from different genera on two tropical forage legumes. Mycorrhiza 8:149–157

Bohlool BB, Ladha JK, Garrity DP, George T (1992) Biological nitrogen fixation for sustainable agriculture: a perspective. Plant Soil 141:1–11

Budi SW, Cordier C, Trouvelot A, Gianinazzi-Pearson V, Gianinazzi S, Blal B, Lemoine M-C (1998) Arbuscular mycorrhiza as a way of promoting sustainable growth of micropropagated plants. Acta Hortic 457: 71–77

Calvet C, Pera J, Barea JM (1993) Growth response of marigold (Tagetes erecta L.) to inoculation with Glomus mosseae, Trichoderma aureoviride and Pythium ultimum in a peat-perlite mixture. Plant Soil 148:1–6

Carpenter AT, Allen MF (1988) Responses of Hedysarum boreale Nutt to mycorrhizas and Rhizobium: plant and soil nutrient changes in a disturbed shrub-steppe. New Phytol 109:125–132

Cervantes E, Rodríguez-Barrueco C (1992) Relationships between the mycorrhizal and actinorhizal symbioses in non-legumes. In: Norris JR, Read DJ, Varma AK (eds) Methods in microbiology, vol 24. Academic Press, London, pp 417–432

Clapp JP, Young JPW, Merryweather JW, Fitter AH (1995) Diversity of fungal symbionts in arbuscular mycorrhizas from a natural community. New Phytol 130: 259–265

Colonna JP, Thoen D, Ducousso M, Badji S (1991) Comparative effects of Glomus mosseae and P fertilizer on foliar mineral composition of Acacia senegal seedlings inoculated with Rhizobium. Mycorrhiza 1:35–38

Cooper KM, Tinker PB (1978) Translocation and transfer of nutrients in vesicular- -arbuscular mycorrhizas II Uptake and translocation of phosphorus, zinc and sulphur. New Phytol 81:43–52

Cuenca G, De Andrade Z, Escalante G (1998a) Diversity of glomalean spores from natural communities and revegetated communities growing on nutrient-poor tropical soils. Soil Biol Biochem 30:711–719

Cuenca G, De Andrade Z, Escalante G (1998b) Arbuscular mycorrhizas in the rehabilitation of fragile degraded tropical lands. Biol Fertil Soils 26:107–111

Daniels Hetrick BA, Bloom J (1983) Vesicular-arbuscular mycorrhizal fungi associated with native tall grass prairie and cultivated winter wheat. Can J Bot 61: 2140–2146

Danielson RM (1985) Mycorrhizae and reclamation of stressed terrestrial environments. In: Tate R, Klein DL (eds) Soil reclamation processes. Marcel Dekker, New York, pp 173–201

Danso SKA, Bowen GD, Sanginga N (1992) Biological nitrogen fixation in trees in agroecosystems. Plant Soil 141:177–196

De La Cruz RE, Manalo MQ, Aggangan NS, Tambalo JD (1988) Growth of three legume trees inoculated with VA mycorrhizal fungi and *Rhizobium*. Plant Soil 108:111–115

De La Cruz RE, Lorilla EB, Aggangan NS (1990) Growth of *Acacia auriculiformis* and *Acacia mangium* in a marginal grassland in response to inoculations with VA mycorrhiza and/or *Rhizobium*. In: Werner D, Müller P (eds) Fast-growing trees and nitrogen-fixing trees. Gustav Fischer, Stuttgart, 321 pp

Dhillion SS, Roy J, Abrams M (1996) Assessing the impact of elevated CO_2 on soil microbial activity in a Mediterranean model ecosystem. Plant Soil 187:333–342

Di HJ, Condron LM, Frossard E (1997) Isotope techniques to study phosphorus cycling in agricultural and forest soils: A review. Biol Fertil Soils 24:1–12

Díaz G, Azcón-Aguilar C, Honrubia M (1996) Influence of arbuscular mycorrhizae on heavy metal (Zn and Pb) uptake and growth of *Lygeum spartium* and *Anthyllis cytisoides*. Plant Soil 180:241–249

Di Bonito R, Elliot ML, Des Jardin EA (1995) Detection of an arbuscular mycorrhizal fungus in roots of different plant species with the PCR. Appl Environ Microbiol 61:2809–2810

Dodd JC, Jeffries P (1986) Early development of vesicular-arbuscular mycorrhizas in autumn-sown cereals. Soil Biol Biochem 18:149–154

Dodd JC, Burton CC, Burns RG, Jeffries P (1987) Phosphatase activity associated with the roots and rhizosphere of plants infected with vesicular-arbuscular mycorrhizal fungi. New Phytol 107:163–172

Dodd JC, Arias I, Koomen I, Hayman DS (1990a) The management of populations of vesicular-arbuscular mycorrhizal fungi in acid-infertile soils of a savanna ecosystem I The effect of pre-cropping and inoculation with VAM-fungi on plant growth and nutrition in the field. Plant Soil 122:229–240

Dodd JC, Arias I, Koomen I, Hayman DS (1990b) The management of populations of vesicular-arbuscular mycorrhizal fungi in acid-infertile soils of a savanna ecosystem II The effects of pre-cropping on the spore populations of native and introduced VAM-fungi. Plant soil 122:241–247

Edwards SG, Fitter AH, Young JPW (1997) Quantification of an arbuscular mycorrhizal fungus, *Glomus mosseae*, within plant roots by competitive polymerase chain reaction. Mycol Res 101:1440–1444

Elliott ET, Coleman DC (1988) Let the soil work for us. Ecol Bull 39:23–32

Estaún V, Savé R, Biel C (1997) AM inoculation as a biological tool to improve plant revegetation of a disturbed soil with *Rosmarinus officinalis* under semiarid conditions. Appl Soil Ecol 6:223–229

Evans DG, Miller MH (1988) Vesicular-arbuscular mycorrhiza and the soil-disturbance-induced reduction of nutrient absorption in maize. New Phytol 110:75–84

Evans DG, Miller MH (1990) The role of the external mycelial network in the effect of soil disturbance upon vesicular-arbuscular mycorrhizal colonization of maize. New Phytol 114:65–71

Fairchild GL, Miller MH (1988) Vesicular-arbuscular mycorrhizas and the soil-disturbance-induced reduction of nutrient absorption in maize II Development of the effect. New Phytol 110:75–84

Fairchild GL, Miller MH (1990) Vesicular-arbuscular mycorrhizas and the soil-disturbance-induced reduction of nutrient absorption in maize III Influence of P amendments to soil. New Phytol 114:641–650

Fitter AH (1985) Functioning of vesicular-arbuscular mycorrhizas under field conditions. New Phytol 99:257–265

Fitter AH, Graves JD, Watkins NK, Robinson D, Scrimgeour C (1998) Carbon transfer between plants and its control in networks of arbuscular mycorrhizas. Funct Ecol 12:406–412

Francis CF, Thornes JB (1990) Matorral: erosion and reclamation. In: Albaladejo J, Stocking MA, Díaz E (eds) soil degradation and rehabilitation in Mediterranean environmental conditions. CSIC, Murcia, pp 87–115

Francis R, Read DJ (1984) Direct transfer of carbon between plants connected by vesicular-arbuscular mycorrhizal mycelium. Nature 307:53–56

Francis R, Finlay RD, Read DJ (1986) Vesicular-arbuscular mycorrhiza in natural vegetation systems IV Transfer of nutrients in inter- and intra-specific combinations of host plants. New Phytol 102:103–111

Gange AC, Brown VK, Farmer LM (1990) A test of mycorrhizal benefit in an early successional plant community. New Phytol 115:85–91

GAO/RCED (1992) GAO/RCED 92-233 Sustainable agriculture. United States Congress Food, Agriculture Conservation, and Trade Act of 1990, P. L. no 101-624, Washington.

Gardezi AK, Ferrera-Cerrato R (1989) The effect of four levels of phosphorus on mycorrhizal colonization, dry root weight, and nitrogen and phosphorus content of *Acacia saligna* inoculated with *Rhizobium* sp. and endomycorrhiza in a Mexican andisol. Nitrogen Fixing. Tree Res Rep 7:43–45

Gavito ME, Miller MH (1998) Changes in mycorrhiza development in maize induced by crop management practices. Plant Soil 198:185–192

Giller KE, Beare MH, Lavelle P, Izac A-MN, Swift MJ (1997) Agricultural intensification, soil biodiversity and agroecosystem function. Appl Soil Ecol 6:3–16

Glick BR (1995) The enhancement of plant growth by free-living bacteria. Can J Microbiol 41:109–117

Gould AB, Hendrix JW (1998) Relationship of mycorrhizal activity to time following reclamation of surface mine land in western Kentucky. II. Mycorrhizal fungal communities. Can J Bot 76:204–212

Grime JP, Mackey JML, Hillier SH, Read DJ (1987) Floristic diversity in a model system using experimental microcosms. Nature 328:420–422

Gueye M, Diop T, Ndao B (1992) *Acacia albida*, une legumineuse arborescente à fort potentiel mycorhizien et fixateur d'azote. In: IFS (ed) Interactions between plants and microorganisms. International Foundation for Science, Dakar, Senegal, pp 452–461

Gupta VVSR, Germida JJ (1988) Distribution of microbial biomass and its activity in different soil aggregate size classes as affected by cultivation. Soil Biol Biochem 21:777–786

Habte M, Fox RL, Aziz T, El-Swaify SA (1988) Interaction of vesicular-arbuscular mycorrhizal fungi with erosion in an oxisol. Appl Environ Microbiol 54:945–950

Hamel C, Dalpé Y, Furlan V, Parent S (1997) Indigenous populations of arbuscular mycorrhizal fungi and soil aggregate stability are major determinants of leek (*Allium porrum* L.) response to inoculation with *Glomus intraradices* Schenck & Smith or *Glomus versiforme* (Karsten) Berch. Mycorrhiza 7:187–196

Harney SK, Edwards FS, Allen MF (1997) Identification of arbuscular mycorrhizal fungi from *Artemesia californica* using the polymerase chain reaction. Mycologia 89:547–550

Haselwandter K, Bowen GB (1996) Mycorrhizal relations in trees for agroforestry and land rehabilitation. For Ecol Manage 81:1–17

Hayman DS, Barea JM, Azcón R (1976) Vesicular-arbuscular mycorrhiza in southern Spain: its distribution in crops growing in soil of different fertility. Phytopathol Mediterr 15:1–6

Helgason T, Daniell TJ, Husband R, Fitter AH, Young JPW (1998) Ploughing up the wood-wide web? Nature 394: 431

Hepper CM, Azcón-Aguilar C, Rosenthal S, Sen R (1988) Competition between three species of *Glomus* used as spacially separated introduced and indigenous mycorrhizal inocula for leek (*Allium porrum* L.). New Phytol 110:207–211

Herrera MA, Salamanca CP, Barea JM (1993) Inoculation of wooody legumes with selected arbuscular mycorrhizal fungi and rhizobia to recover desertified mediterranean ecosystems. Appl Environ Microbiol 59:129–133

Hetrick BAD, Wilson GWT, Hartnett DC (1989) Relationship between mycorrhizal dependence and competitive ability of two tallgrass prairie grasses. Can J Bot 67:2608–2615

Ibijbijen J, Urquiaga S, Ismaili, M, Alves BJR, Boddey RM (1996) Effect of arbuscular mycorrhizas on uptake of nitrogen by *Brachiaria arrecta* and *Sorghum vulgare* from soils labelled for several years with N^{15}. New Phytol 133:487–494

Ikram A, Jensen ES, Jakobsen I (1994) No significant transfer of N and P from *Pueraria phaseoloides* to *Hevea brasiliensis* via hyphal links of arbuscular mycorrhiza. Soil Biol Biochem 26:1541–1547

Jakobsen I, Nielsen NE (1983) Vesicular-arbuscular mycorrhiza in field grown crops 1. Mycorrhizal infection in cereals and peas at various times and soil depths. New Phytol 93:401–413

Jasper DA (1992) Management of mycorrhizas in revegetation. Abst Int Symp Management Mycorrhizas in Agric Hort Forestry, Perth, Australia, 28th Sept- 2nd Oct 1992. p 135

Jasper DA, Abbott LK, Robson AD (1989a) Soil disturbance reduces the infectivity of external hyphae of vesicular-arbuscular mycorrhizal fungi. New Phytol 112:93–99

Jasper DA, Abbott LK, Robson AD (1989b) Hyphae of a vesicular-arbuscular mycorrhizal fungus maintain infectivity in dry soil, except when the soil is disturbed. New Phytol 112:101–107

Jasper DA, Abbott LK, Robson AD (1989c) Acacias respond to additions of phosphorus and to inoculation with VA mycorrhizal fungi in soils stockpiled during mineral sand mining. Plant Soil 115: 99–108

Jasper DA, Abbott LK, Robson AD (1991) The effect of soil disturbance on vesicular-arbuscular mycorrhizal fungi in soils from different vegetation types. New Phytol 118:471–476

Jastrow JD, Miller RM (1991) Methods for assessing the effects of biota on soil structure. Agric Ecosyst Environ 35:279–303

Jeffries P (1987) Use of mycorrhizae in agriculture. CRC Crit Rev Biotechnol 5:319–358

Jeffries P, Barea JM (1994) Biogeochemical cycling and arbuscular mycorrhizas in sustainability of plant-soil systems. In: Gianinazzi S, Schuepp H (eds) Impact of arbuscular mycorrhizas on sustainable agriculture and natural ecosystems. Birkhäuser, Basel, pp 101–115

Jeffries P, Dodd JC (2000) Molecular ecology of mycorrhizal fungi. In: Priest F, Goodfellow M (eds) Applied microbial systematics, Chapman and Hall, London (in press)

Jeffries P, Spyropoulos T, Vardavarkis E (1988) Vesicular-arbuscular mycorrhizal status of various crops in different agricultural soils in northern Greece. Biol Fertil Soils 5:333–337

Johansen A, Jensen ES (1996) Transfer of N and P to barley interconnected by an arbuscular mycorrhizal fungus. Soil Biol Biochem 28:73–81

Johansen A, Jakobsen I, Jensen SE (1992) Hyphal transport of ^{15}N labelled nitrogen by a vesicular arbuscular mycorrhizal fungus and its effect on depletion of inorganic soil N. New Phytol 122:281–288

Johnson CR, Copeland PJ, Crosskston RK, Pfleger FL (1992) Mycorrhizae: a possible explanation for yield decline associated with continuous cropping of corn and soybean. Agron J 84:387–390

Johnson NC (1998) Responses of *Salsola kali* and *Panicum virginatum* to mycorrhizal fungi, phosphorus and soil organic matter: implications for reclamation. J Appl Ecol 35:86–94

Johnson NC, Pfleger FL (1992) Vesicular-arbuscular mycorrhizae and cultural stress. In: Bethenfalvay GJ, Linderman RG (eds) Mycorrhizae in sustainable agriculture ASA Spec Publ Madison, WI, pp 71–99

Johnson NC, Wedin DA (1997) Soil carbon, nutrients, and mycorrhizae during conversion of dry tropical forest to grassland. Ecol Appl 7:171–182

Joner EJ, Jakobsen I (1995) Growth and extracellular phosphatase activity of arbuscular mycorrhizal hyphae as influenced by soil organic matter. Soil Biol Biochem 27:1153–1159

Kabir Z, O'Halloran IP, Fyles JW, Hamel C (1998) Dynamics of the mycorrhizal symbiosis of corn (*Zea mays* L.): effects of host physiology, tillage practice and fertilization on spacial distribution of extra-radical mycorrhizal hyphae in the field. Agric Syst Environ 68:151–163

Kahiluoto H, Vestberg M (1998) The effect of arbuscular mycorrhiza on biomass production and phosphorus uptake from sparingly soluble sources by leek (*Allium porrum* L.) in Finnish field soils. Biol Agric Hortic 16:65–85

Kennedy AC, Smith KL (1995) Soil microbial diversity and the sustainability of agriculture soils. Plant Soil 170:75–86

Kiernan JM, Hendrix JW, Maronek DM (1983) Endomycorrhizal fungi occurring on orphan strip mines in Kentucky. Can J Bot 61:1798–1803

Kim KY, Jordan D, McDonald (1998) Effect of phosphate-solubilizing bacteria and vesicular-arbuscular mycorrhizae on tomato growth and soil microbial activity. Biol Fertil Soils 26:79–87

Kling M, Jakobsen I (1998) Arbuscular mycorrhiza in soil quality assessment. Ambio 27:29–34

Kloepper JW (1992) Plant growth-promoting rhizobacteria as biological control agents. In: Blaine F, Metting J Jr (eds) Soil microbial ecology. Applications in agriculture forestry and environmental management. Marcel Dekker, New York, pp 255–274

Kloepper JW, Zablotowick RM, Tipping EM, Lifshitz R (1991) Plant growth promotion mediated by bacterial rhizosphere colonizers. In: Keister DL, Cregan PB (eds) The rhizosphere and plant growth. Kluwer, Dordrecht, pp 315–326

Koomen I, McGrath SP, Giller KE (1990) Mycorrhizal infection of clover is delayed in soils contaminated with heavy metals from past sewage sludge application. Soil Biol Biochem 22:871–873

Koske RE (1975) *Endogone* spores in Australian sand dunes. Can J Bot 53:668–672

Koske RE, Halvorson WL (1981) Ecological studies of vesicular-arbuscular mycorrhizae in a barrier sand dune. Can J Bot 59:1413–1422

Kucey RMN, Paul EA (1983) Vesicular-arbuscular mycorrhizal spore populations in various Saskatchewan soils and the effect of inoculation with *Glomus mosseae* on faba bean growth on greenhouse and field trials. Can J Soil Sci 63:87–95

Ladha JK (1992) Preface: role of biological nitrogen fixation in sustainable agriculture. Plant Soil 141:VII

Lal R (1989) Conservation tillage for sustainable agriculture: tropics versus temperate environments. Adv Agron 42:85–185

Le Tacon F, Harley JL (1990) Deforestation in the tropics and proposals to arrest it. Ambio 19:372–378

Linderman RG (1992) Vesicular-arbuscular mycorrhizal and soil microbial interactions In: Bethlenfalvay GJ, Linderman RG (eds) Mycorrhizae in sustainable agriculture. ASA Spec Publ, Madison, WI, pp 45–70

Linderman RG (1994) Role of VAM fungi in biocontrol. In: Pfleger FL, Linderman RG (eds) Mycorrhizae and plant health. APS Press, St Paul, pp 1–26

López-Bermúdez F, Albaladejo J (1990) Factores ambientales de la degradación del suelo en el area mediterránea. In: Albaladejo J, Stocking MA, Díaz E (eds) Soil degradation and rehabilitation in mediterranean environmental conditions. CSIC, Murcia, pp 15–45

López-Sánchez ME, Díaz G, Honrubia M (1992) Influence of vesicular-arbuscular mycorrhizal infection and P addition on growth and P nutrition of *Anthyllis cytisoides* L. and *Brachypodium retusum* (Pers.) Beauv. Mycorrhiza 2:41–45

Manjunath A, Bagyaraj DJ, Gopala Gowda HS (1984) Dual inoculation with VA mycorrhiza and *Rhizobium* beneficial to *Leucaena*. Plant Soil 78:445–448

McNaughton SJ, Oesterheld M (1990) Extramatrical mycorrhizal abundance and grass nutrition in a tropical grazing ecosystem, the Serengeti National Park, Tanzania. Oikos 59:92–96

Merryweather J, Fitter A (1998) The arbuscular mycorrhizal fungi of *Hyacinthoides non-scripta* – I. Diversity of fungal taxa. New Phytol 138:117–129

Meyer JR, Linderman RG (1986) Response of subterranean clover to dual inoculation with vesicular-arbuscular mycorrhizal fungi and a plant growth promoting bacterium, *Pseudomonas putida*. Soil Biol Biochem 18:185–190

Michelsen A, Rosendahl S (1990) The effect of VA mycorrhizal fungi, phosphorus and drought stress on the growth of *Acacia nilotica* and *Leucaena leucocephala* seedlings. Plant Soil 124:7–13

Miller RM, Jastrow JD (1992a) The application of VA mycorrhizae to ecosystem restoration and reclamation. In: Allen MF (ed) Mycorrhizal functioning an integrative plant fungal process. Chapman and Hall, New York, pp 438–467

Miller RM, Jastrow JD (1992b) The role of mycorrhizal fungi in soil conservation. In: Bethlenfalvay GJ, Linderman RG (eds) Mycorrhizae in sustainable agriculture. ASA Spec Publ, Madison, WI, pp 29–44

Morgan RPC, Rickson RJ, Wright W (1990) Regeneration of degraded soils. In: Albaladejo J, Stocking MA, Díaz E (eds) Soil degradation and rehabilitation in Mediterranean environmental conditions. CSIC Murcia, pp 69–85

Mosse B (1986) Mycorrhiza in a sustainable agriculture. Biol Agric Hortic 3:191–209

Oades JM (1984) Soil organic matter and structural stability: mechanisms and implications for management. Plant Soil 76:319–337

Olivares J, Herrera MA, Bedmar EJ (1988) Woody legumes in arid and semi-arid zones: the *Rhizobium-Prosopis chilensis* symbiosis. In: Beck DP, Materon LA (eds) Nitrogen fixation legumes in mediterranean agriculture. ICARDA, Martinus Nijhoff, Dordrecht, pp 65–72

Osonubi O, Mulongoy K, Awotoye OO, Atayese MO, Okali DUU (1991) Effects of ectomycorrhizal and vesicular-arbuscular mycorrhizal fungi on drought tolerance of four leguminous woody seedlings. Plant Soil 136:131–143

Osunde AO (1992) Response of *Gliricidia sepium* to *Rhizobium* and VA-mycorrhizal fungi inoculation on an acid soil. InIFS (ed): Interactions between plants and microorganisms. Dakar, Senegal, pp 156–164

Pace NR (1997) A molecular view of microbial diversity and the biosphere. Science 276:734–740

Peoples MB, Craswell ET (1992) Biological nitrogen fixation: investments, expectations and actual contributions to agriculture. Plant Soil 141:13–39

Read DJ (1989) Mycorrhizas and nutrient cycling in sand dune ecosystems. Proc R Soc Edinb 86B:89–110

Read DJ (1991) Mycorrhizas in ecosystems. Experientia 47:376–391

Read DJ, Koucheki HH, Hodgson J (1976) Vesicular-arbuscular mycorrhiza in natural vegetation systems. I The occurrence of infection. New Phytol 77:641–653

Redecker D, Thierfelder H, Walker C, Werner D (1997) Restriction analysis of PCR-amplified internal transcribed spacers of ribosomal DNA as a tool for species identification in different genera of the order Glomales. Appl Environ Microbiol 63:1756–1761

Remy W, Taylor TN, Haas H, Kerp H (1994) Four hundred-million-year-old vesicular-arbuscular mycorrhizae. Proc Natl Acad Sci 91:11841–11843

Requena N, Jeffries P, Barea JM (1996) Assessment of natural mycorrhizal potential in a desertified semiarid ecosystem. Appl Environ Microbiol 62:842–847

Requena N, Jimenez I, Toro M, Barea JM (1997) Interactions between plant-growth-promoting rhizobacteria (PGPR), arbuscular mycorrhizal fungi and *Rhizobium* spp. in the rhizosphere of *Anthyllis cytisoides*, a model legume for revegetation in mediterranean semiarid ecosystems. New Phytol 136:667–677

Roldan A, Garcia C, Albaladejo J (1997) AM fungal abundance and activity in a chronosequence of abandoned fields in a semiarid Mediterranean site. Arid Soil Res Rehabil 11:211–220

Rosendahl S, Sen R (1992) Isozyme analysis of mycorrhizal fungi and their mycorrhiza. In: Norris JR, Read DJ, Varma AK (eds) Methods in microbiology, vol 24. Academic Press, London, pp 169–194

Roskoski JP, Pepper I, Pardo E (1986) Inoculation of leguminous trees with rhizobia and VA mycorrhizal fungi. For Ecol Manage 16:57–68

Sanders IR, Alt M, Groppe K, Boller T, Wiemken A (1995) Identification of ribosomal DNA polymorphisms among and within spores of the Glomales: application to studies on the genetic diversity of arbuscular mycorrhizal fungal communities. New Phytol 130:419–427

Sanders IR, Clapp JP, Wiemken A (1996) The genetic diversity of arbuscular mycorrhizal fungi in natural ecosystems – a key to understanding the ecology and functioning of the mycorrhizal symbiosis. New Phytol 133:123–134

Sexerud MH, Funke BR (1991) Effects on plant growth of inoculation of stored strip-mining topsoil in North Dakota with mycorrhizal fungi contained in native soils. Plant Soil 131:135–141

Schenck NC, Kinloch RA (1980) Incidence of mycorrhizal fungi on six field crops in monoculture on a newly cleared woodland site. Mycologia 72:445–456

Schreiner RP, Bethlenfalvay GJ (1995) Mycorrhizal interactions in sustainable agriculture. In: Varma A (ed) Arbuscular Mycorrhizae: biotechnological applications: an environmental sustainable biological agent. Crit Rev Biotechnol 15:271–285

Schreiner RP, Bethlenfalvay GJ (1997) Plant and soil response to single and mixed species of arbuscular mycorrhizal fungi under fungicide stress. Appl Soil Ecol 7:93–102

Sieverding E, Howeler RH (1985) Influence of species of VA mycorrhizal fungi on cassava yield response to phosphorus fertilization. Plant Soil 88:213–221

Simon L (1996) Phylogeny of the Glomales: deciphering the past to understand the present. New Phytol 133:95–101

Simon L, Lalonde M, Bruns TD (1992) Specific amplification of 18S fungal ribosome genes from vesicular-arbuscular endomycorrhizal fungi colonizing roots. Appl Environ Microbiol 58:291–295

Simon L, Lévesque RC, Lalonde M (1993) Identification of endomycorrhizal fungi colonizing roots by fluorescent single-strand conformation polymorphism-polymerase chain reaction. Appl Environ Microbiol 59:4211–4215

Siqueira JO, Saggin-Júnior OJ, Flores-Aylas WW, Guimarães PTG (1998) Arbuscular mycorrhizal inoculation and superphosphate application influence plant development and yield of coffee in Brazil. Mycorrhiza 7:293–300

Skujins J, Allen MF (1986) Use of mycorrhizae for land rehabilitation. Mircen J 2:161–176

Smith SE, Read DJ (1997) Mycorrhizal symbiosis. Academic Press, London, 605pp

Stahl PD, Williams SE, Christensen M (1988) Efficacy of native vesicular-arbuscular mycorrhizal fungi after severe soil disturbance. New Phytol 110:347–354

Stutz JC, Morton JB (1996) Successive pot cultures reveal high species richness of arbuscular endomycorrhizal fungi in arid ecosytsems. Can J Bot 74:1883–1889

Sylvia DM (1990) Inoculation of native woody plants with vesicular-arbuscular mycorrhizal fungi for phosphate mine land reclamation. Agric Ecosyst Environ 31:253–261

Sylvia DM, Williams SE (1992) Vesicular-arbuscular mycorrhizae and environmental stresses. In: Bethlenfalvay GJ, Linderman RG (eds) Mycorrhizae in sustainable agriculture. ASA Spec Publ, Madison, WI, pp 101–124

Tarkalson DD, Von Jolley D, Robbins CW, Terry RE (1998) Mycorrhizal colonization and nutrition of wheat and sweet corn grown in manure-treated and untreated topsoil and subsoil. J Plant Nutr 21:1985–1999

Thingstrup I, Rozycka M, Jeffries P, Rosendahl S, Dodd JC (1995) Detection of the arbuscular mycorrhizal fungus Scutellospora heterogama within roots using polyclonal antisera. Mycol Res 99:1225–1232

Tisdall JM (1991) Fungal hyphae and structural stability of soil. Aust J Soil Res 29:729–743

Tisdall JM, Oades JM (1980) The effect of crop rotation on aggregation in a red-brown earth. Aust J Soil Res 18:423–433

Tisserant B, Brenac V, Requena N, Jeffries P, Dodd JC (1998) The detection of Glomus spp. (arbuscular-mycorrhizal fungi) forming mycorrhizas at different stages of seedling development, using mycorrhiza-specific isozymes. New Phytol 138:225–239

Toro M, Azcón R, Barea JM (1997) Improvement of arbuscular mycorrhizal development by inoculation with phosphate-solubilizing rhizobacteria to improve rock phosphate bioavailability (^{32}P) and nutrient cycling. Appl Environ Microbiol 63:4408–4412

Toro M, Azcón R, Barea JM (1998) The use of isotopic dilution techniques to evaluate the interactive effects of Rhizobium genotype, mycorrhizal fungi, phosphate-solubizing rhizobacteria and rock phosphate on nitrogen and phosphorus acquisition by Medicago sativa. New Phytol 138:265–273

Val del C, Barea JM, Azcón-Aguilar C (1999) Assessing the tolerance to heavy metals of arbuscular mycorrhizal fungi isolated from sewage sludge-contaminated soils. Appl Soil Ecol 11:261–269

van der Heijden MGA, Boller T, Wiemken A, Sanders IR (1998a) Different arbuscular mycorrhizal fungal species are potential determinants of plant community structure. Ecology 79:2082–2091

van der Heijden MGA, Klironomos JN, Ursic M, Moutoglis P, Streitwolf-Engel R, Boller T, Wiemken A, Sanders ER (1998b) Mycorrhizal fungal diversity determines plant biodiversity, ecosystem variability and productivity. Nature 396:69–72

Van Tuinen D, Jacquot E, Zhao B, Gollotte A, Gianinazzi-Pearson V (1998) Characterization of root colonization profiles by a microcosm community of arbuscular mycorrhizal fungi using 25S rDNA-targeted nested PCR. Mol Ecol 7:879–887

Veenendaal EM, Monnaapula SC, Gilika T, Magole IL (1992) Vesicular-arbuscular mycorrhizal infection of grass seedlings in a degraded semiarid savanna in Botswana. New Phytol 121:477–485

Verma RK, Arya ID (1998) Effect of arbuscular mycorrhizal fungal isolates and organic manure on growth and mycorrhization of micropropagated Dendrocalamus asper plantlets and on spore production in their rhizosphere. Mycorrhiza 8:113–116

Weinbaum BS, Allen MF, Allen EB (1996) Survival of arbuscular mycorrhizal fungi following reciprocal transplanting across the Great Basin, USA. Ecol Appl 6:1365–1372

Weissenhorn I, Leyval C, Belgy G, Berthelin J (1995) Arbuscular mycorrhizal contribution to heavy metal uptake by maize (*Zea mays* L.) in pot culture with contaminated soil. Mycorrhiza 5:245–251

Williams SCK, Vestberg M, Uosukainen M, Dodd JC, Jeffries P (1992) Effects of fertilizers and arbuscular mycorrhizal fungi on the post-vitro growth of micropropagated strawberry. Agronomie 12:851–857

Wright SF, Upadhyaya A (1998) A survey of soils for aggregate stability and glomalin, a glycoprotein produced by hyphae of arbuscular mycorrhizal fungi. Plant Soil 198:97–107

Wright SF, Morton JB, Sworobuk JE (1987) Identification of a vesicular-arbuscular mycorrhizal fungus by using monoclonal antibodies in an enzyme-linked immunosorbent assay. Appl Environ Microbiol 53:2222–2225

Yano K, Yamauchi A, Iijima M, Kono Y (1998) Arbuscular mycorrhizal formation in undisturbed soil counteracts compacted soil stress for pigeon pea. Appl Soil Ecol 10:95–102

Zak JC, Parkinson D (1983) Effects of surface amendation of two mine spoils in Alberta, Canada on vesicular-arbuscular mycorrhizal development of slender wheatgrass: a 4-year study. Can J Bot 61:798–803

Zapata F, Axmann H (1995) ^{32}P isotopic techniques for evaluating the agronomic effectiveness of rock phosphate materials. Fertil Res 41:189–195

Zeze A, Hosny M, Gianinazzi-Pearson V, Dulieu H (1996) Characterization of a highly repeated DNA sequences (SC1) from the arbuscular-mycorrhizal fungus *Scutellospora castanea* and its detection in planta. Appl Environ Microbiol 62:2443–2448

7 Exchange of Carbohydrates Between Symbionts in Ectomycorrhiza

U. Nehls, J. Wiese, and R. Hampp

CONTENTS

I. Introduction

The nutritive relationships in ectomycorrhizas are of dual nature. Fungal hyphae explore the soil for areas not already depleted of mobile nutrients by the plant roots and take up poorly mobile nutrients such as inorganic phosphate or ammonium via local patches. These nutrients are then either transported via conducting structures (rhizomorphs) toward the host root directly or after assimilation (ammonium). In return, the host plant delivers photoassimilates which are used for fungal growth, for delivering metabolic energy for active uptake, and as a source of carbon skeletons for ammonium assimilation. Thus, supply of photoassimilates to the symbiotic structure is the basis for a successful plant-fungus interaction. Carbon supply is, however, not only a one-way traffic from host to symbiont. Several investigations show that the fungal network can connect different host plants and that a net transfer of carbon is also possible from photosynthetically more active (sun) to

Universität Tübingen, Botanisches Institut, Physiologische Ökologie der Pflanzen, Auf der Morgenstelle 1, 72076 Tübingen, Germany

less productive (shade) hosts (Smith and Read 1997) which could result in a net transfer of carbon (sugars or amino acids) from the fungus to the root.

II. Carbon Supply by the Host Plant

One of the first attempts to assay carbon flow in a mycorrhizal plant was performed by Melin and Nilsson (1957), who could show that after feeding ^{14}C-labelled CO_2 to leaves labelled carbon appeared within 1 day in the hyphal mantle.

In general, the direction of carbon flow in the host is controlled by gradients between production (*source* organs such as leaves) and consumption (*sink* organs such as non-green tissues) of photoassimilates. Carbon will thus always be directed to the most active sink area in a plant and the sink strength has been shown to control the rate of photoassimilate production (Stitt 1991; Koch 1996). This regulatory mechanism should be especially effective in mycorrhizal roots, where the fungal partner can consume up to 30% of the net CO_2 assimilation of the host (Söderström 1992). Indeed, there is ample evidence that mycorrhization can upregulate the rate of net photosynthesis of the host (e.g., Nylund and Wallander 1989; Dosskey et al. 1991; Loewe et al. 1998).

The rate of photosynthesis, on the other hand, affects the degree of mycorrhization. Citrus plants grown under long-day photoperiods had greater rates of root-directed transport of photoassimilates in parallel to improved (endo)mycorrhizal infection (Johnson et al. 1982). Generally, the increased carbon drain due to mycorrhiza formation reduces plant growth in comparison to non-mycorrhizal controls, as long as nutrient availability is not limiting.

We know that the response of plants to elevated CO_2 (global change) depends on nutrient availability and the degree of utilization of photoassimilate. Plants with limited sink capacities do not take advantage of an increased supply of CO_2 due to feedback regulation of carbon fixation by accumulation of products (Stitt 1991; Koch 1996). On the other hand, increased availability of CO_2 should favour the growth of source-limited plants, i.e., mycorrhizal plants which cannot meet the needs of both their own growth and that of the fungal partner. This assumption is confirmed in experiments with seedlings. Under elevated CO_2, the rate of photosynthesis, mycorrhiza formation, and fungal growth were all increased (Norby et al. 1987; Ineichen et al. 1995; Lewis and Strain 1996; Loewe et al. 1998).

III. The Interactive Space

In ectomycorrhiza the zone of solute exchange is the Hartig net. Here, the fungal hyphae form highly branched structures in the cell walls of the root cortex (Fig. 1) which creates a large increase in surface area, facilitating exchange. Experimental evidence supports the idea, that host photoassimilates (preferably sucrose) are delivered to this apoplastic space (Fig. 2). Labelling experiments confirmed that labelled sucrose was the main labelled sugar in roots, but this disaccharide could not be detected in fungal hyphae. Here, mannitol and trehalose carried the main proportion of the label (Bevege et al. 1975). Thus, sucrose cleavage has to precede carbon uptake by the fungal partner. Of the sucrose splitting enzymes, sucrose synthase and invertases, cell wall-bound acid invertase dominates in this area in spruce/fly agaric mycorrhizas, while minor activities of sucrose synthase could detected in the bundle area only (Schaeffer 1995). The invertase is obviously a host enzyme as ectomycorrhiza-forming fungi such as *Hebeloma crustuliniforme* or *Amanita muscaria* (Salzer and Hager 1991; Schaeffer 1995) show no invertase activity and are not able to grow on sucrose.

A longitudinal distinction of different zones of root-fungus-interaction in a single ectomycorrhiza (Rieger et al. 1992; Hampp et al. 1995) showed that in the area of probably most intense interaction (highest ergosterol contents) sucrose was lowest while trehalose, the fugus-specific disaccharide, dominated. In this zone cell wall-bound acid invertase peaked in fine roots (Schaeffer 1995).

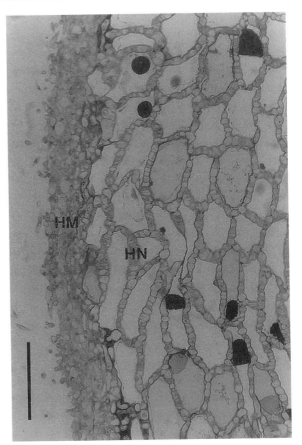

Fig. 1. Tangential cross-section of an ectomycorrhiza formed between a wild-type hybrid aspen and fly agaric. The photograph shows the hyphal mantle (*HM*), covering a fine root, and hyphae, filling the cortex cell walls (*HN* Hartig net). *Bar* 50 μm

IV. The Exchange of Carbohydrates

A. Sugar Uptake by *A. muscaria* Hyphae

From the above it is evident that monosaccharides should be taken up by the fungus. Using protoplasts from *A. muscaria* mycelia, evidence for a hexose-specific uptake system, obviously driven by proton motive force was obtained (Chen and Hampp 1993). The affinity of this monosaccharide uptake system was different for glucose and fructose. The K_M value for the uptake of glucose was 1.25 mM, while that for fructose uptake was one order of magnitude higher (11.3 mM). Furthermore, this monosaccharide uptake system revealed inhibition of fructose uptake by concentrations of glucose as low as 0.5 mM. As the hydrolysis of sucrose at the fungus/plant interface

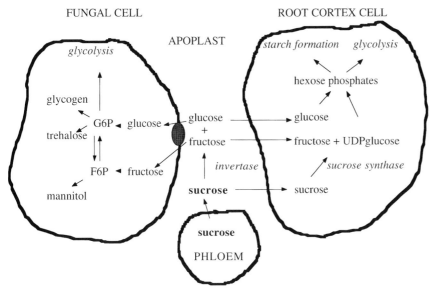

FUNGAL CELL   ROOT CORTEX CELL

Fig. 2. The plant fungus interface in the Hartig net where solute exchange most probably takes place. Of the postulated transport systems involved, only a fungal monosaccharide transporter has been characterized so far (AmMst1 ●); *G6P, F6P* hexose phosphates

of mycorrhizas should yield a 1:1 ratio of glucose and fructose, the obvious preference for glucose uptake by fungal hyphae would lead to an increase of fructose in the mycorrhizal apoplast which, in turn, could inhibit apoplastic acid invertase (Salzer and Hager 1993; Schaeffer et al. 1997). Thus, regulation of glucose supply to the fungus could also be by fructose uptake by the cortex cells of the host. This could prevent apoplastic accumulation of fructose and thus feedback inhibition of invertase.

B. Isolation and Characterization of a Gene Coding for an *A. muscaria* Monosaccharide Transporter

For the identification and characterization of systems for monosaccharide uptake in *A. muscaria*, a molecular approach was used (Nehls et al. 1998). Primers, designed against conserved regions of known fungal monosaccharide transporters, were used to amplify cDNA fragments from *P. abies/A. muscaria* mycorrhizas. These DNA fragments, in turn, were used to isolate full-length cDNA clones from a mycorrhizal cDNA library. Until now one fungal gene coding for a monosaccharide uptake system has been identified. The full-length cDNA clone (AmMST1) contains an open reading frame which codes for a protein of 520 amino acids with a molecular mass of 56455 Da. The protein sequence revealed 12 putative transmembrane domains, arranged in two groups of six each, separated by a hydrophilic spacer of 65 amino acids (Fig. 3, hydrophobicity plot). This is a pattern typical for the superfamily of transmembrane facilitators (Marger and Saier 1993). The best homology for AmMST1 (Fig. 4) was obtained with the *RCO3* monosaccharide transporter gene of *N. crassa* (Madi et al. 1997) the second best with transporters from yeast, i.e., *SNF3* (Celenza et al. 1988) and *RGT2* (Özcan et al. 1996).

Rescreening of the mycorrhizal cDNA library using the AmMST1 cDNA as a probe, as well as using a different set of primers, resulted in the identification of the same AmMST1 gene only. Furthermore, also by Southern blot analysis, no additional monosaccharide transporters homologous to AmMst1 could be identified under either high or medium stringency conditions of hybridization. We thus conclude that AmMST1 represents the main, if not the only, monosaccharide import system for *A. muscaria* in symbiosis and when cultured at elevated external glucose concentrations. This does not necessarily exclude

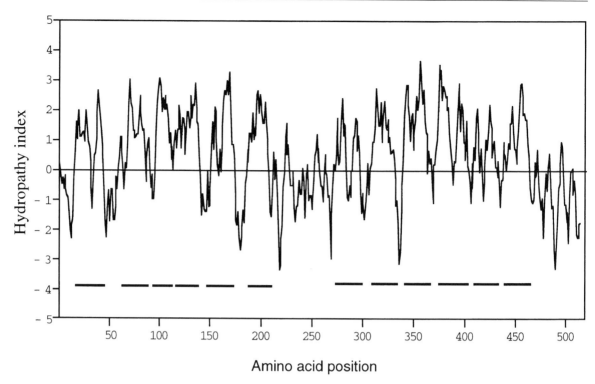

Fig. 3. Hydropathy plot of the deduced amino acid sequence of AmMST1. The hydropathy index of a window of six amino acids was plotted against the amino acid position according to Kyte and Doolittle (1982). Putative membrane spanning α-helical amino acid stretches are *underlined*

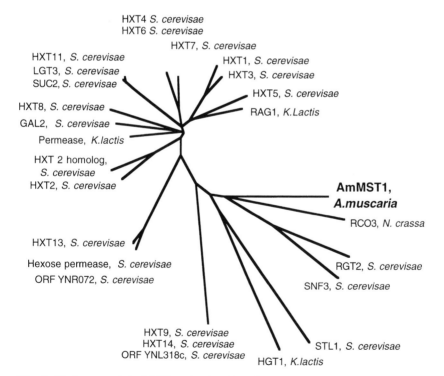

Fig. 4. Cladogram of AmMST1 and other fungal monosaccharide transporters. The relationship of the deduced protein sequence of AmMST1 to other fungal monosaccharide transporters was determined by multiple alignment using ClustalW. (Thompson et al. 1994)

the presence of further monosaccharide transporters which could be expressed at low levels in mycorrhizas or at low external monosaccharide concentrations, and which could be of only limited homology with AmMST1. In comparison, in yeast, more than 20 monosaccharide transporters are encoded by the genome, and a number of these genes were identified by using heterologous yeast probes (Reifenberger et al. 1995). Monosaccharide uptake systems are also present in other fungi (for a review see Jennings 1995), but in most cases the actual number of monosaccharide transporters is unknown.

C. External Monosaccharide Concentration Regulates AmMST1 Expression

A. muscaria hyphae grown at glucose concentrations up to 5 mM expressed the AmMST1 gene at a basal level, while monosaccharide concentrations above 5 mM triggered a fourfold increase of the amount of the AmMST1 transcript, and this could not be further enhanced (Fig. 5; Nehls et al. 1998). Thus, we assume that the change in AmMST1 expression is regulated in a threshold manner, depending on the extracellular concentration of monosaccharides.

An increase in AmMST1 expression, comparable to that found in fungal mycelia when cultivated at elevated monosaccharide concentrations, was also observed in symbiosis of *Amanita muscaria* with both the gymnosperm *Picea abies* (Fig. 5) and the angiosperm *Populus tremula* x *tremoloides*. We therefore assume that the increase of AmMST1 expression in mycorrhizas is also sugar-regulated and depends on the in vivo concentration of monosaccharides at the fungus-plant interface, which should be in mM range. For the apoplast of barley leaves, glucose and fructose concentrations of approx. 5 mM each were reported (Lohaus et al. 1995). From our own data on the amount of glucose of spruce/fly agaric mycorrhizas (about 25 nmol mg^{-1} dry weight; Rieger et al. 1992), we expect apoplastic glucose concentrations of well above 5 mM when an even distribution and a dry weight:fresh weight ratio of about 1:3 is assumed.

The increase in AmMST1 expression is a slow process. A transition from constitutive to maximal content of AmMST1 transcripts occurred between 18 and 24 h of fungal culture, which is in sharp contrast to the time needed for the induction of the high affinity glucose transport system Hxt2 in yeast (90 min; Bisson and Fraenkel 1984). This is possibly due to the fact that soil-growing mycelia are exposed to low concentrations of carbohydrates (Wainwright 1993). Only locally and for a limited time were higher monosaccharide concentrations found. Under such conditions, the basal rate of AmMST1 expression and/or that of other, yet unknown, monosaccharide uptake systems, preferentially expressed at low external monosaccharide concentrations, could be sufficient for the uptake of limited monosaccharide quantities. In contrast to soil-growing hyphae, hyphae at the symbiotic interface are exposed to a continuous supply of plant-derived carbohydrates, which can represent up to 30% of the photosynthetically fixed carbon (Söderström 1992). In order to adapt the flux of carbohydrates into the fungus, monosaccharides must be taken up quickly and either metabolized or converted into compounds for intermediate or long-term storage (e.g., trehalose, mannitol, glycogen). It can thus be assumed that both the extended lag phase for enhanced AmMST1 expression, and its threshold response to elevated monosaccharide concentrations, are adaptations of the ectomycorrhizal fungus to the conditions found at the symbiotic interface. Interestingly, at the threshold value of 5 mM glucose, the uptake capacity of the *Amanita* monosaccharide transporter ($K_M = 1.25$ mM; Chen and Hampp

1 2 3 4 5 6 7 8 9 10 11 12

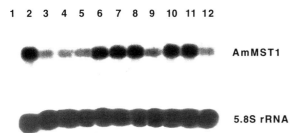

AmMST1

5.8S rRNA

Fig. 5. AmMST1 expression in mycorrhizas and in mycelia grown at different monosaccharide concentrations. After precultivation of fungal mycelia in the absence of glucose, mycelia were cultivated at different glucose or fructose concentrations (0.5 to 50 mM) for 2 days. Total RNA was extracted, separated on an agarose gel, transfered to nylon membranes and hybridized with AmMst1 cDNA or 5.8 S rRNA. (*1*) non-inoculated *P. abies* lateral root; (*2*) *P. abies/A. muscaria* mycorrhizas; (*3*) no glucose; (*4*) 1 mM glucose; (*5*) 2 mM glucose; (*6*) 5 mM glucose; (*7*) 40 mM glucose; (*8*) 95 mM glucose; (*9*) 2 mM fructose; (*10*) 40 mM fructose; (*11*) 20 mM trehalose (hydrolyzed into glucose by fungal extracellular trehalase); (*12*) 40 mM gluconate

1993) is nearly saturated. Obviously, the fungus can sense saturation of monosaccharide transport and, in due course, enhances AmMST1 expression. This results in additional transporter proteins with a concommittant increase of the monosaccharide import capacity, obviously an adaptation to the conditions during symbiotic interaction. The enhanced photoassimilate sequestration by the fungus should then trigger photoassimilate supply by the host with the consequence of increased rates of photosynthesis (see above).

D. Kinetic Properties of the AmMST1 Protein

The function of the AmMST1 protein as active monosaccharide transporter could be confirmed by heterologous expression of the full-length cDNA in a yeast mutant lacking a functional endogeneous monosaccharide uptake system (Fig. 6; A. Wiese, unpubl.). The uptake properties of the AmMST1 monosaccharide transporter resemble those obtained with *A. muscaria* proto-plasts (Chen and Hampp 1993; A. Wiese, unpubl.).

The K_M value for glucose was determined to 0.4 mM, and for fructose to 4 mM (A. Wiese, unpubl.). Also, a strong inhibition of fructose uptake in the presence of glucose was found. The finding that the AmMST1-dependent monosaccharide uptake resembles that of fungal protoplasts again corrob-orates the assumption that AmMST1 is the major monosaccharide uptake system in *Amanita mus-caria*. Interestingly, the *Amanita muscaria* mono-saccharide uptake system exhibits an affinity for glucose which is in between the high (K_M 10 μM) and the low affinity (K_M 10 mM) monosaccharide uptake systems of yeast. This further supports the view that sugar transport by the ectomycorrhizal fungus *Amanita muscaria* is highly adapted to the symbiotic fungus/plant interface.

V. Carbon Metabolism in Fungal Cells

The above studies show that glucose is preferen-tially taken up by the basidiomycete *A. muscaria*. The preference for glucose has also been shown

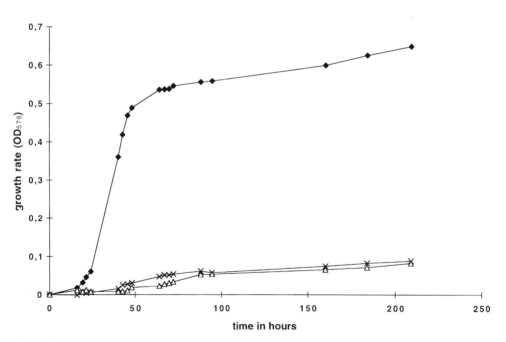

Fig. 6. Heterologous expression of AmMST1 cDNA in *S. cerevisiae*. A monosaccharide transporter-deficient *S. cere-visae* mutant was transformed with constructs containing the AmMST1 cDNA in sense (♦) and antisense (Δ) orien-tation under the control of the yeast PGK-promoter and the control (X) (vector only). The AmMST1 cDNA in sense orientation was expressed under its own internal start ATG. The transformants were precultivated on medium containing gluconate for 1 week and were then transfered to a medium containing 100 mM glucose as carbon source. The growth rates (OD_{578}) of the transformed yeast strains on glucose are shown

for other ectomycorrhiza-forming fungi such as *Cenococcum geophilum* (Hampp et al. 1995), *Cenococcum graniforme* (Martin et al. 1985) or *Sphaerosporella brunnea* (Martin et al. 1988). Labelling studies with ^{13}C showed that glucose taken up by the fungus is rapidly converted as indicated by the fast randomization of ^{13}C after cycling through the mannitol pool. In addition, after 6h of continuous labelling nearly 40% of the soluble ^{13}C was recovered from the amino acid pool (Martin et al. 1988). It would be interesting to know how much of this pool is returned to the host in the symbiotic interaction.

VI. Conclusion

The carbohydrate requirements of ectomycorrhiza-forming fungi clearly affect carbon allocation in the host plant with the result of increased photoassimilate supply (preferably sucrose) toward the area of symbiotic interaction. In order to become available for the fungus, hexoses have to be formed. This is part of the plant's duties. Hexose uptake by the fungus (*A. muscaria*) is mainly performed by a single monosaccharide transporter. The affinity of this transporter for glucose is between that of the high and the low affinity transport systems reported for saprophytic fungi. The regulation of its expression in a threshold manner, as well as the long lag-phase before enhanced gene expression becomes visible, are obviously adaptations of the fungus to the homeostatic, monosaccharide-rich environment which is only available in the space of symbiotic interaction, but not in soil.

A number of questions remain, however, which have to be addressed in future research:

1. What generates the signal for enhanced AmMST1 expression when hyphae are exposed to higher external concentration of monosaccharides?
2. What are the properties of the monosaccharide uptake system of the plant partner in comparison to that of the fungus?
3. Does the external monosaccharide concentration generate a general environment-indicating signal for hyphal physiology (low hexose concentrations in the soil versus high hexose availability within the plant-fungus interface) and could thereby regulate fungal functions in the

different parts of an ectomycorrhiza (Hartig net, mantle structure, rhizomorphs, plaques)?

Acknowledgments. We thank Magret Ecke and Elke Klenk for excellent technical assistance. This work was supported by the Deutsche Forschungsgemeinschaft (Graduiertenkolleg Organismische Interaktion in Waldökosystemen).

References

Bevege DI, Bowen GD, Skinner MF (1975) Comparative carbohydrate physiology of ecto- and endomycorrhizas. In: Sanders FE, Mosse B, Tinker PB (eds) Endomycorrhizas. Academic Press, London, pp 149–174

Bisson LF, Fraenkel DG (1984) Expression of kinase-dependent glucose uptake in *Saccharomyces cerevisiae*. J Bacteriol 159:1013–1017

Celenza JL, Marshall-Carlson L, Carlson M (1988) The yeast SNF3 gene encodes a glucose transporter homologous to the mammalian protein. Proc Natl Acad Sci USA 85:2130–2134

Chen X-Y, Hampp R (1993) Sugar uptake by protoplasts of the ectomycorrhizal fungus *Amanita muscaria*. New Phytol 125:601–608

Dosskey MG, Boersma L, Linderman RG (1991) Role for the photosynthate demand of ectomycorrhizas in the response of Douglas fir seedlings to drying soil. New Phytol 117:327–334

Hampp R, Schaeffer C, Wallenda T, Stülten C, Johann P, Einig W (1995) Changes in carbon partitioning or allocation due to ectomycorrhiza formation: biochemical evidence. Can J Bot 73 (Suppl 1):S548–S556

Ineichen K, Wiemken V, Wiemken A (1995) Shoots, roots and ectomycorrhiza formation of pine seedlings at elevated atmospheric carbon dioxide. Plant Cell Environ 18:703–707

Jennings DJ (1995) The physiology of fungal nutrition. Cambridge University Press, Cambridge

Johnson CR, Menge JA, Schwab S, Ting IP (1982) Interaction of photoperiod and vesicular-arbuscular mycorrhizae on growth and metabolism of sweet orange *Citrus sinensis*. New Phytol 90:665–670

Koch KE (1996) Carbohydrate-modulated gene expression in plants. Annu Rev Plant Physiol Plant Mol Biol 47:509–540

Kyte J, Doolittle RF (1982) A simple method for displaying the hydropathic character of a protein. J Mol Biol 157:105–132

Lewis JD, Strain BR (1996) The role of mycorrhizas in the response of *Pinus taeda* seedlings to elevated CO_2. New Phytol 133:431–443

Loewe A, Einig W, Shi L, Dizengremel R, Hampp R (2000) Mycorrhization and elevated CO_2 both increase the capacity for sucrose synthesis in source leaves of spruce and aspen. New Phytol 145:565–574

Lohaus G, Winter H, Riens B, Heldt HW (1995) Further studies of the phloem loading process in leaves of barley and spinach. The comparison of metabolite concentrations in the apoplastic compartment with those in the cytosolic compartment and in the sieve tubes. Bot Acta 108:270–275

Madi L, McBridge S, Bailey LA, Ebbole DJ (1997) Rco-3, a gene involved in glucose transport and conidiation in *Neurospora crassa*. Genetics 146:499–508

Marger MD, Saier MH (1993) A major superfamily of transmembrane faciliators that catalyze uniport, symport and antiport. Trends Biol Sci 18:13–20

Martin F, Canet D, Marchal JP (1985) Carbon-^{13}NMR study of mannitol cycle and trehalose synthesis during glucose utilization by the ectomycorrhizal ascomycete *Cenococcum graniforme*. Plant Physiol 77:499–502

Martin F, Ramstedt M, Söderhäll K, Canet D (1988) Carbohydrate and amino acid metabolism in the ectomycorrhizal ascomycete *Sphaerosporella brunnea* during glucose utilization. A ^{13}C NMR study. Plant Physiol 86:935–940

Melin E, Nilsson H (1957) Transport of ^{14}C-labelled photosynthate to the fungal associate of pine mycorrhiza. Sven Bot Tidskr 51:166–186

Nehls U, Wiese A, Guttenberger M, Hampp R (1998) Carbon allocation in ectomycorrhiza: identification and expression analysis of an *A. muscaria* monosaccharide transporter. Mol Plant Microb Interact 11:167–176

Norby RJ, O'Neill EG, Hood WG, Luxmore RBJ (1987) Carbon allocation, root exudation and mycorrhizal colonization of *Pinus echinata* seedlings grown under CO$_2$ enrichment. Tree Physiol 3:203–210

Nylund JE, Wallander H (1989) Effects of ectomycorrhiza on host growth and carbon balance in a semi-hydroponic cultivation system. New Phytol 112:389–398

Özcan S, Dover J, Rosenwald AG, Wölfl S, Johnston M (1996) Two glucose transporters in *Saccharomyces cerevisiae* are glucose sensors that generate a signal for induction of gene expression. Proc Natl Acad Sci USA 93:12428–12432

Reifenberger E, Freidel K, Ciriacy M (1995) Identification of novel HXT genes in *Saccharomyces cerevisiae* reveals the impact of individual hexose transporters on glycolytic flux. Mol Microbiol 16:157–167

Rieger A, Guttenberger M, Hampp R (1992) Soluble carbohydrates in mycorrhized and non-mycorrhized fine roots of spruce seedlings. Z Naturforsch 47c:201–204

Salzer P, Hager A (1991) Sucrose utilization of the ectomycorrhizal fungi *Amanita muscaria* and *Hebeloma crustuliniforme* depends on the cell wall-bound invertase activity of their host, *Picea abies*. Bot Acta 104:439–445

Salzer P, Hager A (1993) Characterization of cell wall-bound invertase isoforms of *Picea abies* cells and regulation by ectomycorrhizal fungi. Physiol Plant 88:52–59

Schaeffer C (1995) Untersuchung des Kohlenhydratstoffwechsels von Ektomykorrhizen (Pilz-Baumwurzel-Symbiosen). Einfluß der Mykorrhizierung von *Picea abies* mit *Amanita muscaria* und *Cenococcum geophilum* auf Enzymaktivitäten und Metabolite des Saccharosestoffwechsels und der Glykolyse. PhD Thesis, Tübingen University, Tübingen, Germany

Schaeffer C, Wallenda T, Hampp R, Salzer P, Hager A (1997) Carbon allocation in mycorrhizae. In: Rennenberg H, Eschrich W, Ziegler H (eds) Trees – contributions to modern tree physiology. Backhuys, Leiden, The Netherlands, pp 393–407

Smith SE, Read DJ (1997) Mycorrhizal symbiosis, 2nd edn. Academic Press, San Diego

Söderström B (1992) The ecological potential of the ectomycorrhizal mycelium. In: Read DG, Lewis DH, Fitter AH, Alexander IJ (eds) Mycorrhizas in ecosystems. CAB International, Wallingford, pp 77–83

Stitt M (1991) Rising CO$_2$ levels and their potential significance for carbon flow in photosynthetic cells. Plant Cell Environ 14:741–762

Thompson JD, Higgins DG, Gibson TJ (1994) ClustalW: improving the sensitivity of progressive multiple sequence alignment through sequence weighting, position-specific gap penalties and weight matrix choice. Nucleic Acids Res 22:4673–4680

Wainwright M (1993) Oligotrophic growth of fungi – stress or natural state. In: Jennings DJ (ed) Stress tolerance of fungi. Marcel Dekker, New York, pp 127–144

Further Fungal Interactions

8 *Piriformospora indica*: An Axenically Culturable Mycorrhiza-Like Endosymbiotic Fungus

A. Varma[1], A. Singh[1], Sudha[1], N.S. Sahay[1], J. Sharma[1], A. Roy[1], M. Kumari[1], D. Rana[1], S. Thakran[1], D. Deka[1], K. Bharti[1], T. Hurek[2], O. Blechert[3], K.-H. Rexer[3], G. Kost[3], A. Hahn[4], W. Maier[5], M. Walter[5], D. Strack[5], and I. Kranner[6]

CONTENTS

[1] School of Life Sciences, Jawaharlal Nehru University, New Delhi, 110067, India
[2] Philipps-University Marburg, Max-Planck-Institute for Terrestrial Microbiology, Karl-von-Frisch-Straße, 35032 Marburg, Germany
[3] Philipps-University Marburg, Fachbereich Biologie, Systematic Botany and Mycology, Karl-von-Frisch-Straße, 35032 Marburg, Germany
[4] Technical University of Muenchen at Weihenstephan, Department of Botany, Alte Akademie 12, 85350 Freising, Germany
[5] Institute of Plant Biochemistry, Weinberg 3, 06120 Halle (Saale), Germany
[6] Institute of Plant Physiology, Karl-Franzens University of Graz, Schubertstraße 51, 8010 Graz, Austria

I. Introduction

Soil microflora influence plant growth and health both beneficially and adversely. The microorganisms used as biofertilisers stimulate plant growth response by providing necessary nutrients as a result of their colonisation of the rhizosphere (*Azotobacter*, *Azospirillum*, phosphate-solubilising bacteria and cyanobacteria) or by symbiotic association (*Rhizobium*, mycorrhizae fungi and *Frankia*). They may also regulate the physiological processes in the ecosystems by involvement in the decomposition of organic matter, fixation of atmospheric nitrogen, secretion of growth-promoting substances, increase in the availability of mineral nutrients, immobilisation of these assimilable nutrients and protection of plants from pathogens (Mukerji et al. 1998). Thus, rhizosphere effects through microbial activities, modify the plants by providing plant growth substances and increasing the availability of nutrients at the root zone. Plant root anatomy and tissue articulation play a significant role in the symbiotic processes (Lynch 1995).

The mycorrhizal symbiosis is an intimate association which exists between the plant root system and certain groups of soil fungi (Sahay et al. 1998). The name mycorrhiza, literally translated root fungus, was coined by Frank in 1885. This relationship benefits the plant growth by enabling a greater proportion of the nutrients available in the soil to be absorbed by the plants. Possible benefits include increased crop yield, protection against certain root pathogens, increased tolerance to environmental stress and perhaps, most importantly, a reduction in the input of chemical fertilisers (Linderman 1994). The efficient use of mycorrhiza has been found to be more meaningful in tropical countries (Bagyaraj and Varma 1995; Hindav et al. 1998). The mycorrhiza-plant partnership is the basic, essential and integral part of plant survival and growth (Varma and Schuepp

The Mycota IX
Fungal Associations
Hock (Ed.)
© Springer-Verlag Berlin Heidelberg 2001

1995). In situations where native mycorrhizal fungal inoculum potential is low or ineffective, providing the appropriate fungi for plant production systems is worth considering (Varma 1995a,b; Read 1999). With the current state of technology, inoculation is the best method for transplanted crops and in areas where soil disturbance has reduced native inoculum potential (Sahay et al. 1998; Varma 1998, 1999a). Inoculum production of arbuscular mycorrhizal fungi (AMF) presents a very different problem (Varma 1995b). These fungi will not grow like any other fungi apart from their hosts (obligate symbionts). Although genetic engineering of these fungi may, in future, provide the possibility of their production in the absence of the host plants, the classical technique will still be in use for a considerable time. Obligate symbiotic mode of growth, non-availability of pure culture and expensive means of production and their unreliability for the beneficial effects have greatly jeopardised/undermined the mycorrhizal science, and may be appropriately called a biological enigma. Because of the absence of an authentic pure culture, commercial production is the greatest bottleneck in the use and application of mycorrhizal biotechnology. The lack of authentic culture also slows down molecular analysis to understand the basis for the fungal interaction with plant cross-talk and their biotechnological applications (Varma 1998, 1999a). However, now there is a silverlining for the scientists dealing with the mycorrhizal science, as Varma and his colleagues have discovered a new root endophyte designated *Piriformospora indica*, belonging to the Hymenomycetes (Basidiomycota). *P. indica* drastically improves plant growth and overall biomass and can be easily cultivated on a variety of synthetic media. Hyphae colonise the root and show inter- and intracellular structures (vesicles and hyphal coils). Chlamydospores are formed both inside the root tissues and externally into the environment (Verma et al. 1998).

The fungus grows on a wide range of synthetic simple and complex media, e.g., on minimal media (MM1) normally used for *in vitro* germination of AM fungi with 2% sugar or glucose as carbon source, on different media for *Aspergillus* sp. (CM1 and MM2) and on Moser b medium (Fig. 1a). Mass cultivation of the fungus can be easily achieved on simplified broth culture (Fig. 1b). The temperature range of the fungal growth is 20–35 °C; the optimum temperature and pH being 30 °C and 5.8, respectively.

Fig. 1a–c. Cultivation of *Piriformospora indica* on agar and liquid broth. **a** Mycelial growth on agar disc transferred onto minimal (without organic complexes), CM, Moser *b* and MM 2 media (clockwise). Petri plates were incubated at 30 °C for 28 days. **b** Broth culture in aspergillus medium. **c** Highly interwoven mycelial mat. (cf. Verma et al. 1998)

By now, we have established beyond any doubt that the new fungus *P. indica* promotes plant growth. Interestingly, it has a wide host range, monocots and dicots including legumes. In addition, the tissue culture-raised plantlets, on treatment with the fungus, showed early root differentiation. When such plantlets (after biological hardening) were transferred to the pots and later to the field, the survival rate ranged from 90 to 100%. In contrast, the untreated plantlets had a low survival rate, ranging from 10 to 60%. Early callus differentiation into roots and shoots was also seen (Sahay 1999; Sudha 1999).

Recently, we have also found that *P. indica* shows potential as an agent for biological control of disease against soil-borne pathogens. The fungus was coinoculated with a potent plant pathogen, *Gaeumannomyces graminis*, on a pathogen medium typical for the latter, containing inorganic nutrients and yeast extract with saccharose as the carbon source. It was found that, in contrast to the normal growth of *P. indica*, the growth of the pathogen was completely inhibited. A similar observation was made for *P. indica* when used in combination with several other established root pathogenic fungi.

To sum up, the new fungus acts both as a plant stimulator and a pathogenic inhibitor. This opens up new perspectives for its application in agriculture, flori-horticulture, viticulture and reclamation of degraded and heavily mined soils.

II. Fungal Morphology and Taxonomy

Significant quantitative and morphological changes were detected when the fungus was challenged to grow on different media. Incubation under shaking retarded the growth in MM1 liquid cultures, whereas no such negative effect was ever observed during cultivation on any other substrate. In fact, the fungal biomass was considerably enhanced on shaking cultures with aspergillus medium (CM) (Fig. 1b). On Moser b medium, the colonies appeared compact, wrinkled with furrows and constricted. The mycelium produced fine zonation and a great amount of white aerial hyphae.

The hyphae were highly interwoven, often adhered together and gave the appearance of simple intertwined cord (Fig. 1c). Young mycelia were white and almost hyaline, but inconspicuous zonations were recorded in other cultures. Hyphae were thin-walled and of different diameters ranging from 0.7 to 3.5 μm. The septate hyphae often showed anastomosis. New branches emerged irregularly and the hyphal walls showed some external deposits at regular intervals, perhaps polysaccharides and/or some hydrophobic proteins, which stained deeply with toluidine blue. Since septation was irregular, the single compartments could contain more than one nucleus. The chlamydospores appeared singly or in clusters and were distinctive due to their pear-shaped structure (Fig. 2a,b). The chlamydospores were (14–) 16–25(–33) μm in length and (9–) 10–17 (–20) μm in width. Very young spores had thin, hyaline walls. At maturity, these spores had walls upto 1.5 μm thick, which appeared two-layered, smooth and pale yellow. The cytoplasm of the chlamydospores was densely packed with granular material and usually contained 8–25 nuclei (Fig. 2c). Neither clamp connections nor sexual structures could be observed.

The new fungus produced chlamydospores at the apex of undifferentiated hyphae. Different kinds of substrates were tested to induce sexual development, like young and mature leaves of hemp, young leaves of *Cynodon dactylon* and pollen grains, oat meal, potato, carrot or tomato dextose agar. Since all these efforts did not lead to the desired results, there were only a few features to characterize the fungus morphologically to group it according to the classical species concept. According to the ultrastructure of the septal pore (see **IV**) and the molecular data (see **V**), the fungus was placed within the Hymenomycetes (Basidiomycota).

III. Geographical Occurrence

The fungus was originally found in soil samples from the rhizosphere of the woody shrubs *Prosopsis juliflora* and *Zizyphus nummularia* growing in the western part of Rajasthan, a typical desert region of the Indian subcontinent. Several strains of glomaceous fungi were screened (Mathew et al. 1991; Neeraj et al. 1991). Later, some spores were cultivated on minimal agar medium (Varma 1995a). Following the serial dilution technique, a pure axenic culture was obtained which was later named *P. indica*. In a communication, Professor

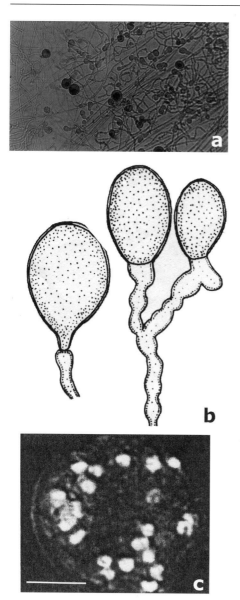

Fig. 2a–c. Morphology of *P. indica*. **a** Intertwined hyphae with spores. **b** Pear-shaped chlamydospores at the tip of the hyphae. **c** Nuclei in chlamydospores. Chlamydospores were stained with DAPI and observed in epifluorescence. Different optical planes were assembled in one picture using improvision software package (IMPROVISION, Govenny, UK). *Bar* 10 μm

isons of 18S rDNA sequences obtained after PCR with specific primers indicated that a similar kind of fungus apparently occurs in the rhizosphere soil of *Leptochloa fusca* or *Oryza minuta* in Pakistan or in the Phillipines, respectively. Evaluations of 18S rDNA clone libraries obtained with general primers suggested that the fungus was not abundant in any of these rhizospheres. However, none of the European soils tested were found to contain this fungus.

IV. Ultrastructure of the Hyphae

The cell walls were very thin and showed multi-layered structures. The septa consisted of dolipores with continuous parenthosomes (Fig. 3). The dolipores were very prominent with a multi-layered cross-wall and a median swelling mainly consisting of electron-transparent material. The parenthosomes were always straight and had the same diameter as the corresponding dolipore. No pores of any kind could be detected, thereby showing that they are flat discs without perforation (Verma et al. 1998). The parenthosomes consisted of an electron-dense outer layer and a less dense inner layer, which showed an inconspicuous dark line in the median region. These studies formed the basis for the systematic position within the Hymenomycetes.

V. Molecular Studies

Molecular sequence data are useful for systematics, where morphological characters, like the taxonomically decisive sexual states, are missing. A neighbour–joining analysis on comparisons of partial 18S rDNA sequences (525 nucleotide positions) placed *P. indica* within the Basidiomycota close to the *Rhizoctonia solani* group (*Ceratobasidiales*) (Verma et al. 1998). A maximum likelihood analysis on almost complete 18S rDNA sequences (1550 nucleotide positions) confirmed this assignment (Fig. 4). Similar results were obtained by distance and parsimony methods. A comprehensive phylogenetic analysis of *Rhizoctonia* using sequences from mitochondrial and nuclear rDNA on more representatives should provide an insight into the evolution of this important group and clarify the evolutionary rela-

A.G. Khan, and his coworkers in New South Wales, Australia, presented a fungus similar to the description to *P. indica*. Although they considered their isolate as one of the species of *Glomus*, after careful examination of the description of the figures and photographs, the present authors are inclined to categorise the Australian isolate as a prototype similar to *P. indica* (pers. comm. described, see Ph.D. thesis, Saif 1997). Compar-

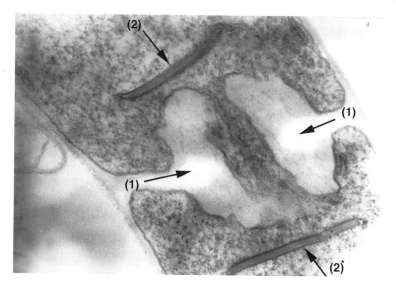

Fig. 3. Dolipore and parenthosomes of *P. indica.* Sections of hyphae were observed by TEM. Arrows indicate the dolipore (1) and the continuous parenthosomes (2). This septal pore type is typical for Hymenomycetes

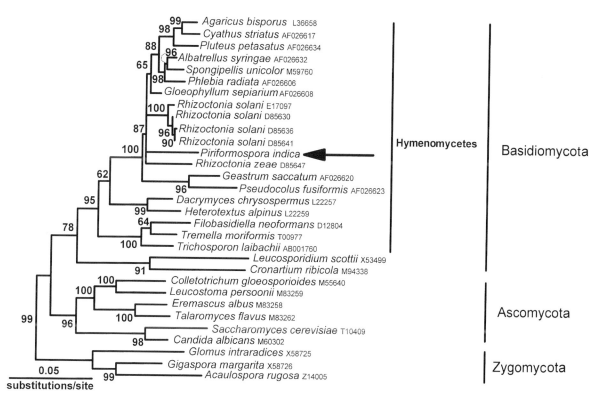

Fig. 4. Maximum-likelihood tree estimated by the quartet puzzling method (Strimmer and Haeseler, 1996) as implemented by PAUP 4.0b2a (Swofford, 1998) on 18S rDNA sequences showing phylogenetic relationships between *Piriformospora indica* and other representatives of the Basidiomycetes. Branches with support values below 55% were collapsed. Puzzling support indices are shown at each branch

tionship of *P. indica* and *Rhizoctonia* within the Hymenomycetes.

There is no existing genus which covers all the characters of the new fungus. Other groups of Basidiomycota which may be related according to the molecular or ultrastructural data are not known to be able to produce chlamydospores or have a different way of life. Therefore, a new genus was erected and the fungus was designated *Piriformospora indica* (Verma et al. 1998).

VI. Root Colonization

Most of the plants (mono- and dicots, including legumes) tested were colonised by *P. indica*. The hyphae first colonised the surface and produced structures similar to appressoria. Subsequently, they entered and traversed through the cortical cells and produced vesicles and also differentiated into intracellular highly coiled structures (Fig. 5a, b, c, d). At maturity, external and internal spores were formed. Like AM fungi, hyphae multiplied within the cortical tissues and never traversed through endodermis. Likewise, also did not invade the aerial portion of the plant (stem and leaves). Protein profile of the colonised root showed some qualitative and very little quantitative changes (Fig. 6, Table 1) but the protein profile of the leaf showed no qualitative but some quantitative changes (Fig. 7, Table 2).

VII. Axenic Culture

Williams (1992) provided the term axenic, which is used to refer to growth of a single species in the absence of whole, living organisms or living cells of any other species. The growth of arbuscular mycorrhizae in axenic culture is a matter of global concern. If these fungi could be grown in culture, many more characteristics might be considered and classification facilitated; in particular, their genetic modification, selection to obtain superior strains, and biotechnological applications would

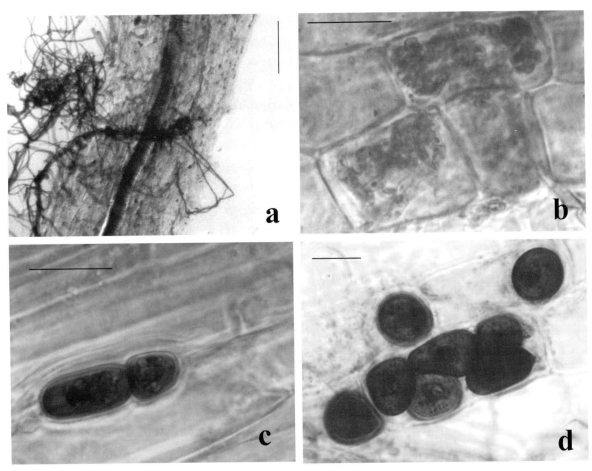

Fig. 5a–d. Root colonisation. Maize plants were inoculated with *Piriformospora indica* and harvested 5 weeks after inoculation. Root samples were stained with Chlorazol black E and investigated by light microscopy. Roots were heavily colonised with the fungus (**a**) in the root cortex, development of intracellular hyphal coils (**b**) and round bodies. Round bodies could be observed (**c**) intercellularly or (**d**) intracellularly. Bars **a** 50 μm; **b, c, d** 10 μm

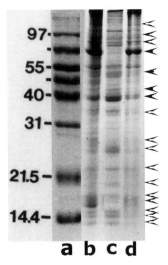

Fig. 6. Protein profile (SDS-PAGE) of *P. indica* colonised root biomass of tobacco plant. *Lane a* shows the standard marker (molecular weight kDa); *lane b* untreated; *lane c P. indica* treated biomass; *lane d G. mosseae*-treated biomass. Qualitative and quantitative changes were observed

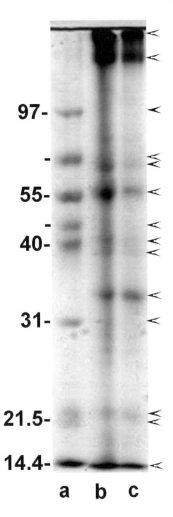

Fig. 7. Protein profile (SDS-PAGE) of *P. indica* colonised leaf biomass of tobacco plant. *lane a* shows the standard marker (molecular weight kDa); *lane b* untreated control; *lane c P. indica*-treated biomass

become feasible. Axenic culture of AMF has now become a subject of interest to commercial firms, who foresee the possibilities of producing high-quality inoculum under controlled conditions. Despite the large amount of hyphal extension which can occur on defined media, AMF growth has always been found to be dependent on the attachment of the hyphae to the parent spore, unless root infection has taken place. Azcon-Aguilar et al. (1986) suggested that over 90% germination can be obtained axenically on water agar in the absence of any mineral or organic nutrient supply.

There are many possible reasons why AM fungi fail to show extensive and continuous growth unless they are part of a symbiotic partnership with a host root. Hepper (1984) postulated that they may have a simple nutritional requirement which, due to our lack of knowledge, has not yet been fulfilled, or it may be necessary to supply some nutrients continuously at a low concentration, or else these fungi may have lost a considerable part of their genetic material and this necessitates their interaction with the host's metabolism. It is also postulated that as these organisms have no saprophytic phase, under normal ecological conditions some part of the genome is in the repressed condition and that the host supplies the inducer to allow nucleic acid transcription. If this is true, then the postulated

inducer must be almost universally present in plant roots and might be extractable. In this particular area of research on AMF, more than in any other, there is probably a pool of unpublished data relating to negative results or to the effect of nutrients or conditions which provided only marginal improvements in hyphal growth. All results of this type are valuable, even if only to prevent repetition of unprofitable lines of approach.

The cultivation of AMF under axenic conditions continues to be one of the challenging goals in mycorrhizal research. However, following the discovery of the new isolate, *P. indica*, which mimics AMF in many respects and can be easily cultivated axenically on synthetic medium, may present a model system to understand the

Table 1. Quantitative protein profile of *Piriformospora indica* colonised tobacco root

Root	Control	*P. indica*	*G. mosseae*
Total bands	17	18	21
Weakly expressed	5 (kDa 12, 14, 16, 23, 28)	4 (kDa 12, 14, 16, 74)	10 (kDa 12, 14, 16, 18, 20,23, 26, 34, 36, 45)
Strongly expressed	12 (kDa 3 > 97, 26, 33, 36, 40, 43, 47, 49, 55, 74)	14 (kDa 4 > 97, 23, 26, 34, 36, 41, 43, 45, 47, 49, 55)	11 (kDa 4 > 97, 16, 47, 55, 80, 84, 95, 97)

Water-soluble proteins separated on SDS-PAGE from root samples depict protein bands of weakly and strongly expressed nature which are represented in tabular form with their approximate molecular weights to compare with *G. mosseae* and control.

Table 2. Qualitative protein profile in leaves of *Piriformospora indica* colonised tobacco

Leaves	Control	*P. indica*
Total bands	9	10
Weakly expressed	4 (kDa 17, 21.5, 31, 47)	5 (kDa 21.5, 31, 47, 74, 97)
Strongly expressed	5 (kDa 2 > 97, 14.4, 34, 55)	5 (kDa 2 > 97, 14.4, 34, 55)

Water-soluble proteins separated on SDS-PAGE from leaf samples depict protein bands of weakly and strongly expressed nature which are enumerated in the table with their approximate molecular weights.

molecular basis of photo- and mycobiont interaction and would be helpful in the achievement of the above-mentioned goals.

VIII. Root Organ Culture

Plant factors involved in the growth of arbuscular-mycorrhizal fungi remain to be discovered. Even before symbiosis has been established, germinating spores need certain factors to promote hyphal growth (Bécard and Piché 1989a,b). Typical mycorrhizal fungal interaction must be produced under controlled biological conditions for critical study of the effects of AMF on plant growth.

Mosse and Hepper (1975) were the first to produce a simplified *in vitro* system for the study of AM development using excised roots in place of whole plants. Mugnier and Mosse (1987) modified the technique by using Ri T-DNA transformed roots (hairy roots) as host tissue. Transformed root organ culture of carrot together with *P. indica* showed luxuriant root growth promotion compared to the control (Fig. 8a,b).

This technique allows standardised quantitative measurement of fungal growth and can be successfully used to outline the factors necessary for mycorrhizal development (Bécard and Piché 1992). Normal root organ culture of tobacco along with the inoculation of *P. indica* shows root induction and proliferation (Fig. 8c,d).

Root organ culture and seedling experiments in association with the *P. indica* show more surface colonisation than inter- and intracellular hyphae (Fig. 9). Characteristic pear-shaped spore formation occurs in the cortical regions of the root as well as in the extramatrical environment of rhizosphere.

IX. Photosymbiont Promotional Features

A. Impact of Fungal Hyphae

P. indica improves the growth and overall biomass production of different grasses, trees and herbaceous species. More recently, we have identified a model plant, *Setaria italica* which has, a small

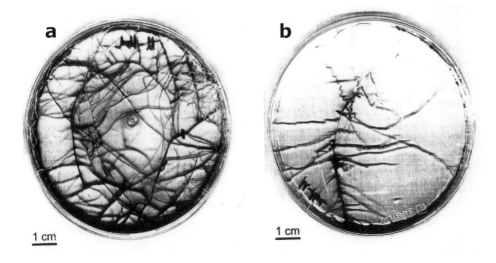

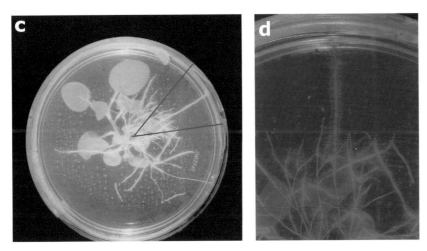

Fig. 8a–d. Effect of *P. indica* on transformed carrot roots after 20 days of inoculation. **a** Heavy growth of root system in the presence of the fungus. **b** Control (without fungus) showing poor development of root system. **c** Root initiation by *P. indica* in tobacco plantlets. **d** An enlarged view of a sector

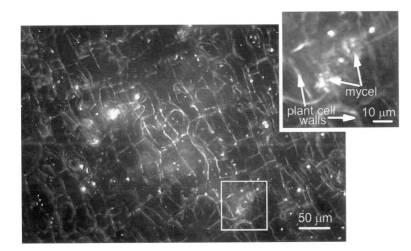

Fig. 9. Colonisation of carrot root organ culture with *P. indica*. Top view of an optical section through roots stained with live/dead (Viability Kit, Molecular Probes, Europe BV). Cells heavily colonized intracellularly (insert)

control   P. indica

Fig. 10. Growth of *Setaria italica* in pots containing expanded clay. Inoculum at the rate of $2\,gl^{-1}$ was placed at the early stages of seed germination. Photographs were made after 4 weeks

genome next to rice. The roots were colonised and the plants highly promoted by interacting with *P. indica* (Fig. 10). The new fungus also promotes several tropical legumes tested (*Cicer arietinum*, *Phaseolus aureus*, *Ph. mungo*, *Pisum sativum* and *Glycine max*). Interestingly, like AMF, the new fungus did not invade the root of two strains of *Brassica* (Cruciferae) and the myc⁻ mutants of *Glycine max* and *Pisum sativum*, obtained from Prof. Peter Gresshoff, Knoxwille, USA and Dr. V. Gianninazzi-Pearson, Dijon, France, respectively.

B. Impact of Culture Filtrate

P. indica is a potent plant growth-promoting fungus. It is not only the mycelium in association with roots but also the culture filtrate of the mycelium containing fungal exudates (hormones, enzymes, proteins, etc) which exerts this positive effect. In the culture experiment a very small amount of this filtrate $(50\,\mu l)$ is sufficient to promote root and shoot growth (Fig. 11).

Fifteen-day-old germinated seedlings were transferred to disposable plastic pots containing vermiculite (autoclaved) and sand (acid-washed) at the ratio of 3:1. Experiments were designed to see the effect of culture filtrate of *P. indica* on plant growth. Fifteen ml of freshly eluted culture filtrate was applied to each pot for this purpose. Minimal media (on which the fungus was growing) and equal amounts of double-sterilised distilled water served as effective control for this purpose. There was an increase in root length, shoot length and plant biomass in the *P. indica*-treated hosts, maize, *Bacopa monnieri* and tobacco. Similar

results were obtained in culture tube experiments with the induction of secondary roots.

X. Interaction with Orchids

The interaction of *P. indica* with orchids differs fundamentally from the described interactions with other higher plants. Blechert et al. (1999) inoculated two years old asymbiotically grown *Dactylorhiza maculata* with *P. indica*. They found that pelotons as in typical orchid mycorrhiza were formed and that the growth of the plants was significantly promoted.

Blechert et al. (1999) also tested the influence of *P. indica* on the germination of *D. purpurella* seeds *in vitro*. They found that the germination rates of the seeds were not influenced by *P. indica*, but the growth rates of the developing protocorms were increased by the fungus. They also described the initial stages of the interaction between *D. purpurella* protocorms and *P. indica*.

In further studies, *D. majalis* and *D. fuchsii* were inoculated with *P. indica* and a natural mycorrhizal partner of *D. majalis*, identified as *Rhizoctonia* sp., to compare the interactions (Rexer et al, ms in prep.). The inoculations were carried out according to Clements et al. (1986).

The interaction of *P. indica* with the protocorms started with the colonisation of the rhizoid. The hyphae formed small appressoria-like structures on the rhizoid surface and penetrated into the cell by a narrow penetration neck. Within the cells, the hyphae expanded to a normal diameter and grew towards the protocorm. From the basal

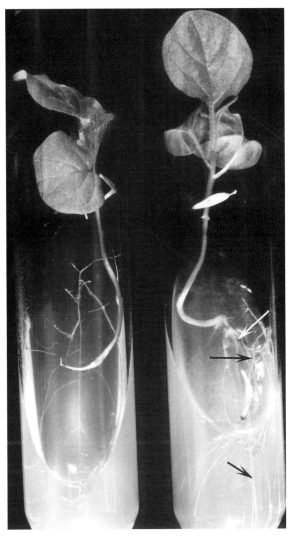

Fig. 11. Effect of culture filtrate on brinjal (*Solanum melongena* var. pusa purple cluster). The seeds were germinated on half-strength MS medium. After 15 days of germination, they were transferred to the culture tubes containing MS media; 15 µl of *P. indica* culture filtrate obtained from 10-day-old fungal culture was applied on the surface of the medium during transfer. The *arrows* indicate the root promotional effect. In the control, an equal volume of the culture medium was included (*left* control; *right* treated)

interaction of *P. indica* with *D. majalis* and *D. fuchsii,* respectively, was identical to the findings of Blechert et al. (1999), who studied the early stages of development in *D. purpurella.* The morphological structures formed during the interaction between the orchids and the fungi tested were almost identical (Fig. 12). The similarity between the two fungi concerned also the ontogeny and chronology of the interaction development. So *P. indica* acts as a specific orchidaceous mycorrhizal fungus in all *in vitro* tested *Dactylorhiza* sp. The development of the interaction of *D. majalis* with *P. indica* was almost identical to the findings of Rasmussen (1990).

XI. Stimulatory Factors

Freshly harvested *P. indica* hyphae were washed with water and aliquots (1 g fresh weight) were further washed twice in 5 ml of 80% aqueous methanol. The supernatant was used for analytical HPLC separation. The analysis showed seven peaks in the hyphae and one main peak in the culture filtrate (Fig. 13a). Preparative HPLC analyses of hyphal and culture filtrate showed a major peak identified as benzoic acid (Fig. 13b–e). The function of this compound is not clear. Compounds identical to benzoic acid and their analogues (benzoic acid, α-hydroxybenzoic acid, 3–4 di-hydroxybenzoic acid, vanillic acid, cinnamic acid, p-coumaric acid, caffeic acid, ferulic acid) showed no stimulation on the plants tested. At this stage of our knowledge, we do not know the stimulating factor which promotes the plant growth.

Freshly harvested intact roots of different plants were washed with water and aliquots were treated twice in 5 ml of 80% aqueous methanol. The mixture was centrifuged and the supernatant was used for HPLC analysis. The cyclohexomone derivative blumenin, which accumulated in roots of cereals and other members of the *Poaceae* colonised by arbuscular mycorrhizal fungi (Maier et al. 1995; Fester et al. 1998), was also present here. The exact function of accumulated cyclohexomones is still not known. It is speculated that these secondary compounds might be involved in the regulation of mycorrhizal colonization (Fester et al. 1998). HPLC analysis of methanolic extracts from 6-week-old infected and non-infected plants showed quantitative but no qualitative changes in the root samples as a result of interaction of

swelling of the rhizoids, inter- and intracellular hyphae were spreading. After the penetration of the cortical cells, especially in the basal region of the protocorm, typical pelotons were formed (Fig. 12c, d). In the epidermal cells, hyphal coils never appeared. The first digested pelotons were found 17 days after inoculation.

The natural symbiont of *D. maculata, Rhizoctonia* sp., developed identical interaction structures and showed the same ontogeny of these structures as *P. indica* (Fig. 12a, b). The mode of

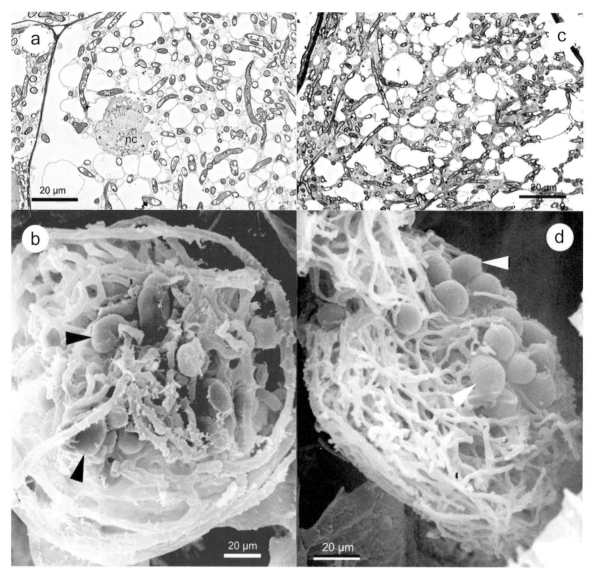

Fig. 12a–d. Semithin sections (**a,c**) and SEM pictures (**b,d**) of pelotons formed during the interaction of *Dactylorhiza majalis* with *Rhizoctonia* sp. (**a,b**) and *P. indica* (**c,d**), respectively. Dense hyphal coils were formed in the orchid tissues. On the SEM photographs, starch grana (*arrowheads*) were visible. The symbiotic cellular structures of both fungi are almost identical; *nc* nucleus of the plant

P. indica with barely, maize and fox tail millet (*Setaria italica*) roots (Fig. 13d). No changes were recorded for the hosts, rice and wheat. The UV spectra obtained from HPLC photodiode array detector showed a cluster of peaks between 7.5 and 12.5 min of the HPLC chromatograph on interaction of maize, barley, rice and foxtail millet with *P. indica*, indicating the presence of indole derivatives, e.g., tryptophane, tryosine and tyramine or their derivatives (unpubl. results).

Root extracts of maize showed a significant presence of cyclic hydroxamine acids like DIMBOA (2,4-dihydroxy-7-methoxy-1, 4-benzoxazin-3-one) in wheat but none in rice, barley and *Setaria*. As a result of the maize root colonisation by *P. indica*, the authors obtained an enhancement of different benzoxazinone levels. HPLC analysis of methanol extracts of infected maize roots showed eight different peaks with a photodiode array detector with typical UV spectra from benzoxazinone derivatives. None of these

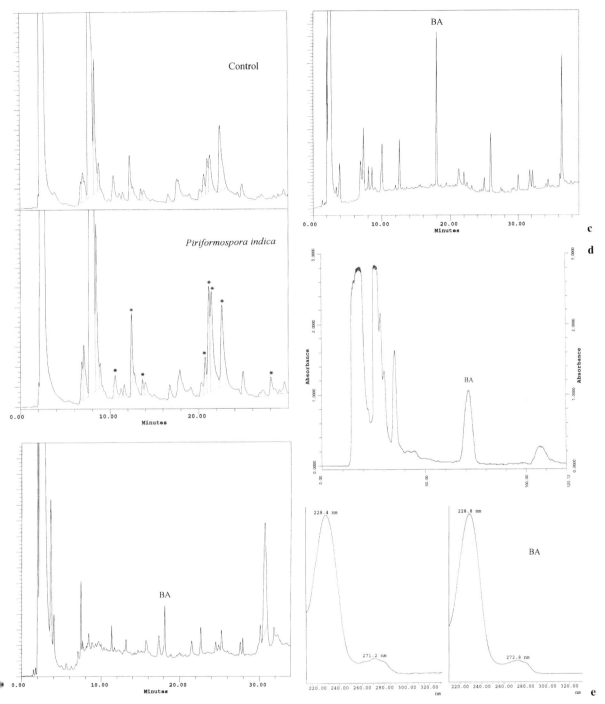

Fig. 13. a HPLC traces of methanolic extracts from 6-week-old non-infected (control) and the respective roots of maize infected with *P. indica*. For each 5 ml of methanolic extraction, (volume from 1 g fresh weight of root material), 20 μl was injected to the HPLC column. Components were traced by maxplot detection between 210 and 450 nm; the *asterisks* mark the eight peaks corresponding to different cyclic hydroxamic acids. Note the dramatic increase in three compounds (as yet unknown) after infection with *P. indica*. **b** Analytical HPLC traces of methanolic extracts from *P. indica* hyphae. With spectroscopic methods, the structure of the main component was identified as benzoic acid (*BA*). **c** Analytical HPLC traces of culture filtrate of *P. indica*; *BA* benzoic acid. **d** Preparative HPLC from the methanolic extracts from 15.5 g *P. indica* with BA as one of the main components. The extract was separated in seven fractions: 1: 13, 3–24, 6 min, 2: 24, 6–33, 0 min, 3: 33, 0–59, 0 min, 4: 59, 0–64, 2 min, 5: 64, 2–80, 0 min (BA), 6: 80, 0–97, 6 min, 7: 97, 6–117, 6 min. **e** UV spectra (max plot detection between 210–450 nm) of BA (*right*) and BA from the culture filtrate (*right*)

peaks was identical to DIMBOA or DIBOA. The chemical structures of these compounds are to be identified by spectrophotometric methods.

XII. Cell Wall-Degrading Enzymes

Cellular actions leading to reciprocal morpho-functional integration between symbionts during mycorrhiza establishment are complex. There is now evidence that the penetration of roots by endomycorrhizal fungi involves low and regulated production of a combination of cell wall-degrading hydrolytic enzymes leading to an organised colonisation of the plant root (Blilou et al. 1996). Investigations have demonstrated the production of pectinase, cellulase, xylanase and chitinase (Garcia-Romera et al. 1990, 1991a,b, 1996; Varma and Bonfante 1994; Perotto et al. 1995a,b, 1997) from the hyphae and mycorrhized roots. It seems that mycorrhizal fungi colonise the root tissues of their host plant by a combination of mechanical and enzymatic mechanisms (Bonfante and Perotto 1995; Gianinazzi-Pearson et al. 1996). Very weak and localised production of enzymes might ensure that the viability of the host is main-

tained, defence responses are not triggered and a high degree of compatibility is reached (Varma 1999b). Despite their potential significance to symbiosis, only a limited number of isolates of a small number of endomycorrhizal fungal species have been screened for enzyme production under a limited range of conditions.

Hyphae of *P. indica* were able to intrude through the cells of the host root. Cell wall-degrading enzymes were tested following the conventional assay methods. Carboxymethylcellulase (CMCase), xylanase and polygalacturonase were found in significant quantities in both the mycelial culture filtrate and the root samples which were treated with *P. indica* (Table 3). It is possible that the cell wall penetration may be due to dual action of hydrolytic enzymes and the mechanical pressure and/or either of them (Fig. 14).

XIII. Serological Characterization

Mycorrhizal associations are the most important mutualistic biotrophic interactions (Harley 1991). To thoroughly understand these complex systems, more information is required at the cellular level. Antibodies, the main tool of immuno-

Blow up

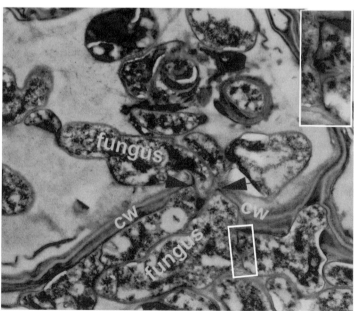

Overview

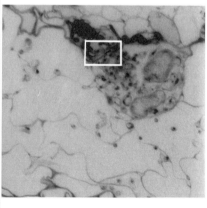

Fig. 14. TEM photograph of transformed carrot root organ culture. Note the entry of *P. indica* hyphae through the cell wall and establishment within the cortical cells. At the point of entry, the hypha broadens and then penetrates, indicating the formation of appressorium-like structure

Table 3. Hydrolytic enzymes in tissue culture (derived from tobacco roots), colonised by *Piriformospora indica*

CMCase

Treatment	IU
Control	0.003
Glomus mosseae	0.002
P. indica (10 days)	0.007
P. indica (45 days)	0.013

1% CMC-Na salt (Sigma) was used as the substrate

Xylanase

Treatments	IU
Control	0.006
Glomus mosseae	0.016
P. indica (10 days)	0.005
P. indica (45 days)	0.062

1% oat spelt xylan (Sigma) was used as the substrate

One unit activity (IU) of CMCase and xylanase was defined as the amount of enzyme which releases 1 µmol of reducing sugar as equivalent to glucose or xylose per min and ml

Polygalacturonase (PG)

Treatments	IU
Control	0.003
Glomus mosseae	0.007
P. indica (10 days)	0.007
P. indica (45 days)	0.017

0.5% Na-polypectate (Sigma) was used as the substrate. One unit activity (IU) of PGase is defined as the amount of enzyme which releases 1 µmol of carboxyl group as equivalent to the amount of Na-thiosulphate added to neutralise the residual iodine

Polymethylgalacturonase (PMG) was completely absent

Table 4. Cross-reaction of *Piriformospora indica* with authentic AMF

Fungal species	ELISA signal (OD 405 nm) Ab diluted	
	1:800	1:1600
Piriformospora indica	0.751	0.577
Glomus mosseae	0.676	0.305
G. intraradices	0.765	0.405
Gigaspora margarita	0.757	0.383
Gi. gigantea	0.697	0.437
Scutellospora margarita	0.713	0.371
SD	0.001–0.008	0.002–0.03

chemical techniques, can be used to characterise extra- and intracellular surfaces and to quantify cellular constituents. Immunochemistry provides elegant tools for these tasks (Goebel et al. 1998; Hahn et al. 1998). The bottleneck is the availability of antibodies which are provided by immunological means. Knowledge of immunological mechanisms allows the researcher to generate efficient probes for physiological, taxonomic and ecological experiments to study the formation, interaction and processes of mycorrhizas.

Antibodies (Ab), immunoglobulins, are produced in response to an antigenic stimulus by plasma cells derived from B-lymphocytes. In analogy to the definition of an antibody, any molecule that elicits a response by the immune system is called an antigen. When the population of antibodies is derived from a set of different B-lymphocytes and usually contains different Ig classes with different affinities and specificities,

they are called polyclonal antibodies (pAb). After their harvest and purification, they can be directly coupled, by chemical means, to detection molecules such as fluorescent dyes for fluorescence microscopic analysis, enzymes for enzyme-linked immunoassays, or heavy metals for immunocytochemical analysis with an electron microscope (Hahn et al. 1999).

Peters (1988) points out that despite modern methods of Ab production such as monoclonal antibodies (mAb) and, more recently, recombinant antibody technologies, pAb are still valuable tools, due to their low cost and ease of production. In addition, a well-designed immunisation scheme in most cases yields an antiserum containing a substantial amount of antibodies with very high specificity and affinities, which can then be used for analysis at a very high dilution. In contrast, the search for Ab of such qualities with mAb technology may be very cost- and labour-intensive.

Immunofluorescence and ELISA tests were carried out after making the polyclonal antibodies against *P. indica*. Antibodies were raised from crushed mycelium and spores. Immunofluorescence studies indicated the recognition of *P. indica* and an affinity with the spores of *Glomus intraradices*. The antigenic similarities were shown with most of the members of Glomales tested so far (Table 4), with lesser cross-reactivity with ectomycorrhizal fungi from Basidiomycota and the least with other fungi belonging to Ascomycetes and Deuteromycetes (Sharma and Varma 2000a). Immunoblot studies further confirmed the observations which were recorded by ELISA. The observations of sero-characterization of *P. indica* and *G. mosseae*, following ELISA, immunoblotting, immuno-fluorescence and immuno-gold labelling, are summarized in Table 5 and Table 6.

Table 5. Performance of the antiserum raised against *P. indica* tested with various endo-, ecto- and non-mycorrhizal antigens using various immunological techniques

Antigens	Techniques employed			
	ELISA	WB	IF	IgL
Piriformospora indica	+	+	+	+
Glomus mosseae	+	+	+	+
Amanita muscaria	+	–	–	–
Pisolithus tinctorius	+	+	–	–
Lentinus edodes	+	–	–	–
Ustilago maydis	+	–	–	–
Schizophyllum commune				
Agaricus bisporus	+	–	–	–

ELISA, Enzyme Linked Immunosorbant Assay; WB, Western Blotting; IF, Immunofluorescence; IgL, Immunogold labelling; +: significant reaction of the antigen with the antiserum; –: no reaction

These results further confirmed the observations recorded earlier for ELISA (Sharma et al 2000a, b)

XIV. Nutrient (P³²) Transport

Phosphorus is an essential mineral nutrient for the plant growth. Plants acquire this mineral from the environment either directly by their roots or indirectly from mycorrhizal fungi which form inter- or intracellular symbiotic associations with the roots. VAM colonisation may increase the rate of phosphorus accumulation beyond the limits which can be currently utilised, thus reducing the rate of phosphorus utilisation efficiency. Such momentary "luxury consumption" of phosphorus may, however, serve a storage function and be utilised subsequently, allowing mycorrhized plants ultimately to outperform non-mycorrhized plants (Koide 1991). Recent P^{32} experiments suggest that *P. indica* is important for P acquisition by the root especially in the arid and semiarid regions (ms in prep.).

Acid phosphatases have been observed to be active in *P. indica* mycelium (Sharma and Varma 2000b). The fungus could utilize a variety of inorganic and organic phosphate sources which is in accordance with the broad range of substrates utilized by the acid phosphatases of many fungi. Besides this, phosphate starvation of *P. indica* led to an overall 27 % increase in the intracellular acid phosphatase activity. This increase was probably due to the appearance of a P-repressible isoform of acid phosphatase in addition to the constitutive

Table 6. Summary of the performance of antiserum raised against *G. mosseae* against various endo–, ecto– and non–mycorrhizal antigens employing various immunological techniques

Antigens	Techniques employed			
	ELISA	WB	IF	IgL
Glomus mosseae	+	+	+	+
G. intraradices	+	–	–	–
Piriformospora indica	+	+	+	+
Amanita muscaria	+	+	–	–
Pisolithus tinctorius	+	+	–	–
Rhizopus sp	+	–	–	–
Ustilago maydis	+	–	–	–

one observed in the enzyme staining of the native polyacrylamide gels. The significance of these enzymes in the phosphate transport needs to be further substantiated by the studies on the plant roots colonized with *P. indica*.

XV. Comparative Characteristics with Arbuscular Mycorrhizae

The characters of *P. indica* were compared with the established arbuscular mycorrhizal fungi, namely, *Glomus* and *Gigaspora*. (Table 7). It seems that the new fungus is similar to AMF in morphology, function and serological features. The data of 18S rDNA and septal pore ultrastructure were different, warranting us to place the new fungus in Basidiomycetes instead of Zygomycetes. *P. indica* possesses undulating hyphae with intermittent spores located at the terminal end and/or at the intercalary position. Spores are invariably pear-shaped. Both the spores and the hyphae are at least four to five times thinner and smaller in diameter that those of AMF (Fig. 15). Like AM fungi, *P. indica* is a wide host range fungus colonizing the roots of mono- dicots including legumes.

XVI. Biotechnological Applications

A. Transformation Studies

1. Protoplast Fusion

Somatic hybridisation by protoplast fusion is a powerful tool to bring together the desirable char-

Table 7. Comparison of *Piriformospora indica* with AMF

	P. indica	AMF
Geographical distribution	India, Pakistan, Philippines Australia	Ubiquitous
Axenic culture	Yes	No
Morphology		
Hyphal strands	often undulating	Straight
Hyphae diameter	0.7–3.5 µm	10–20 µm
Spore shape	Pear-shaped	Globose
Spore colour	Faint yellow to golden	Golden yellow
No. of nuclei/spores	8–25	>1000
Dolipore	Present	Absent
Parenthosomes	Present	Absent
Sexual stage	Absent	Absent
Root colonisation	Yes	Yes
Aerial portion colonisation	Absent	Absent
Extramatrical hyphae	Present	Present
Appressorium	Present	Present
Vesicle	Yes	Yes
Arbuscule	?	Yes
Sporulation	Yes	Yes
Enzymes		
Acid phosphatase	Detected	Detected
Alkaline phosphatase	Detected	Detected
Nitrate reductase	Detected	Detected
Chitinase	Detected	Detected
Mannase	Not known	Detected
Glucanase	Detected	Not known
Ferulase	Detected	Not known
Laccase	Detected	Not known
Tyrosinase	Detected	Not known
Amylase	Detected	Not known
Proteinase	Detected	Detected
Peroxidase	No	Detected
Catalase	No	Detected
Polyphenol oxidase	No	Detected
Polymethylgalacturonase	No	Detected
Phytohormone(s)	Yes	Yes
Plant promotional effect	Yes	Yes
Rooting (in root organ culture)	Profuse	Moderate
Biocontrol agent for plant disease(s)	Yes	Yes
Plant defense compounds	Hydroxy amino acid related to DIMBOA, DIBOA	Callose, phenolic compounds, PR-proteins, silicon
Yellow pigment in roots	Yes	Yes
Mycelial HPLC	Analysed	Not analysed
Benzoic acid	Yes	Not known
Hydroxamicacid	Yes	Not known
Colonisation in crucifers	No	No
Root colonisation of myc⁻ mutants		
Glycine max	Absent	Absent
Pisum sativum	Absent	Absent
Biohardening agent for tissue culture-raised plants	Positive	Positive
Orchid mycorrhiza	Yes	No

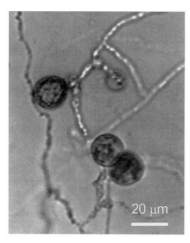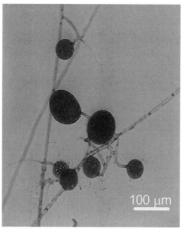

Fig. 15. Fungal discs of *P. indica* (left) and *Glomus intraradices* (right) were grown on transformed carrot root organ cultures. Sporulation occurred after 4 weeks. Note the highly undulating thin hyphae and small spores in the former fungus. In contrast, the latter formed straight thick hyphae and large spores. *Bars* 20 and 100 μm, for *P. indica* and *G. intraradices*, respectively

acteristics of two organisms which are reproductively isolated. The removal of hyphal wall at protoplast formation may permit more extensive interaction between them (Kevei and Peberdy 1977; Baltz 1978). Protoplast fusion overcomes the problem of low recombination frequencies typically attained with natural systems (Hopwood 1981) and creating novel strains that display enhanced biotechnological potential (Kavanagh and Whittaker 1996). The advantages of protoplast fusion which make it an excellent genetic tool (Holt and Saunder 1985) are the following:

- protoplast formation and fusion are universally applicable (Alfoldi 1982);
- recombinants can be obtained in microorganisms in which no natural form of gene transfer has been demonstrated;
- both parents appear to play an equal role in recombinant formation. However, protoplast fusion can be polarised by UV irradiation or SDS treatment to select against one parental population, if desired (Hopwood 1981);
- protoplast fusion between more than two sets of parental strains can be utilised to generate multiparental recombinants;
- protoplasts can be transformed by plastids as well as fused with a wide range of other cells and organelles to promote interspecific hybrid formation and uptake of organelles;
- protoplast fusion allows the transfer of relatively large segments of genomic DNA (Sathe et al. 1992);

Protoplast fusion technique essentially involves isolation of sufficient quantities of viable protoplasts, their fusion, regeneration and selection of fusion products (Sathe et al. 1992).

2. Protoplasts

Protoplasts can be used for various purposes, including the investigation of cell wall synthesis and synchronised cell growth (Davis 1985), for the isolation of cytoplasmic organelles (Billich et al. 1988), for cell fusion (Minuth and Esser 1983), for the preparation of high molecular weight DNA for cloning and karyotyping (Barrett et al. 1989), and as a source material for mutant induction as well as for transformation experiments designed for strain improvement (Hynes 1986; Hampp et al. 1998).

Until recently, all these experiments seemed a distant possibility for manipulating AM fungi due to the inability to culture them axenically. However, following the discovery of *P. indica*, these obstacles would no longer hinder our advances to better understand the basis and outcome of the plant-fungus relationship. We were successful in isolating protoplasts from *P. indica* (Fig. 16), which opens important possibilities to improve symbiosis through transgenic manipulation of the fungal component through the introduction of desirable genes. It would be interesting to see the effect on symbiosis caused by the introduction of genes coding, for example, for cell wall-degrading enzymes. In this case, the probes already isolated from pathogenic fungi or an organism producing high amounts of cell wall-degrading enzymes could prove very useful and would help us to understand to what extent the production of hydrolytic enzymes plays a role in the compatibility between the plant and the mycorrhizal fungi. Another rationale could be to engineer a novel genotype of *P. indica* for a more efficient mobilisation of insoluble phosphates. The

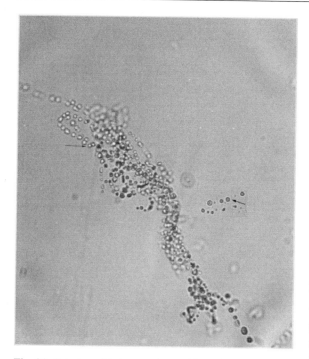

Fig. 16. Intact purified protoplasts of *P. indica* after hyphal wall degradation with Novozym 234, as seen under the bright field microscope

conversion of insoluble phosphates into soluble ones and their subsequent transport to the host is a well-known mycorrhizal function.

In recent years, green fluorescent protein (GFP) has been developed as a reporter for gene expression, a marker of subcellular protein localisation, a tracer of cell lineage and a label to follow development of pathogens within the host plants (Chalfie et al. 1994; Spellig et al. 1996). As a marker system, GFP, which accepts excitation energy from luciferases or photoproteins (Prashar 1995), has several advantages (Cubitt et al. 1995; Niedenthal et al. 1996) over existing reporters (Sheen et al. 1995). Thus, it would now be possible to transform *P. indica* protoplasts and to apply such strains to rhizosphere to investigate population processes such as colonisation patterns, dispersal of marked populations released to nature, the microenvironmental relationships (Caldwell et al. 1992) and the demographic factors (Andrews et al. 1987).

B. Immobilisation of Spores and Hyphae

The concept of using microbial enzymes, attached to solid surfaces, as stabilised reusable catalysts, commonly called immobilised enzymes, was proposed in the early 1900s by Nelson and Griffin (1916). Immobilisation may be interpreted as the presence of organisms or individual animal or plant cells either on the surface or within the particles. This phenomenon has been attributed to increased cell-to-cell contact, induced stress due to concentration gradients and partial differentiation.

In the past, it has been difficult to obtain large amounts of AMF inoculants for agronomic use as the spores of these obligate symbionts produce mycelium with limited independent growth (Hepper 1984). Mycelia regenerated in pure culture from intraradical vesicles have the ability to infect root cortices (Strullu and Romand 1987). Ectomycorrhizal fungi, in contrast, can now be produced in a fermenter and by entrapping the mycelium in sodium/calcium alginate beads. Several workers (Le Tacon et al. 1985; Mauperia et al. 1987) have found this to be an efficient tool for the production of inoculants of AM fungi. Immobilisation can preserve the physiological properties of mycorrhizal roots, promote the regeneration of mycelium and the formation of mycorrhizas (Strullu and Plenchette 1990).

In our experiments, in which hyphae and spores of *P. indica* were immobilised in alginate beads, the former were found to retain the germinating capacity even after two months when used as an inoculum in pot culture experiments using *Sorghum* sp. as the host. We have achieved success in preparing the coloured alginate beads by including low amounts of safranin (0.2–0.5%) and trypan blue (0.1–0.4%) without affecting the viability of the fungus (Fig. 17). Now the effort is to make the alginate beads with dual culture loaded with *P. indica* and seeds of terrestrial orchids. The combination is intended to make a breakthrough for the commercial micropropagation of terrestrial orchids, as the survival rate of these orchid seeds is normally extremely poor.

C. Tools for Biological Hardening of Micropropagated Plants

The rate of survival of micropropagated plants during field trials has been found to be extremely low. Micropropagation provides optimal growth conditions to plants which differ from those available in field pot environment, notably a weak vapour pressure deficit (VPD), reduced light and presence of sucrose, inducing somaclonal

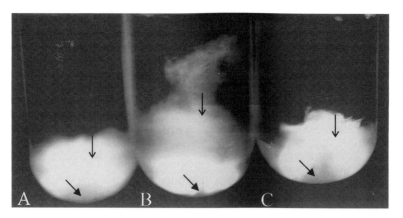

Fig. 17A–C. Immobilised beads containing three to five spores of *P. indica* were coloured with malachite green, safranin, and trypan blue at the rate of 1–2%. **A–C** Spore germination after transfer to fresh minimal medium after a storage of 60 days at room temperature. The beads containing **A** malachite green, **B** safranin, and **C** trypan blue

variations like non-functional stomata and less developed cuticle that renders the plants sensitive to water stress. Inoculation of micropropagated plantlets with active cultures of AM fungi appear to be critical for their survival and growth (Varma and Schuepp 1995, 1996; Lovato et al. 1999). Although the mycorrhizal response of different species varies, this avoids transient transplantation shock and stunted growth on transfer to the field and promotes rooting in difficult plant strains, thereby increasing the rate of outplanting performance of micropropagated plants.

1. Selection of Microbes for Biohardening

Micropropagation involves mycorrhizal inoculation after roots are formed. The process of biohardening (acclimatisation) of micropropagated plants using an endophyte depends on the following factors:

- high specific efficiency (phytohormone formation, acquisition of nutrients, protection from pathogens and stress);
- high affinity to a broad spectrum of host plants;
- competitivity towards other deleterious rhizosphere microorganisms;
- good adaptation to different ecological conditions;
- reproducible stimulation of plant development and effective technological solution for inoculum production.

2. Fungus-Host Specificity

AMF are generally regarded as non-host-specific. However, the mycorrhizal dependency of plant may vary with the fungus species or with the medium, which is regarded as a kind of functional host specificity (Clarke and Mosse 1981; Fortuna et al. 1992). Guillemin et al. (1992) found that certain cultivar-fungus associations were more efficient than others in enhancing plant growth. The influence of fungal strain on the efficiency of symbiosis varies with the plant growth media and with the fertilisation rates applied.

3. Growth Substrate

Growth substrate differs with timing of inoculation. *In vitro* inoculation can be carried out on a synthetic medium having a composition crucial not only for the development of symbiosis, but also for overall plant growth.

The substrates favoured for mycorrhization are mostly peat-based. We have established a mixture of vermiculite (autoclaved) and sand (acid-washed) in the ratio of 3:1, which is an important growth substrate for *in vitro* inoculation of *P. indica* and *Glomus* sp. with micropropagated tobacco, brinjal and *Bacopa monnieri* plantlets (Fig. 18a,b). The survival rate of biologically hardened micropropagated tobacco plants in the case of live mycelium of *P. indica* was maximum (Table 8) as compared to control, *G. mosseae* and dead mycelium of *P. indica*. Inoculum in each case was added at the rate of 1% of the substratum.

Transfer of the tissue culture raised plantlets to field conditions after biological hardening with either established AMF or *P. indica* showed a good growth and higher survival rate. More details on plant promotional effects have been described recently (Varma et al. 1999a,b). *P. indica* colonised the tissue culture raised plantlets of brahmi (*Bacopa monnieri*, a medicinal plant), tobacco (*Nicotiana tabacum cult Xanthi*), brinjal (*Solanum melongena*) and neem (*Azadirachta indica*). In general, the rate of plantlet survival was much higher than for the untreated controls. Higher biomass was also recorded in the treated plants (Sahay and Varma 1999, 2000).

Micropropagation is an effective tool to multiply agricultural and flori-horticultural plants and those used in regeneration and revegetation programmes. The problems with the plants during this process could be overcome, at least to some extent, by AM inoculation, as seen in the case of woody plants difficult to root (including oak, apple, plum, hortensia and pear). Similarly, the dormancy of certain *Prunus* and *Malus* root stocks could be resolved with AMF symbiosis (Varma and Schuepp 1994a,b; Vidal et al. 1992). *P. indica* has been found to behave in an identical manner with the hosts tested.

Fig. 18. Biological hardening of tobacco plantlets raised from tissue culture. Well-developed plantlets were transferred to plastic pots containing sterile sand (acid-washed) and soil mixture in the ratio of 3 : 1 (v/v). To the second and fourth pots (*from left*), live inocula of *G. mosseae* and *P. indica*, respectively, were added at the rate of 1%. The extreme left pot represents control and the third pot from the left was fortified with an equal quantity of dead fungal biomass (second control). Photographs taken **a** after 4 weeks and **b** after 8 weeks of treatment

4. AM-Induced Bioactive Principles in Cell Culture

Currently, there has been an increasing awareness towards plant-derived herbal antimicrobial agents and naturally occurring phytochemicals. The methods developed for the *in vitro* production of secondary metabolites are economically viable and commercially advantageous, micropropagation being one of them. It has the following benefits:

– production of desired product without adversely affecting the plants;
– increased quality and quantity of products, like bioactive pigments;
– easy inoculation and purification of the product.

The dual biotechnology, micropropagation and mycorrhization, promises the production of secondary metabolites efficiently. Research concerning micropropagation and mycorrhization of *Bacopa monnieri* and *Spilanthes calva* is in progress. *B. monnieri* is widely used in various ayurvedic (herbal and natural medicine) preparations due to its tremendous potential for increasing memory and its anti-ageing properties. Due to the pressing need for the metabolite, mass multi-

Table 8. Biological hardening of tissue culture-raised plants

	No. of plants used	Biomass (g) after 8 weeks	Percent colonisation	Percent survival
Glomus mosseae	7	4.2	65	70
Piriformospora indica, live	7	5.9	90	95
Piriformospora indica, dead	7	6.8	–	60
Control	7	3.9	–	60

Percent colonisation, biomass and percent survival after 8 weeks of biological hardening with *G. mosseae* and *P. indica* live and dead biomass. Added inoculum at the rate of 1% (w/w). Statistical analysis was done through RM ANOVA on Ranks test shows, chi-square = 8.40 with 3 degrees of freedom. P (est.) = 0.0384; P (exact) = 0.0190, the differences in the median value among the treatment groups are greater than would be expected by chance, there is statistically significant difference (P = 0.0190).

plication of both the plants is gaining momentum and may open up new ways for commercial production of phytochemical-based industries.

D. Biological Control Agent of Soil-Borne Diseases

AMF are the major components of the rhizosphere (mycorrhizosphere). Literature suggests that they reduce soil-borne diseases and/or the severity of diseases caused by phytopathogens (Dehne 1982).

P. indica was challenged with a virulent root and seed pathogen Gaeumannomyces graminis (Fig. 19). In another experiment, when the fungus was allowed to grow earlier and the pathogen was inoculated later in the centre of the solidified nutrient medium, pathogen growth was completely blocked. Culture filtrate of P. indica also completely stopped the growth of the pathogen. Similar inhibitory effects were seen with interaction of P. indica and Aspergillus sydowii. These experiments indicate that the new fungus has the potential to act as a biological agent for the control of root diseases; however, the chemical nature of the inhibitory factor is still not known.

XVII. Future Perspectives

Piriformospora indica is a new category of axenically culturable root colonising, broad host-range

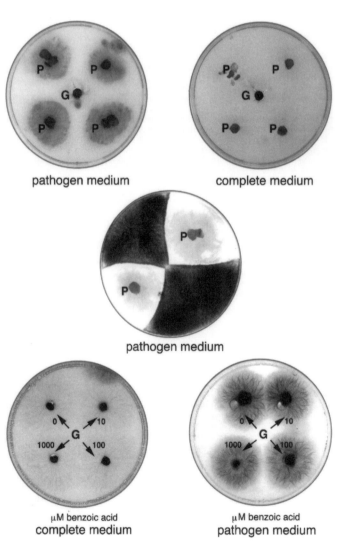

Fig. 19. Interaction of root pathogen *Gaeumannomyces graminis* (G) with *P. indica* (P). The pathogen was placed in the centre of the nutrient agar plates and *P. indica* discs were placed equidistantly on the four sides across the pathogen disc. Growth on pathogen medium (*top left*) and on complete medium (*top right*). Note the significant suppression of the pathogen. In the middle plate, pathogen and mycosymbiont discs were placed in an alternate manner. The former did not suppress the latter. *Bottom plates* show the growth of the pathogen in the presence of different concentrations of benzoic acid

plant-promoting fungus and also an amicable agent for the control of plant root diseases. This fungus also helps in the establishment and survival of micropropagated rootlets.

The immediate questions to be asked are regarding its symbiotic nature, scenario for pre- and postroot colonisation, molecular basis of root penetration (mechanical or enzymatic and/or both), signalling between the plant and the fungus, and ecological distribution.

After obtaining first insights into the metabolism of *P. indica*, it is a challenge to study the stress physiology of this fungus. A better understanding of the properties of *P. indica* could even lead to the following results, of potential importance for agriculture. An approach to using *P. indica* could be made to improve drought resistance in crop plants. In higher plants, the proficiency to survive desiccation is restricted to a number of poikilohydric plants and, within fairly tight limits, to dormant seeds and spores. By contrast, desiccation tolerance is widespread among algae, cyanobacteria, fungi, and particularly among bryophytes and lichenised fungi. The majority of higher plants, however, cannot survive desiccation, and this is one of the major problems of agriculture in arid zones.

The capacity to cope with the injurious effects of water removal requires a complex interplay of several adaptation mechanisms. Indeed, desiccation-tolerant plants have some remarkable features that allow them to survive desiccation without losing their capability to regenerate and grow. Three major mechanisms allowing desiccation-tolerant organisms to overcome desiccation are found on the level of proteins, non-reducing sugars and antioxidants. The ability to cope with free radicals, the formation of which is well documented in desiccated tissues, is one of the basic requirements of desiccation-tolerant organisms. Indeed, desiccation tolerance is correlated with maintenance and/or synthesis of antioxidants such as glutathione (γ-glutamyl-cysteinyl-glycine, GSH), ascorbic acid and tocopherols, and with activated oxygen-processing enzymes such as superoxide dismutase, catalase or peroxidase. Glutathione, in particular, may play a crucial role for overcoming periods of drought (for review see Kranner and Lutzoni 1999).

In theory, there are two ways of enhancing resistance to drought in higher plants. If we knew the mechanisms allowing those highly specialised desiccation-tolerant organisms to function, we could, using methods of genetic engineering, increase the drought resistance of crop plants in future. However, another approach could be made to use *P. indica* for improving the drought resistance of plants. *P. indica* has been isolated from desert soil and therefore is a promising candidate for an attempt to improve the drought resistance of crop plants. If tests confirm that *P. indica* itself is desiccation-tolerant, the next step will be to test whether or not plants inoculated with *P. indica* are better adapted to dry conditions than plants without this fungus. This would be a milestone in the development of agriculturally important plants, of special significance for the arid zones of the earth. For assessing the mechanisms of adaptation to drought, the knowledge about reactions of antioxidants such as GSH and other free radical scavengers can be used.

Moreover, it will be interesting to study two other aspects of the GSH metabolism in *P. indica*. First, the enzyme glutathione-S-transferase (GST) catalyses the GSH-dependent detoxification of xenobiotics, including many pesticides. Secondly, in some plant families, the so-called phytochelatins [(γ-glutamyl-cysteinyl)$_n$-glycine] are involved in the GSH-dependent detoxification of heavy metals. If *P. indica* has a high potency to detoxify xenobiotics and/or heavy metals, this knowledge could be used to grow plants in symbiosis with *P. indica* in areas where the contamination of the soil leads to environmental problems. In this way, excessive xenobiotics or heavy metals could be removed from the soil, thus contributing to the health of a sustainable environment.

Acknowledgments. The authors from JNU, New Delhi, are thankful to the Department of Biotechnology and UGC for partial financial support. Ajit Varma is thankful to DAAD, Bonn, Germany, for conducting some of the experiments described in this chapter. The senior author also acknowledges with thanks the facilities received at the Institute of Plant Physiology, Karl-Franzens University, Graz, for completing the proofreading and changes incorporated. IK acknowledges financial support from the Austrian Academy of Science (APART 428) and the Austrian Science Foundation (grant P12690-BIO). K.-H. Rexer is thankful to DAAD, Bonn for organising a visit to JNU, New Delhi.

References

Alfoldi L (1982) Fusion of microbial protoplasts: problems and perspectives. In: Hollander A (ed) Genetic engineering of microorganisms for chemicals. Plenum Press, New York, pp 59–71

Andrews JH, Kinkel LL, Berbee FM, Nordheim EV (1987) Fungi, leaves and theory of island biogeography. Microb Ecol 14:277–290

Azcon-Aguilar C, Diaz-Rodriguez RM, Barea JM (1986) Effect of soil micoorganisms on spore germination and growth of the vesicular arbuscular fungus *Glomus mosseae*. Trans Br Mycol Soc 86:337–340

Bagyaraj DJ, Varma A (1995) Interaction between arbuscular mycorrhizal fungi and plants and their importance in sustainable agriculture in arid and semi-arid tropics. Adv Microb Ecol 14:119–142

Baltz RH (1978) Genetic recombination of *Streptomyces fradiae* by protoplast fusion and cell regeneration. J Gen Microbiol 107:93–102

Barrett V, Lamke PA, Dixon RK (1989) Protoplast formation from selected species of ectomycorrhizal fungi. Appl Microb Biotechnol 30:381–387

Becard G, Piche Y (1989a) New aspects on the acquisition of biotrophic status by a vesicular arbuscular mycorrhizal fungus, *Gigaspora margarita*. New Phytol 112:77–83

Becard G, Piche Y (1989b) Fungal growth stimulation by carbondioxide and root exudates in VAM symbiosis. Appl Environ Microbiol 55:2320–2325

Becard G, Piche Y (1992) Establishment of vesicular arbuscular mycorrhiza in root organ culture: Review and proposed methodology. In: Norris JR, Read DJ, Varma AK (eds) Methods in microbiology. Academic Press, London, pp 89–108

Billich A, Kellar U, Kleinhauf H, Zocher R (1988) Production of protoplast from *Fusarium sciripi* by lytic enzymes from *Streptomyces tsusimaensis*. Appl Microbiol Biotechnol 28:442–444

Blechert O, Kost G, Hassel A, Rexer K-H, Varma A (1999) First remarks on the symbiotic interactions between *Piriformospora indica* and terrestrial orchids. In: Varma A, Hock B (eds) Mycorrhizae: Structure, function, molecular biology and biotechnology, 2nd edn. Springer, Berlin Heidelberg New York, pp 683–688

Blilou I, Martin J, Ocampo JA (1996) Influence of cellulase on the susceptibility of non-host cabbage to colonization by *Glomus intraradices*. In: Azcon-Aguilar C, Barea JM (eds) Mycorrhizas in integrated systems from genes to plant development. Official Publications of the European Communities, Luxemburg, pp 215–217

Bonfante P, Perotto S (1995) Strategies of arbuscular mycorrhizal fungi when infecting host plants. New Phytol 130:3–21

Caldwell DE, Krober DR, Lawrence JR (1992) Confocal laser microscopy and digital image analysis in microbial ecology. Adv Microb Ecol 12:1–67

Chalfie M, Tu Y, Euskirchen G, Ward WW, Prasher DC (1994) Green fluorescent protein as a marker of gene expression. Science 263:802–805

Clarke C, Mosse B (1981) Plant growth response of vesicular arbuscular mycorrhiza XII. Field inoculation responses of barley at two soil P levels. New Phytol 87:695–703

Clements MA, Muir H, Cribb PJ (1986) A preliminary report on the symbiotic germination of European terrestrial orchids. Kew Bull 41:437–445

Cubitt AB, Heim R, Adams SR, Boyd AE, Gross LA, Tsien RY (1995) Understanding, improving and using green fluorescent proteins (GFP). Gene 173:33–38

Davis B (1985) Factors influencing protoplast isolation. In: Peberdy JF, Ferenczy L (eds) Fungal protoplasts. Dekker, New York, pp 45–71

Dehne HW (1982) Interaction between vesicular-arbuscular mycorrhizal fungi and plant pathogens. Phytopathology 72:1115–1119

Fester T, Maier W, Strack D (1998) Accumulation of secondary compounds in barley and wheat roots in response to inoculation with an arbuscular mycorrhizal fungus and co-inoculation with rhizosphere bacteria. Mycorrhiza 8:241–246

Fortuna P, Citernisi S, Morini S, Giovanetti M, Loreti F (1992) Infectivity and effectiveness of different species of arbuscular mycorrhizal fungi in micropropagated plants of plum root stock. Agronomie 12:825–830

Frank AB (1885) Über die auf Wurzelsymbiose beruhende Ernährung gewisser Bäume durch unterirdische Pilze. Ber Dtsch Bot Ges 3:128–145

Garcia-Romera I, Garcia-Garrido JM, Martinez-Molina E, Ocampo JA (1990) Possible influence of hydrolytic enzymes on vesicular arbuscular mycorrhizal infection in alfalfa. Soil Biol Biochem 22:149–152

Garcia-Romera I, Garcia-Garrido JM, Ocampo JA (1991a) Pectolytic enzymes in the vesicular-arbuscular mycorrhizal fungus *Glomus mosseae*. FEMS Microbiol Lett 78:343–346

Garcia-Romera I, Garcia-Garrido JM, Ocampo JA (1991b) Pectinase activity in vesicular-arbuscular mycorrhiza during colonization of lettuce. Symbiosis 12:189–198

Garcia-Romera I, Garcia-Garrido JM, Ocampo JA (1996) Hydrolytic enzymes in arbuscular mycorrhizae. In: Azcon-Aguilar C, Barea JM (eds) Mycorrhizas in integral systems from genes to plant development. Official Publications from the European Communities, Luxemburg, pp 234–237

Gianinazzi-Pearson V, Dumas-Gaudot E, Armelle G, Tahiri A, Gianinazzi S (1996) Cellular and molecular defence-related root responses to invasion by arbuscular mycorrhizal fungi. New Phytol 133:45–58

Goebel C, Hahn A, Giersch T, Hock B (1998) Monoclonal antibodies for the identification of arbuscular mycorrhizal fungi. In: Varma A (ed) Mycorrhizamanual. Springer, Berlin Heidelberg New York, pp 271–287

Guillemin JP, Gianinazzi S, Trouvelots A (1992) Screening of arbuscular mycorrhizal fungi for establishment of micropropagated pineapple plants. Agronomie 12:831–836

Hahn A, Goebel C, Hock B (1998) Polyclonal antibodies for detection of AM fungi. In: Varma A (ed) Mycorrhiza manual. Springer, Berlin Heidelberg New York, pp 255–287

Hahn A, Goebel C, Hock B (1999) Immunochemical properties of mycorrhizas. In: Varma A, Hock B (eds) Mycorrhizae: structures, function, molecular biology and biotechnology, 2nd edn. Springer, Berlin Heidelberg New York, pp 177–201

Hampp R, Stülten C, Nehls U (1998) Isolation and regeneration of protoplasts from ectomycorrhizal fungi. In:

Varma A (ed) Mycorrhiza manual. Springer, Berlin Heidelberg New York, pp 115–126

Harley JL (1991) Introduction: The state of art. In: Norris JR, Read DJ, Varma AK (eds) Methods in microbiology, vol 21. Academic Press, London, pp 1–24

Hepper CM (1984) Isolation and culture of vesicular arbuscular mycorrhizal (VAM) fungi. In: Powell CL, Bagyaraj DJ (eds) VA mycorrhiza. CRC Press, Boca Raton, Florida, pp 95–112

Hindav R, Kumari M, Mondal N, Paul J, Sahay N, Sarma J, Singh A, Sudha, Varma A (1998) One kilo tropical soil is equal to one kilo gold: this is microbial science. In: Varma A (ed) Microbes for Health, Wealth and Sustainable Environment. MPH, New Delhi, pp 1–23

Holt G, Saunder G (1985) Genetic manipulation of the industrial microorganisms. In: Bull AT, Dalton H (eds) Comprehensive biotechnology: the principles, applications and regulations of biotechnology in industry, agriculture and medicine, vol 1. Pergamon Press, pp 51–76

Hopwood DA (1981) Genetic studies with bacterial protoplasts. Annu Rev Microbiol 35:237–273

Hynes MG (1986) Transformation of filamentous fungi. Exp Mycol 10:1–8

Kavanagh K, Whittaker PA (1996) Application of protoplast fusion to the non-conventional yeast. Enzyme Microb Technol 18:45–51

Kevei F, Peberdy JF (1977) Interspecific hybridization between *Aspergillus nidulans* and *A. rugulosus* by fusion of somatic protoplasts. J Gen Microbiol 102:255–262

Koide RT (1991) Nutrient supply, nutrient demand and plant response to mycorrhizal infection. New Phytol 117:365–386

Kranner I, Lutzoni F (1999) Evolutionary consequences of transition to a lichen symbiotic state and physiological adaptation to oxidative damage associated with poikilohydry. In: Lerner HR (ed) Plant response to environmental stress: from phytohormones to genome reorganization. Dekker, New York, M Inc pp 591–628 (1991)

Le Tacon F, Jung G, Mugnier J, Michelot P, Mauperia C (1985) Efficiency of a forest nursery of an ectomycorrhizal inoculum produced in a fermentor and entrapped in polymeric gels. Can J Bot 63:1664–1668

Linderman RG (1994) Mycorrhizae for plant health. Symposium Series. In: Fleger FL, Linderman RG (eds), APS Press, St Paul, Minnesota, pp 1–25

Lovato PE, Schuepp H, Trouvelot A, Gianinazzi A (1999) Application of arbuscular mycorrhizal fungi (AMF) in orchard and ornamental plants. In: Varma A, Hock B (eds) Mycorrhiza; structures, function, molecular biology and biotechnology, 2nd edn. Springer, Berlin Heidelberg New York, pp 443–467

Lynch J (1995) Root architecture and plant productivity. Plant Physiol 109:7–13

Maier W, Peipp H, Schmidt J, Wray V, Strack D (1995) Levels of a terpenoid glycoside (blumenin) and cell wall-bound phenolics in cereal mycorrhizas. Plant Physiol 109:465–470

Mathew J, Shankar A, Neeraj, Varma AK (1991) Glomalaceous fungi associated with spineless cacti, a fodder supplement in deserts. Trans Mycol Soc Jpn 32: 225–233

Mauperia C, Mortier F, Garbaye J, Le Tacon F, Carr G (1987) Viability of an ectomycorrhizal inoculum produced in a liquid medium and entrapped in calcium alginate gels. Can J Bot 65:2326–2329

Minuth W, Esser K (1983) Intraspecific, interspecific and intergeneric recombination in β-lactam-producing fungi by protoplast fusion. Appl Microbiol 18: 38–46

Mosse B, Hepper CM (1975) Vesicular arbuscular mycorrhizal infection in root organ cultures. Physiol Plant Phytopathol 5:215–223

Mugnier J, Mosse B (1987) VAM infection in transformed root inducing T-DNA root grown axenically. Phytopathology 77:1045–1050

Mukerji KG, Mandeep, Varma A (1998) Mycorrhizosphere microorganisms: screening and evaluation. In: Varma A (ed) Mycorrhiza manual. Springer, Berlin Heidelberg New York, pp 85–98

Neeraj, Shankar A, Mathew J, Varma AK (1991) Occurrence of VA mycorrhizae within the Indian semi-arid soil. Biol Fert Soils 11:140–144

Nelson JM, Griffin EG (1916) Adsorption of invertase. J Am Chem Soc 38:1109–1111

Niedenthal RK, Riles L, Johnstons, Hegemann JH (1996) Green fluorescent protein as a marker for gene expression and subcellular localization in budding yeast. Yeast 12:773–786

Perotto S, Bettini V, Favaron F, Alghisi P, Bonfante P (1995a) Polygalacturonase activity and location in arbuscular mycorrhizal roots of *Allium porrum*. Mycorrhiza 5:157–165

Perotto S, Perotto R, Schubert A, Varma A, Bonfante P (1995b) Ericoid mycorrhizal fungi: cellular and molecular basis of their interaction with host plant. Can J Bot 73:557–568

Perotto S, Coisson JB, Perugini I, Cometti V, Bonfante P (1997) Production of pectin degrading enzymes by ericoid mycorrhizal fungi. New Phytol 135:151–160

Peters JH (1988) Immunisierung von groeberen Versuchstieren zur Herstellung von Antiserum. In: Peters JH, Baumgartum H (eds) Monoklonale Antikoerper. Springer, Berlin Heidelberg New York

Prasher DC (1995) Using GFP to see the light. Trends Genet 11:320–323

Rasmussen H (1990) Cell differentiation and mycorrhizal infection in *Dactylorhiza majalis* (Rchb. f.) Hunt and Summeh. (Orchidaceae) during germination *in vitro*. New Phytol 116:137–147

Read DJ (1999) Mycorrhiza – the state of art. In: Varma A, Hock B (eds) Mycorrhizae: structure, function, molecular biology and biotechnology, 2nd edn. Springer, Berlin Heidelberg New York, pp 3–34

Rexer K-H, Blechert O, Kost G, Varma A (2000) *Piriformospora indica versus Epulorhiza repens* – A comparison of the interactions with *Dactylorhiza* sp. (Orchidaceae) (in prep.)

Sahay NS (1999) Interaction of *Piriformospora indica* with tissue culture raised plant. PhD Thesis, Jawaharlal Nehru University, New Delhi

Sahay NS, Sudha, Singh A, Varma A (1998) Trends in endomycorrhizal research. Indian J Exp Biol, NISCOM 36:1067–1086

Sahay NS, Varma A (1999) *Piriformospora indica*: a new biological hardening tool for micropropagated plants. FEMS Microbiol Lett 181:297–302

Sahay NS, Varma A (2000) Biological approach towards increasing the survival rates of micropropagated plants. Curr Sci 78:126–129

Saif M (1997) Comparative study of production, infectivity, and effectiveness of arbuscular-mycorrhizal fungi produced by soil-based and soil-less techniques. PhD Thesis, University of Sydney-Macarthur, Campbelltown NSW, Australia

Sathe S, Sivaraman H, Gokhale DV (1992) Protoplast fusion in yeast strain improvement in *Saccharomyces*. Ind J Microbiol 32:15–27

Sharma J, Varma A (2000) *In vitro* and *in vivo* immunological characterization of *Glomus mosseae*. FEMS (Communicated)

Sharma J, Hurek T, Varma A (2000a) Serological characterization of *Piriformospora indica*. Mycol Res, UK (Communicated)

Sharma J, Neeraj, Varma A (2000b) *In vitro* immunocharacterization of *Piriformospora indica*. Mycol Res, UK (communicated)

Sharma J, Varma A (2000) Biochemical studies of acid phosphotase and the effects of phosphatic compounds on *in vitro* growth of a plant growth promoting endophytic fungus, *Pirifomospora indica*. New Phytol (communicated)

Sheen J, Hwang S, Niwa Y, Kobayaski H, Gabraith DW (1995) Green fluorescent protein as a new vital marker in plant cell. Plant J 8:777–784

Spellig T, Bottin A, Kahmann R (1996) Green fluorescent protein (GFP) as new vital marker in the phytopathogenic fungus *Ustilago maydis*. Mol Gen Genet 252:503–509

Strimmer K, Haeseler A (1996) Quartet Puzzling: A quartet maximum likelihood method for reconstructing tree topologies. Mol Biol Evol 13:964–969

Strullu DG, Plenchette C (1990) Encapsulation de la forme intraracinaire de *Glomus* dans l'aiginate et utilization des capsules comme inoculum. C R Acad Sci Paris 310:447–452

Strullu DG, Romand C (1987) Culture axenique de vesicules isolées a partir d'endomycorhizes et réassociation in vitro à des racines de tomates. C R Acad Sci Paris 305:15–19

Sudha (1999) In vitro study of endosymbionts associated with tissue culture-raised medicinal plants. PhD thesis, Jamia Hamdard University, New Delhi

Swofford DL (1998) PAUP*. Phylogenetic analysis using parsimony (*and other methods), Version 4. Sinauer Associates, Sunderland. MA

Varma A (1995a) Arbuscular mycorrhizal fungi: the state of art. Crit Rev Biotechnol 15:179–199

Varma A (1995b) Ecophysiology of arbuscular mycorrhizal fungi. In: Varma A, Hock B (eds) Mycorrhiza: structures, function, molecular biology and biotechnology, 2nd edn. Springer, Berlin Heidelberg New York, pp 561–591

Varma A (1998) Mycorrhizae, the friendly fungi: what we know, what should we know and how do we know? In: Varma A (ed) Mycorrhiza manual. Springer, Berlin Heidelberg New York, pp 1–24

Varma A (1999a) Ecology and physiology of endomycorrhizal fungi in arid soils. In: Varma A, Hock B (eds) Mycorrhiza: structures, function, molecular biology and biotechnology, 2nd edn. Springer Berlin Heidelberg New York, pp 521–556

Varma A (1999b) Hydrolytic enzymes from arbuscular mycorrhizae; the current status. In: Varma A, Hock B (eds) Mycorrhiza: structure, function, molecular biology and biotechnology, 2nd edn. Springer, Berlin Heidelberg New York, pp 373–389

Varma A, Bonfante P (1994) Utilization of cell wall-related carbohydrates by ericoid mycorrhizal endophytes. Symbiosis 16:301–313

Varma A, Schuepp H (1994a) Infectivity and effectiveness of *Glomus intraradices* on micropropagated plants. Mycorrhiza 5:29–37

Varma A, Schuepp H (1994b) Positive influence of arbuscular mycorrhizal fungus on in vitro-raised hortensia plantlets. Angew Bot 68:108–115

Varma A, Schuepp H (1995) Mycorrhizae: their application in micropropagated plantlets. Crit Rev Biotechnol 15:313–328

Varma A, Schuepp H (1996) Influence of mycorrhization on the growth of micropropagated plants. In: Mukerji KG (ed) Concepts in mycorrhizal research. Hand Book Vegetation Sciences Series. Kluwer, Dordrecht pp 113–132

Varma A, Verma S, Sudha, Sahay NS, Franken P (1999a) *Piriformospora indica*, a cultivable plant growth promoting root endophyte with similiarities to arbuscular mycorrhizal fungi. Appl Environ Microbiol 65:2741–2744

Varma A, Sudha, Sahay NS, Singh A, Kumari M, Bharti K, Sarbhoy AK, Maier W, Walter MH, Strack D, Franken P (1999b) Proceedings on the symposium Pollution abatement through biological treatment of industrial effluents March 24, 1998. Central Pollution Control Board (CPCB), Delhi (in press)

Verma S, Varma A, Rexer K-H, Hassel A, Kost G, Sarbhoy A, Bisen P, Buetehorn B, Franken P (1998) *Piriformospora indica*, gen. et sp. nov., a new root-colonizing fungus. Mycologia 90:896–903

Vidal MT, Azcon-Aguilar C, Barea JM (1992) Mycorrhizal inoculation enhances growth and development of micropropagated plants of avocado. Hortic Sci 27:785–787

Williams PG (1992) Axenic culture of arbuscular mycorrhizal fungi. In: Norris JR, Read DJ, Varma AK (eds) Methods in microbiology, vol 24. Academic Press, London, pp 203–220

9 *Geosiphon pyriforme*, an Endocytosymbiosis Between Fungus and Cyanobacteria, and its Meaning as a Model System for Arbuscular Mycorrhizal Research

A. Schüßler and M. Kluge

CONTENTS

I. Introduction

Geosiphon pyriforme (Kütz.) v. Wettstein is a coenocytic soil fungus and until now the only known example of a fungus living in endocytobiotic association with a cyanobacterium, i.e. with *Nostoc punctiforme*. The symbiotic nature of the system was first recognized by F.v. Wettstein (1915), who described it as a symbiosis between a heterotrophic siphonal chlorophyceaen alga and *Nostoc*. The fungal nature of the macrosymbiont was recognized by Knapp (1933). The fungus lives together with the cyanobacterium on the surface and in the upper layer of wet soils poor in inorganic nutrients, particularly in phosphate. When a fungal hypha comes into contact with free-living *Nostoc* cells, the latter are incorporated by the

Institut für Botanik, Technische Universität Darmstadt, 64287 Darmstadt, Germany

fungus at the hyphal tip, which thereafter swells and forms a unicellular "bladder", about 1–2 mm in size and appearing on the soil surface (Fig. 1). Inside this bladder the cyanobacteria are physiologically active and dividing. Life history, ultrastructure and physiological activity of the system will be described in this chapter.

Due to the physiological activities of the endosymbiont the consortium is capable of C- and N-autotrophic life. *Geosiphon* can be considered as a primitive endocytobiotic system, because the photobiont can be experimentally separated and cultured without the fungal partner, which is obligate symbiotic. The system may provide interesting insights into evolutionary steps leading to more derived systems which finally gave rise to the organelles within the eukaryotic cells. In spite of its potential importance for research on endocytobioses and biology of the eukaryotic cell, there are still only few studies on *Geosiphon*, mainly because the organism is quite rare in nature and its cultivation very difficult.

Recently, it has been suggested that *Geosiphon* could provide an important model system for another symbiosis, the arbuscular mycorrhiza (AM). It bears a great potential for the study of many fundamental mechanisms and evolutionary questions concerning AM. The state of knowledge on *Geosiphon* until 1997 was reviewed in Kluge et al. (1994, 1997). In the present chapter we report on new results obtained in our laboratory.

II. Taxonomic Position of the Fungus

Until recently, little was known about the taxonomic position of the fungal partner of the *Geosiphon* consortium. Knapp (1933) recognized it as a phycomycete and Mollenhauer (1992) first proposed that it might belong to the genus *Glomus*. The verification of this assumption appeared to be extremely interesting against the

The Mycota IX
Fungal Associations
Hock (Ed.)
© Springer-Verlag Berlin Heidelberg 2001

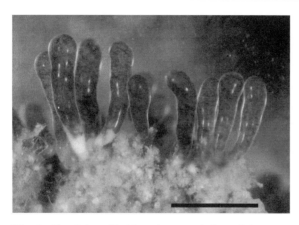

Fig. 1. *Geosiphon* bladders, harvested from laboratory culture on natural substrate. *Bar* 1 mm

background that the members of the Glomales comprise the ecologically and also economically important AM fungi, and it is conceivable that *Geosiphon* is also capable of AM formation.

A. Molecular Phylogeny

Classically the taxonomy of the arbuscular mycorrhizal fungi (AMF) is based mainly on the characteristics of the spores. Thus, Schüßler et al. (1994) performed a detailed study comparing morphological and ultrastructural criteria of *Geosiphon* spores with those of other glomalean fungi. This investigation indeed uncovered striking similarities between *Geosiphon* on the one hand and *Glomus versiforme* on the other. The same holds true for some other species of the genus *Glomus* investigated so far (unpubl. results). Final unequivocal evidence in favor of the view that *Geosiphon* belongs to the Glomales, the fungal order in which the AMF are placed (Morton and Benny 1990), was provided by Gehrig et al. (1996). These authors analyzed and compared the nearly complete small subunit ribosomal RNA (SSU rRNA) genes of *Geosiphon pyriforme* and *Glomus versiforme*. Taking into account the secondary structure (De Rijk et al. 1992), the sequences were aligned with those of other glomalean isolates and a large set of other fungal taxa outside the order Glomales. The phylogenetic trees obtained clearly show that the Glomales include *Geosiphon* and form a distinct branch not clustering with any other group of the Zygomycetes sequenced so far. Within the Glomales, several clades exist, formed by the families

Gigasporaceae and Acaulosporaceae and by the genus *Glomus*, as defined by the usual criteria (Morton and Benny 1990). *Glomus* species show considerable large phylogenetic distances, suggesting that the genus is not well defined from the conventional point of view. This is also indicated by classical criteria (Walker 1992).

As far as *Geosiphon* is concerned, it is now clear that the fungus represents a probably ancestral member of the Glomales, and recent sequence analyses (Redecker et al. 2000; Schüßler 1999) show that it is very closely related to the dimorphic AMF *Acaulospora gerdemannii* (synonym *Glomus leptotichum*), which was recently sequenced independently by two other laboratories (Redecker et al. 2000; Sawaki et al. 1999). The question of whether this phylogenetic clade, containing *Geosiphon* and AMF, is ancestral within the Glomales, as indicated by the sequence analyses, cannot yet be definitively answered. Work is now in progress to analyze the complete SSU rRNA sequences of more glomalean fungi. Our results support the view of Simon et al. (1993) and Walker (1992) that the genus *Glomus* probably has a polyphyletic origin. The question remains open as to how this genus should be finally defined.

B. Implications for the Interpretation of Arbuscular Mycorrhizal Evolution

Sequence analyses (Simon et al. 1993; Gehrig et al. 1996; Simon 1996) indicate that *Glomus* represents the most ancestral genus of the Glomales. Presumably *Glomus*-like fungi represent the mycobionts in the early endomycorrhizal associations which developed in the Paleozoicum in parallel to the appearance of the first land plants. This is in agreement with *Glomus*-like fossil records (Stubblefield et al. 1987) and the discovery of arbuscules in 400-Ma-old *Aglaophyton* fossils (Remy et al. 1994).

AM are formed by more than 80% of all vascular plant species (Smith and Read 1997), including ferns (Peterson et al. 1981) and Lycopodiaceae (Schmid and Oberwinkler 1993). Also all classes of Bryophytes contain species with AM-like associations (Stahl 1949; Parke and Linderman 1980; Ligrone 1988; Ligrone and Lopes 1989). Probably the evolution of land plants was highly promoted by the associations between plants and *Glomus*-like fungi, efficiently supplying the plants with water and nutrients from the soil (Malloch et al.

1980; Pirozynski 1981; Simon et al. 1993; Marschner and Dell 1994). Pirozynski and Malloch (1975) postulated a partnership between two basically aquatic organisms, an alga and a "phycomycetous" fungus, as the initial step of land plant evolution, a hypothesis which became widely accepted but lacks direct evidence. *Geosiphon pyriforme*, representing a symbiotic association between a glomalean fungus and a photoautotrophic prokaryotic alga could reflect an ancestral partnership, and thus, indirectly but substantially, supports the view of Pirozynski and Malloch (1975). Thus, it is very plausible to assume that in the beginning of terrestrial plant life also other associations between glomalean fungi and photoautotrophic organisms existed.

Against the background of our assumption that *Geosiphon* represents a member of a probably ancestral clade within the Glomales, the interesting question arises of whether *Geosiphon* itself can act as fungal partner in forming AM. This question is still open, but experiments are in progress in our laboratory to answer it. For this purpose we are also developing molecular probes to evaluate the occurrence of *Geosiphon* in the natural habitat.

III. Initiation and Development of the Symbiotic Association

A. Development of the Symbiosis

A detailed study on the initiation and further development of the symbiotic association *Geosiphon pyriforme* has been recently provided by Mollenhauer et al. (1996). Initially, the cells of the cyanobacterium *Nostoc punctiforme* live freely together with the future fungal partner in and on the soil. There, the partners come into contact, but a successful interaction of the fungus with *Nostoc* to form the symbiosis depends on the appropriate developmental stage of the cyanobacterium (for terminology of the nostocacean life cycle see Dodds et al. 1995). The life cycle of *Nostoc* starts from germinating akinetes leading to vegetative colonies. These colonies release motile trichomes (hormogonia) which are positively phototactic in dim light and negatively phototactic in strong light. As a consequence, the hormogonia often gather just below the soil surface. Here, they can undergo transformation into an aseriate stage

called primordium. This stage differentiates into vegetative cells which then form gelatinous *Nostoc* colonies (thalli). Only the early primordia of *Nostoc* can interact with the future fungal partner to give rise to the symbiotic consortium.

The life cycle of the fungal partner of the association starts from resting spores which contain a lot of storage compounds (Schüßler et al. 1994) and are formed by the fungus in the upper soil layer. The spores germinate by the protrusion of one or sometimes more than one hypha that branches to form a mycelium of a few centimeters inside the soil. There the young hyphae come into contact and incorporate *Nostoc* primordia. Each single incorporation event leads to the formation of a pear-shaped above-ground bladder up to 2 mm in length at the location where the incorporation took place. As already mentioned, in these bladders the incorporated *Nostoc* cells divide and become physiologically active. From experiments in our laboratory we know that, as is also true for AM, P limitation of the nutrient solution (1–2 μM phosphate) triggers the stable establishment of the symbiosis. N limitation seems not to be a crucial factor.

The incorporation of *Nostoc* into the fungal hypha proceeds by the following steps. Upon contact of the hyphal tip with a developing *Nostoc* primordium, a portion of fungal cytoplasm bulges out just below the apex of the hypha. The bulging process is repeated several times so that finally the hyphal tip forms an irregularly shaped mantle which surrounds the contacted *Nostoc* primordium and thus incorporates it into the fungal cytoplasm. Afterwards, the fungal bladder develops from this *Nostoc*-containing structure. It is important to note that heterocysts (specialized N_2-fixing cells), differentiating within the *Nostoc* primordia, are never enclosed by the fungal cytoplasm (see below) and therefore remain outside the hypha during the incorporation process (Mollenhauer et al. 1996).

Within the first hours after incorporation into the fungal cytoplasm, the *Nostoc* filaments are heavily affected and the photosynthetic pigments of the cells bleach considerably. These alterations and significant changes in ultrastructure suggest that during the initial state of endocytotic life the incorporated cyanobacteria suffer serious stress. During the following maturation of the *Geosiphon* bladder, the enclosed *Nostoc* cells recover gradually, i.e., the cells multiply, grow to a volume considerably larger than that of free-living cells, and arrange in filaments in which heterocysts are formed with the same frequency as in the filaments outside the bladders. Details of this process are also documented in a scientific film which is available in English and German (Mollenhauer and Mollenhauer 1997).

B. Specificity of Partner Recognition

The initial reaction between the partners leading to the establishment of the symbiotic *Geosiphon* association is to a large extent specific. Cells of certain strains of *Nostoc punctiforme* can be incorporated by *Geosiphon* and lead to the formation of functional fungal bladders. With cells of other strains, although being incorporated, the formation of bladders is blocked in an early stage of development, and there are, moreover, *Nostoc punctiforme* strains which are not incorporated at all by the fungus. An even more striking argument for a specific recognition process is the fact that, among the various developmental stages of *Nostoc*, exclusively the primordia are incorporated by the fungus. Not only the physiological activity of the primordia is different from the other stages of the *Nostoc* life cycle (Bilger et al. 1994) but also the composition of the gelatinous envelope. When differentiating to primordia, a mannose-containing slime is produced by the cells, whereas other sugars within the extracellular glycoconjugates could be detected only in earlier or later stages of the life cycle. In context with the finding that the heterocysts of *Nostoc* primordia are never incorporated by the fungus, it is interesting to note that these specialized cells are not surrounded by the mannose-containing glycoconjugate (Schüßler et al. 1997). Thus, it is tempting to speculate, but remains to be proven, that the specific partner recognition is based on the glycoconjugate composition of the cyanobacterial envelope.

IV. Structure and Compartmentation of *Geosiphon*

A. The *Geosiphon* Bladders and Spores

The *Geosiphon* bladder, up to 2 mm in size, represents a cell which is coenocytic with the fungal mycelium in the soil. The bladders show a strong polarity with a photosynthetic active region in the apical part, which is exposed to light and air, and a much smaller storage region in the basal part of the bladders, which is in nature embedded in the soil surface. The center of the bladder is highly vacuolized (Schüßler 1995). Schematic drawings of the compartmentation of *Geosiphon* are shown in Fig. 2.

Our ultrastructural observations show *Geosiphon pyriforme* as an endosymbiotic system with very close contact between the partners. It represents a symbiotic consortium of three organisms:

1. The fungus, supplying the consortium with inorganic nutrients like phosphate, trace elements, and water.
2. The cyanobacteria, supplying the consortium with carbohydrates by photosynthesis and nitrogen compounds by N_2 fixation.
3. The bacteria-like organisms (BLOs; see Figs. 4 and 5), found in many glomalean fungi, with their yet unknown physiological function.

Within the bladders, the cyanobacteria are located in a single compartment, the symbiosome, which is cup-shaped and located peripherally. The *Nostoc* cells divide and are physiologically active as endosymbionts. They are much larger than free-living vegetative cells (Fig. 3). The ca. tenfold increased volume of the endosymbiotic *Nostoc* cells could be caused by the high osmotic pressure inside the bladders. The isoosmolar concentration of sorbitol was observed with oil-filled microcapillaries and determined to be 220–230 mM, corresponding to a turgor pressure (P) of about 0.6 MPa (Schüßler et al. 1995). In many symbioses with plants cyanobacteria are known to increase in size, but not to the same extent (see, e.g., Johansson and Bergman 1992; for a review see Grilli Caiola 1992). This is probably a result of the high osmotic pressure of the surrounding medium. High NaCl concentrations are also known to cause an increase in volume of cyanobacteria (Erdmann and Schiewer 1984). Despite the increase in size, the *Nostoc* cells inside the *Geosiphon* bladder show a normal ultrastructural appearance (Fig. 3). They contain a high number of thylakoids and carboxysomes. Heterocysts also look normal and, in a relation to the vegetative cells, occur in the same ratio as in the free-living stage. It can be deduced from these observations that in the case of the *Geosiphon* symbiosis the major role of the cyanobacteria is photosynthesis. In contrast, in

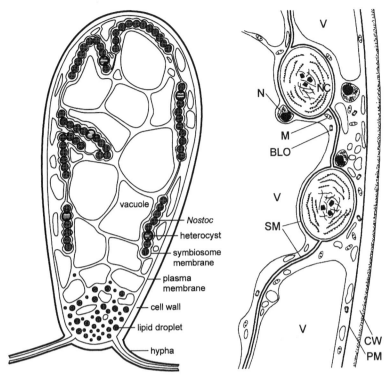

Fig. 2. Schematic drawings of *Geosiphon* bladder compartmentation. (Schüßler 1995). Overview (*left*) and detail (*right*); *BLO* bacteria-like organism; *CW* cell wall; *M* mitochondrium; *N* nucleus; *NC Nostoc* cell; *PM* plasma membrane; *SM* symbiosome membrane; *V* vacuole

symbioses with plants, e.g., with bryophytes or in the endosymbiosis with *Gunnera*, the main role for the cyanobacteria usually is N_2 fixation, which is reflected by a great enlargement of the relative heterocyst number.

Within the cytoplasm of the fungus, glycogen granules occur as storage compounds. Many lipid droplets are found in the basal part of the bladder. No dictyosomes exist, and microtubules can rarely be observed. In general, the fixation of the cells during preparation of the specimens for electron microscopy is often not sufficient and is best performed by microwave acceleration. This is due to the low cell wall permeability, which is discussed below.

The ultrastructure of the spores of *Geosiphon* is described in detail by Schüßer et al. (1994). Preparation of these spores for electron microscopy is even more difficult than of the bladders. This problem also exists with other glomalean species (Maia et al. 1993). The spores show a thick cell wall, which is laminated and very similar to that of some *Glomus* species. Two main storage compounds exist inside the spores: lipid droplets of different size, and structured granules, which

occupy about 25% of the spore volume. The latter are proposed to be storage compounds and are discussed below with respect to element analysis. They show paracrystalline inclusions similar to spores of some other glomalean fungi. Small vacuoles are found in germinating spores as well as in hyphae, which often contain dark deposits (Fig. 4). These look like the deposits in AMF and probably represent polyphosphate granules. In general, the ultrastructural appearance of the fungal symbiosis partner of *Geosiphon* represents that of AMF.

B. The Symbiotic Interface Between the Partners

The ultrastructure of *Geosiphon* was first studied by Schnepf (1964) and was the crucial investigation leading to the theory of the compartmentation of the eukaryotic cell. Recent ultrastructural studies (Schüßler et al. 1996) showed that inside the fungal bladder the symbiotic *Nostoc* cells are located in a single peripheral, cup-shaped compartment, the symbiosome. The space between the

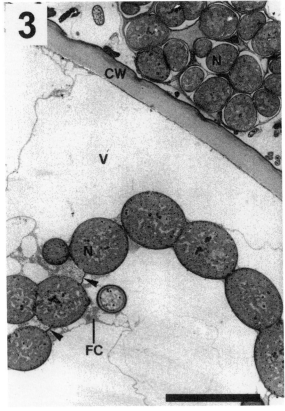

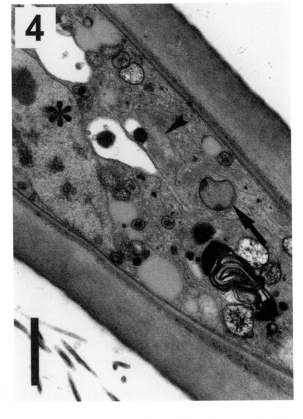

Figs. 3, 4. Electron micrographs of *Geosiphon*. Glutaraldehyde/OsO₄ fixation; *CW* cell wall; *FC* fungal cytoplasm; *N* *Nostoc* cell; *V* vacuole. *Bars* **3** 10 μm; **4** 1 μm. **Fig. 3.** Part of a *Geosiphon* bladder, cross-section. A vegetative colony of *Nostoc* (*upper right*) is attached to the outside of the bladder. Endosymbiotic *Nostoc* cells are much larger than the free-living ones. Parts of the symbiosome where no endosymbiotic cyanobacteria are located are marked by *arrowheads*; cf. also Fig. 8. (Schüßler et al. 1996). **Fig. 4.** Tangentially cut *Geosiphon* hypha. Nucleus (*asterisk*), BLO (*arrow*) and a hard to recognize mitochondrium (*arrowhead*) are marked. *Dark deposits* can be seen within small vacuoles. (Schüßler et al. 1994)

symbiosome membrane and the wall of the *Nostoc* cells enclosed by it is only 30–40 nm thick and contains a layer of electron microscopically opaque and amorphous-appearing material (Figs. 5, 6, 7) which was formerly assumed to be slime produced by the endosymbiont (Schnepf 1964). However, confocal laser scanning microscopical (CLSM) studies by means of affinity techniques with six fluorescence-labeled lectins with different specificity revealed by the occurrence of chitin that this amorphous layer inside the symbiosome represents a rudimentary fungal cell wall. Ultrastructural studies with WGA gold conjugates (Fig. 6) confirmed this results (Schüßler et al. 1996). Thus, the fungal cell wall as well as the electron-opaque layer within the symbiosome contain chitin, meaning that the symbiosome membrane surrounding the *Nostoc*

cells is homologous to the fungal plasma membrane.

Striking similarities exist between the fungal cell wall material present in the *Geosiphon* symbiosome and the thin cell wall bordering the symbiotic fungus from the penetrated plant cell in AM: both are electron-dense after OsO₄ fixation, amorphous in structure, about 30–40 nm thick, and show the same appearance. Considering also the results on the phylogenetic position of *Geosiphon* and the known or proposed nutrient flows between the symbiotic partners, it was suggested that the symbiotic interface in the AM and *Geosiphon* symbioses are homologous (Schüßler et al. 1996). The main differences are the relations of macro- and microbiont: in *Geosiphon*, the photoautotrophic partner (cyanobacterium) is the microsymbiont, in the AM the photoautotrophic

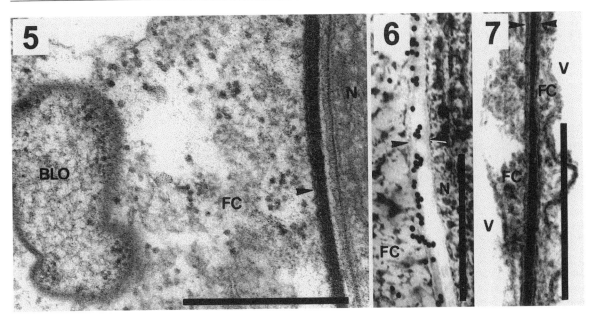

Figs. 5–7. Electron micrographs of the symbiotic interface of *Geosiphon*; *BLO* bacteria-like organism; *FC* fungal cytoplasm; *N Nostoc* cell; *V* vacuole. *Bars* 0.5 µm. (Schüßler et al. 1996). **Fig. 5.** Conventional fixation (glutaraldehyde/OsO₄), the symbiosome membrane is marked by an *arrowhead*. **Fig. 6.** Material freeze substituted without OsO₄. The thin layer attached to the symbiosome membrane is labeled by WGA gold. The plasma membranes of the symbiosis partners, enclosing the symbiosome space, are marked by *arrowheads*. **Fig. 7.** Conventional fixation (glutaraldehyde/OsO₄), part of the symbiosome where no cyanobacteria are located (cf. also Fig. 3). The symbiosome membrane, enclosing the osmiophilic layer, is marked by *arrowheads*

partner (plant) is the macrosymbiont. For a better understanding, Fig. 8 shows a scheme summarizing these assumptions.

V. Metabolism and Nutrient Acquisition

A. Photosynthetic Carbon Acquisition and Nitrogen Fixation

^{14}C-tracer experiments by Kluge et al. (1991) revealed that due to the endosymbionts *Geosiphon* is capable of photosynthetic carbon acquisition. Phosphorylated compounds and oligosaccharides, among them trehalose and raffinose, are the main photosynthetic end products. In light, but never in darkness, there was also a quick transfer of carbon into insoluble compounds, presumably polyglucans. Thus, it is conceivable that the glycogen granula which are found in the fungal cytoplasm derive from the photosynthetic activity of the endosymbionts. The bladders show also dark CO_2 fixation, but the rates are much lower than that of

CO_2 fixation in light. As is to be expected, the end products of dark CO_2 fixation were organic acids (mainly malate) and amino acids.

Compared with free-living cells of the same strain, the *Nostoc* cells in the bladder show considerably higher photosynthetic activity (Bilger et al. 1994). This is indicated by higher quantum yields in photosystem II, higher quantum flux density required to saturate the photosynthetic electron transport rates, and lower susceptibility

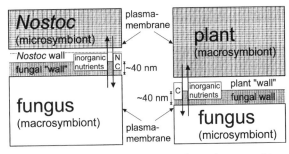

Geosiphon pyriforme   arbuscular mycorrhiza

Fig. 8. Schematic drawing comparing the interfaces of the *Geosiphon* and AM symbiosis

to photoinhibition. The reason for this different photosynthetic behavior is not yet known, but it is reasonable to assume that CO_2 concentration, and thus the availability of the major photosynthetic substrate for the photobionts, is better inside than outside the bladder.

From the fact that the *Nostoc* filaments in the fungal bladders develop heterocysts it can be assumed that *Geosiphon* is capable of nitrogen fixation. This was supported by Kluge et al. (1992), who showed substantial nitrogenase activity in the bladders.

B. Uptake of Nutrients from the Outside

As shown by electrophysiological experiments (unpubl. results), inorganic ions (nitrate, chloride) and small organic molecules (e.g., glycin, cystein) lead to rapid, transient depolarization of the plasma membrane potential of the *Geosiphon* bladders, suggesting that these substances are actively taken up from the outside. On the other hand, there were no changes in membrane potential if hexoses (e.g., glucose) and larger amino acids were applied, suggesting that such molecules could not be taken up from the environment. Since metabolism of radioactively labeled hexoses by the bladders also could not be detected after usual incubation times, we speculated about the cell wall being responsible for these negative results. This was supported by observations of Schüßler et al. (1995), showing that the presence of solutes having large molecule radii leads to irreversible cytorrhysis, i.e., collapse of the whole bladder including the cell wall, occurred, whereas in presence of small solutes plasmolysis occurred, or cytorrhysis was quickly reversed. This different transport behavior was presumably due to the selective permeability of the bladder wall.

By using solutes with known molecular radii, it was shown that the limiting pore radius of the *Geosiphon* bladder wall is about 0.5 nm, which, compared with other cell walls, is very low. Provided that such a small pore size holds true also for the hyphal wall of *Geosiphon*, the fungus could have difficulties with saprophytic acquisition of organic molecules such as glucose, sucrose, glutamate, etc., and thus might depend largely on the photoautotrophic endosymbiont. For the acquisition of inorganic hydrated ions the pore radius of 0.5 nm should not be a problem. However, cell wall

permeability is a complicated topic and, dependent on their differentiation, hyphae might show different cell wall permeability. We will try to answer this question in the future.

C. Element Composition and Distribution Within *Geosiphon*

At present it is not known why AMF cannot be cultured axenically, and there is little information available about the trace element requirements and general element composition of the AMF. For this and other reasons, considering the fact that these fungi supply the majority of land plants with inorganic elements, it is necessary to perform further studies on element composition. Therefore PIXE (proton induced X-ray emission; Johansson et al. 1995) measurements on *Geosiphon pyriforme* were initiated to obtain first indications on the macro- as well as the microelement composition of the spores and bladders. Moreover, element content of some subcellular compartments could be quantitatively analyzed and, by a differential approach, that of others calculated. PIXE, combined with STIM (scanning transmission ion microscopy) allows elemental concentrations to be absolutely quantified with a lateral resolution in the $1\,\mu m$ range and with high accuracy and precision (Maetz et al. 1999b).

First results on the *Geosiphon* bladders were published in Maetz et al. (1999a). They show (all values given here are related to dry weight) that the fungal partner of the symbiosis, grown on a poor nutrient solution (e.g., containing only $1\,\mu M$ phosphate), accumulates high amounts of Cl (about 2.5%) and K (about 8%), which both seem to play a major role in osmoregulation of the fungus. The symbiosome (including the cyanobacteria) contains only small amounts of these elements. P also is accumulated by the fungus in high concentrations (about 2%), but not within the symbiosome; it is probably stored in the vacuoles. Mg, S, and Ca occur in concentrations comparable to that found in plants. The same holds true for the microelements Fe, Mn, Cu, and Zn. Se concentration is below 1 ppm. Some interesting results were shown for microelements playing a role in N metabolism. Mo is present within the symbiosome at much lower concentration compared to the rest of the bladder. This is a little surprising, since Mo is a constituent of nitrogenase, required for N_2 fixation. Reasons could be that nitrogenase is present also in the fungal part of the bladders, e.g., in the BLOs, that other Mo enzymes (e.g., nitrate reductase, sulfite oxidase) occur in relevant amounts, or that Mo is located in the fungal vacuole. The most striking result regarding the microelements was that Mn and Ni are present in the symbiosome in high amounts. In future we plan to investigate in detail which compartments contain these elements. Candidates for Mn are, e.g., Mn superoxide dismutase or the water-cleaving Mn protein of photosystem II. N_2-fixing *Rhizobium* is known to contain MnSOD in significant amounts. Ni could be present in urease (the only known Ni enzyme in plants) or prokaryotic NiFe hydrogenases.

Unpublished and preliminary results on the element composition of the *Geosiphon* spores show that the structured granules (SGs) which are 4–6 μm in diameter, are included within a vesicle, occupy about 25% of the spore volume, and contain most of the total P, K, and S present within the spores. We have not yet investigated the chemical storage form of P. The spore cell wall shows a relatively high S concentration (0.25%), probably because of a high protein content, which was shown for a *Glomus* species by Bonfante and Grippiolo (1984). Cl and K are concentrated within the spores in much lower amounts, compared to the bladders.

VI. Conclusions and Future Research

Geosiphon pyriforme is the only known fungal endocyanosis, and the fungal partner of the symbiosis doubtless belongs to the Glomales. Because of its size and the exposition of the fungal membrane in the symbiotic stage, it has considerable advantages for the study of particular physiological features of glomalean fungi, e.g., by electrophysiological measurements. A further advantage of the *Geosiphon* symbiosis is that only one eukaryote is present. This means that this organism is predestined for molecular biological studies dealing with the expression of fungal genes. In contrast to the AM symbiosis, where in the symbiotic stage the plant is always present, in *Geosiphon* it is relatively easy to isolate specifically the fungal mRNA because the prokaryotic mRNA of *Nostoc* lacks a poly(A) tail. Thereafter,

the fungal mRNA can be reversably transcribed into cDNA and differentially expressed genes identified. By the method of differential display reverse transcription PCR (DDRT-PCR) or subtractive hybridization it should be possible to identify, for example, specific fungal genes related to the acquisition and metabolism of inorganic nutrients such as phosphate. This method also makes it possible to identify and characterize sugar- or nitrate transporters. Probably a set of genes might be identified which are related to the different metabolic conditions. The obtained DNA fragments could be used for the identification and quantification of the corresponding genes in the AM symbiosis.

Altogether, on the molecular level, *Geosiphon* appears to be an extremely promising model system and could lead to major advances in AM research.

Acknowledgments. We thank the Deutsche Forschungsgemeinschaft for support of our work on *Geosiphon* and AMF phylogeny (SFB199 TPA3, SCHU 1203/1-2).

References

Bilger W, Büdel B, Mollenhauer R, Mollenhauer D (1994) Photosynthetic activity of two developmental stages of a *Nostoc* strain (cyanobacteria) isolated from *Geosiphon pyriforme* (Mycota). J Phycol 30: 225–230

Bonfante P, Grippiolo R (1984) Cytochemical and biochemical observations on the cell wall of the spore of *Glomus epigaeum*. Protoplasma 123:140–151

De Rijk P, Neefs JM, Van de Peer Y, De Wachter R (1992) Compilation of small ribosomal subunit RNA sequences. Nucleic Acids Res 20 (Suppl):2075–2089

Dodds WK, Gudder DA, Mollenhauer D (1995) The ecology of *Nostoc*. J Phycol 31:2–18

Erdmann N, Schiewer U (1984) Cell size changes as indicator of salt resistance of blue green algae. Arch Hydrobiol Suppl (Algol Stud) 67:431–439

Gehrig H, Schüßler A, Kluge M (1996) *Geosiphon pyriforme*, a fungus forming endocytobiosis with *Nostoc* (cyanobacteria), is an ancestral member of the Glomales: evidence by SSU rRNA analysis. J Mol Evol 43:71–81

Grilli Caiola M (1992) Cyanobacteria in symbiosis with bryophytes and tracheophytes. In: Reisser W (ed) Algae and symbiosis: plants, animals, fungi, viruses, interactions explored. Biopress, Bristol, pp 231–253

Johansson C, Bergman B (1992) Early events during the establishment of the *Gunnera/Nostoc* symbiosis. Planta 188:403–413

Johansson SAE, Campbell JL, Malmqvist KG (1995) Particle-induced X-ray emission spectrometry (PIXE). John Wiley, New York

Kluge M, Mollenhauer D, Mollenhauer R (1991) Photosynthetic carbon assimilation in *Geosiphon pyriforme* (Kützing) F.v. Wettstein, an endosymbiotic association of fungus and cyanobacterium. Planta 185: 311–315

Kluge M, Mollenhauer D, Mollenhauer R, Kape R (1992) *Geosiphon pyriforme*, an endosymbiotic consortium of a fungus and a cyanobacterium (*Nostoc*), fixes nitrogen. Bot Acta 105:343–344

Kluge M, Mollenhauer D, Mollenhauer R (1994) *Geosiphon pyriforme* (Kützing) von Wettstein, a promising system for studying endocyanoses. Prog Bot 55:130–141

Kluge M, Gehrig H, Mollenhauer D, Schnepf E, Schüßler A (1997) News on *Geosiphon pyriforme*, an endocytobiotic consortium of a fungus with a cyanobacterium. In: Schenk HEA, Herrmann R, Jeon KW, Müller NE, Schwemmler W (eds) Eukaryotism and symbiosis. Springer, Berlin Heidelberg New York, pp 469–476

Knapp E (1933) Über *Geosiphon pyriforme* Fr.v. Wettst., eine intrazelluläre Pilz-Algen-Symbiose. Ber Dtsch Bot Ges 51:210–217

Ligrone R (1988) Ultrastructure of a fungal endophyte in *Phaeoceros laevis* (L.) Prosk. (Anthocerophyta). Bot Gaz 149:92–100

Ligrone R, Lopes C (1989) Cytology and development of a mycorrhiza-like infection in the gametophyte of *Conocephalum conicum* (L.) Dum. (Marchantiales, Hepatophyta). New Phytol 111:423–433

Maetz M, Przybylowicz WJ, Mesjasz-Przybylowicz J, Schüßler A, Traxel K (1999a) Low dose nuclear microscopy as a necessity for accurate quantitative microanalysis of biological samples. Nucl Instrum Meth B 158:292–298

Maetz M, Schüßer A, Wallianos A, Traxel K (1999b) Subcellular trace element distribution in *Geosiphon pyriforme*. Nucl Instrum Meth B 150:200–207

Maia LC, Kimbrough JW, Erdos G (1993) Problems with fixation and embedding of arbuscular mycorrhizal fungi (Glomales). Mycologia 85:323–330

Malloch DW, Pirozynski KA, Raven PH (1980) Ecological and evolutionary significance of mycorrhizal symbiosis in vascular plants (a review). Proc Natl Acad Sci USA 77:2113–2118

Marschner H, Dell B (1994) Nutrient uptake in mycorrhizal symbiosis. Plant Soil 159:89–102

Mollenhauer D (1992) *Geosiphon pyriforme*. In: Reisser W (ed) Algae and symbiosis: plants, animals, fungi, viruses, interactions explored. Biopress, Bristol, pp 339–351

Mollenhauer D, Mollenhauer R (1997) Endosymbiosis between *Nostoc* and *Geosiphon pyriforme*. Institut für den Wissenschaftlichen Film, Göttingen, Film No C1955

Mollenhauer D, Mollenhauer R, Kluge M (1996) Studies on initiation and development of the partner association in *Geosiphon pyriforme* (Kütz.) v. Wettstein, a unique endocytobiotic system of a fungus (Glomales) and the cyanobacterium *Nostoc punctiforme* (Kütz.) Hariot. Protoplasma 193:3–9

Morton JB, Benny GL (1990) Revised classification of arbuscular mycorrhizal fungi (Zygomycetes): a new order, Glomales, two new suborders, Glomineae and Gigasporineae, and two new families, Acaulosporaceae and Gigasporaceae, with an emendation of Glomaceae. Mycotaxon 37:471–491

Parke JL, Lindermann RG (1980) Association of vesicular-arbuscular mycorrhizal fungi with the moss *Funaria hygrometrica*. Can J Bot 58:1898–1904

Peterson RL, Howarth MJ, Whittier DP (1981) Interactions between a fungal endophyte and gametophyte cells in *Psilotum nudum*. Can J Bot 59:711–720

Pirozynski KA (1981) Interactions between fungi and plants through the ages. Can J Bot 59:1824–1827

Pirozynski KA, Malloch DW (1975) The origin of land plants: a matter of mycotrophism. BioSystems 6:153–164

Redecker D, Morton JB, Bruns TD (2000) Ancestral lineages of arbuscular mycorrhizal fungi (Glomales). Mol Phylogenet Evol 14:276–284

Remy W, Taylor TN, Hass H, Kerp H (1994) Four hundred-million-year-old vesicular arbuscular mycorrhizae. Proc Natl Acad Sci USA 91:11841–11843

Sawaki H, Sugawara K, Saito M (1999) Phylogenetic position of an arbuscular mycorrhizal fungus, *Acaulospora gerdemannii*, and its synanamorph *Glomus leptotichum*, based upon 18S rRNA gene sequence. Mycoscience 39:477–480

Schmid E, Oberwinkler F (1993) Mycorrhiza-like interactions between the achlorophyllous gametophyte of *Lycopodium clavatum* L. and its fungal endophyte studied by light and electron microscopy. New Phytol 124:69–81

Schnepf E (1964) Zur Feinstruktur von *Geosiphon pyriforme*. Arch Mikrobiol 49:112–131

Schüßler A (1999) Glomales SSU rRNA gene diversity. New Phytol 144:205–207

Schüßler A (1995) Strukturelle und funktionelle Charakterisierung der Pilz/Blaualgen Endosymbiose *Geosiphon pyriforme*: Physiologie, Zellbiologie und Taxonomie. PhD Thesis, University of Heidelberg, Heidelberg

Schüßler A, Mollenhauer D, Schnepf E, Kluge M (1994) *Geosiphon pyriforme*, an endosymbiotic association of fungus and cyanobacteria: the spore structure resembles that of arbuscular mycorrhizal (AM) fungi. Bot Acta 107:36–45

Schüßler A, Schnepf E, Mollenhauer D, Kluge M (1995) The fungal bladders of the endocyanosis *Geosiphon pyriforme*, a *Glomus*-related fungus: cell wall permeability indicates a limiting pore radius of only 0.5 nm. Protoplasma 185:131–139

Schüßler A, Bonfante P, Schnepf E, Mollenhauer D, Kluge M (1996) Characterization of the *Geosiphon pyriforme* symbiosome by affinity techniques: confocal laser scanning microscopy (CLSM) and electron microscopy. Protoplasma 190:53–67

Schüßler A, Meyer T, Gehrig H, Kluge M (1997) Variations of lectin binding sites in extracellular glycoconjugates during the life cycle of *Nostoc punctiforme*, a potentially endosymbiotic cyanobacterium. Eur J Phycol 32:233–239

Simon L (1996) Phylogeny of the Glomales: deciphering the past to understand the present. New Phytol 133: 95–101

Simon L, Bousquet J, Lévesque RC, Lalonde M (1993) Origin and diversification of endomycorrhizal fungi and coincidence with vascular land plants. Nature 363:67–69

Smith SE, Read DJ (1997) Mycorrhizal symbiosis, 2nd edn. Academic Press, London

Stahl M (1949) Die Mycorrhiza der Lebermoose mit besonderer Berücksichtigung der thallösen Formen. Planta 37:103–148

Stubblefield SP, Taylor TN, Trappe JM (1987) Fossil mycorrhizae: a case for symbiosis. Science 237:59–60

von Wettstein F (1915) *Geosiphon* Fr. v. Wettst., eine neue, interessante Siphonee. Österr Bot Z 65:145–156

Walker C (1992) Systematics and taxonomy of the arbuscular endomycorrhizal fungi (Glomales) – a possible way forward. Agronomie 12:887–897

Lichens

10 The Symbiotic Phenotype of Lichen-Forming Ascomycetes

R. Honegger

CONTENTS

I. Introduction

A. Peculiarities of the Lichen Symbiosis

Lichens are the symbiotic phenotype of lichen-forming fungi, a polyphyletic, taxonomically heterogenous assembly of nutritional specialists which acquire fixed carbon from a population of minute, extracellularly located green algal or cyanobacterial cells; these are referred to as the photobiont. *Geosiphon pyriforme*, the only known fungal symbiosis with an intracellularly located cyanobacteria, is not normally considered a lichen

Institute of Plant Biology, University of Zürich, Zollikerstr. 107, 8008 Zürich, Switzerland

(see Schüßler and Kluge, Chap. 9, this Vol.). In marked contrast to fungal parasites of algae, lichen-forming fungi do not damage their photoautotrophic partner. The thalli of morphologically advanced taxa are, in fact, very elaborate culturing chambers for photobiont cells, many of which are poor competitors on their own (see below). As many lichen-forming fungi either form symbiotic propagules or disperse successfully by means of thallus fragmentation, the photobiont cells are carried along and invade areas where they would not normally occur in the free-living state. Thus, the lichen symbiosis is regarded as a mutualistic association.

Lichenization is a common and successful nutritional strategy, with more than 25% of fungi being lichenized (Hawksworth et al. 1995; Honegger 1996b). Approximately 98% of lichen-forming fungi are ascomycetes, or about 46% of ascomycetes are lichenized, respectively (Honegger 1997). Lichen-forming fungi are typical representatives of their fungal classes but differ from the other fungi by their various adaptations to the special symbiotic situation, i.e., the cohabitation with a population of minute photobiont cells. As concluded from biogeographic data, the major orders of ascomycetes with exclusively or predominantly lichen-forming taxa are ancient (Galloway 1996). Fossil records are largely missing, but it is hoped that the recent discovery of *Winfrenatia reticulata*, a cyanolichen from the early Devonian Rhinie chert (Taylor et al. 1997), contemporary with the earliest-known arbuscular mycorrhizal fungi (Taylor et al. 1995), might train the eyes and minds of paleomycologists and -botanists.

This chapter focuses on the diversity of symbiotic phenotypes among lichen-forming ascomycetes, features the main building blocks of morphologically advanced foliose and fruticose thalli and the most common types of mycobiont-photobiont interactions therein. Metabolic interactions (mobile carbohydrates, poikilohydric

The Mycota IX
Fungal Associations
Hock (Ed.)
© Springer-Verlag Berlin Heidelberg 2001

water relations, secondary metabolites etc.) are summarized in Honegger (1997).

B. Concentric Bodies

Lichen-forming ascomycetes were thought to differ cytologically from non-lichenized taxa by the possession of concentric bodies, proteinaceous cell organelles of yet unknown origin and function (Fig. 1a–c). Concentric bodies measure approximately 0.3 μm in diameter and are often seen in clusters near the cell periphery. They were found in the majority of lichen-forming ascomycetes collected in the wild; exceptions are aquatic species, which never dry out, and axenic cultures, which are normally kept at constant humidity (review: Ahmadjian 1993). However, concentric bodies are a peculiarity of neither lichen-forming ascomycetes nor of the symbiotic way of life per se as they have been found in non-lichenized, saprotrophic taxa such as litter decomposers or microcolonial ascomycetes of desert rocks, and in plant pathogens (reviews: Ahmadjian 1993; Honegger 1993). The only feature shared by all ascomycetous cells containing concentric bodies seems to be their relative longevity and desiccation tolerance. As concentric bodies are not membrane-bound and, as seen in freeze-fractured preparations (Fig. 1b), have a gas-filled centre, it was hypothesized that they were remains of cytoplasmic cavitation events, e.g., denatured proteins from the surface of drought stress-induced cytoplasmic gas bubbles (Honegger 1995). Reversible cytoplasmic cavitation during desiccation, first observed in ascospores by deBary (1866), is a normal feature in desiccation-tolerant lichen-forming and non-lichenized fungi (Fig. 1c) and is survived unharmed (Honegger 1995, 1997; Scheidegger et al. 1995).

C. Lichen Photobionts

The species names of lichens refer to the fungal partner; photoautotrophic partners of lichen-forming fungi have their own names and phylogenies. In less than 2% of lichens has the photobiont ever been identified at the species level; quite often, not even the generic affiliation is known. Only after isolation and culturing under defined conditions can lichen photobionts be properly identified since many of them change their growth

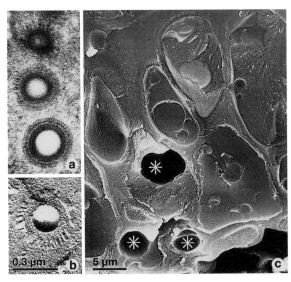

Fig. 1a–c. Concentric bodies, organelles of unknown origin and function, with a proteinaceous core around a gas-filled centre as seen in cortical cells of *Peltigera praetextata* (Flörke ex Sommerf.) Zopf. **a** TEM of concentric bodies in an ultrathin section (2 tangential and 1 median sections, *top to bottom*), and **b** in a replica of a freeze-fractured specimen. **c** LTSEM of a cross-fractured, drought-stressed thallus during the rehydration process. The specimen was cryoimmobilized 60 s after addition of water to the tallus surface. Top cells are partially rehydrated, bottom cells shrivelled and cavitated. *Asterisks* refer to cytoplasmic gas bubbles in cavitated fungal cells. See text for further explanations. Same magnification in **a** and **b**

pattern upon lichenization (Fig. 2e). An estimated 85% of lichen-forming ascomycetes associate with unicellular or filamentous green algae, approximately 10% with cyanobacteria, and 3–4%, the cephalodiate species (see Sect. II.D), simultaneously with both green algal and cyanobacterial partners (Tschermak-Woess 1988). Approximately 100 spp. of lichen photobionts have so far been described. Some are widespread and common elements of aerophilic algal or cyanobacterial communities (e.g., Trentepohliales; Hawksworth 1988a; review: Friedl and Büdel 1996), but the most common green algal lichen photobionts, all representative of the Trebouxiophyceae sensu Friedl (1995), are rarely found outside lichens. This is especially true of the genus *Trebouxia* (approximately 20 spp. described, photobionts of 50% of lichen-forming ascomycetes), which are by far the most common photobionts in lichens of temperate, arctic/alpine, antarctic and desert ecosystems but are very poor competitors outside lichens.

II. Thallus Structure in Microlichens

The distinction between micro- and macrolichens is artificial, and innumerable intermediate forms can be found. However, microlichens comprise morphologically less advanced taxa which may nevertheless reach impressive dimensions (e.g., cover several m² of rock surface as seen in numerous leprose taxa) but are not capable of differentiating large, three-dimensional thalli which rise above the substratum, as is normally the case in macrolichens.

A. Leprose, Microfilamentous and Microglobose Lichens

Leprose lichens, recognizable as powdery crusts over rock or bark surfaces, are morphologically the most simple symbiotic phenotypes. They are formed by representatives of the Caliciales (e.g., *Chaenotheca* spp.), Arthoniales (*Chrysothrix* spp.) and by a range of sterile taxa incertae sedis within the ascomycetes (e.g., *Lepraria*, *Leprocaulon*, *Leproloma* spp.). Fungal hyphae of leprose lichens overgrow groups of unicellular green algae (Fig. 2a) and ensheath them with remarkably hydrophobic cell wall surface material, the chemical composition of which remains to be explored. Many taxa are brightly coloured due to secondary fungal metabolites (white, pale or intense yellow) and thus very conspicuous, especially when covering large areas such as several m² of rock surface. Due to their water repellency, many species of leprose lichens are almost unwettable; these usually develop at dry, often shaded sites such as rock overhangs which are hardly ever wetted by rain; such thalli rehydrate from humid air alone. Thus, only green algal photobionts are compatible partners; these have been shown in other taxa to be capable of rehydrating from humid air whilst cyanobacteriae require liquid water (Lange et al. 1986, 1988).

Various species of lichen-forming ascomycetes overgrow and ensheath either filamentous cyanobacterial colonies (e.g., *Scytonema* sp.; Tschermak-Woess et al. 1983) or filamentous green algae (Trentepohliaceae such as *Trentepohlia* or *Physolinum* spp.; Meier and Chapman 1983); thus, their microfilamentous habit reflects the growth pattern of the photobiont (Figs. 2c–d, 3a).

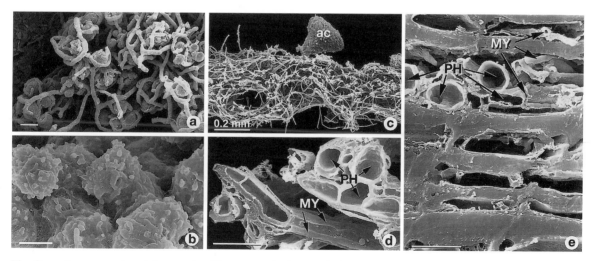

Fig. 2a–e. Leprose, microglobose, microfilamentous and crustose lichens with green algal photobionts. **a** *Lepraria incana* (L.) Ach. (sterile taxa)/*Trebouxia* sp., powdery, leprose crust with very hydrophobic surface. Each of the globose algal cells is contacted by the fungal partner. **b** *Micarea hedlundii* Coppins (Lecanorales)/non-identified photobiont; microglobose thallus, the algal cells are tightly ensheathed by the fungal partner. **c,d** *Coenogonium subvirescens* Nyl. (Gyalectales)/*Trentepohlia* sp., microfilamentous thallus. Loosely interwoven green algal filaments are partially ensheathed by hyphae of the fungal partner. The thick, cellulosic algal cell wall is not penetrated (no haustoria seen in dissected algal cells). **e** *Arthonia ilicina* Taylor (Arthoniales)/*Trentepohlia* sp.; crustose, non-stratified thallus within the bark of *Ilex aquifolium* L. Fungal hyphae grow between bark cells an contact the coccoid cells of the photobiont which, upon isolation and culturing, switches to filamentous growth. Abbreviations: *MY* mycobiont; *PH* photobiont; *ac* ascocarp of the fungal partner. *Bars* 10 µm unless otherwise stated

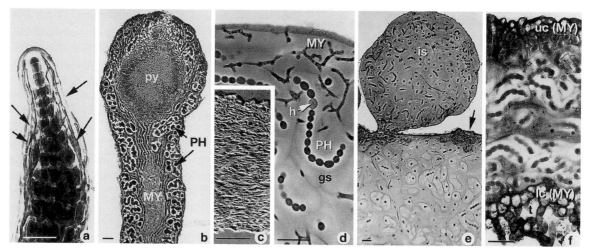

Fig. 3a–f. Gelatinous lichens: compact microfilamentous (**a**), fruticose (**b**) and foliose thalli (**c–f**), all with a cyanobacterial photobiont with an extensive gelatinous sheath. **a** *Ephebe lanata* (L.) Vainio (Lichinales)/*Stigonema* sp. *Arrows* point to the fungal partner which grows within the gelatinous sheath and between the cells of the photobiont. **b** *Lichina pygmaea* (Lightf.) Agardh (Lichinales)/*Calothrix scopulorum* (W. et M.) Agardh. Intertidal, epilithic species, which forms black tufts in association with barnacles and fucoid algae. **c–e** *Collema auriforme* (With.) Coppins & Laundon (Lecanorales)/*Nostoc* sp. in the desiccated state (**c**) with <20% water dw⁻¹, and detail of the top part of the fully hydrated thallus (**d**) with approximately 800% water dw⁻¹. The fungal partner grows mainly between the gelatinous sheaths of the chain-like colonies of the photobiont. **e** Isidium formation is under the control of the mycobiont; the *arrow* points to an isidium initial at the thallus surface. **f** *Leptogium saturninum* (Disks.) Nyl. (Lecanorales)/*Nostoc* sp., with a thin upper and lower fungal cortex and tomentum. As in *Collema*, the mycobiont grows mainly between the gelatinous sheaths of the photobiont. Abbreviations: *MY* mycobiont; *PH* photobiont; *gs* gelatinous sheath of the cyanobacterial photobiont; *h* heterocyst; *is* isidium; *lc* lower cortex; *py* pycnidium of the fungal partner; *uc* upper cortex; *t* tomentum. **a** Whole mount; **b–f** semithin sections, **c** cryofixed, freeze-substituted specimen. Bars 20 μm.

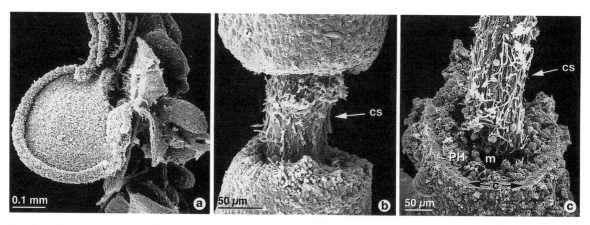

Fig. 4a–c. Squamulose (**a**) and fruticose thalli with internal stratification and unicellular green algal photobionts (**b,c**). **a** *Normandina pulchella* (Borrer)Nyl. (incertae sedis) /*Nannochloris normandinae* Tsch.-Woess, squamulose thallus growing on a liverwort. **b,c** *Usnea hirta* (L.)Weber ex Wigg. (Lecanorales)/*Trebouxia* sp., erect to pendulous, epiphytic beard lichen. *Usnea* spp. derive considerable tensile strength from a prominent, conglutinate central strand (*cs*). The photobiont cells (*PH*) are kept at the periphery of the gas-filled medullary layer (*m*) underneath the peripheral cortex (*c*).

Microglobose lichens are built up by minute, globular elements in which the fungal partner overgrows and ensheaths algal cells (Fig. 2b). The term goniocyst is used in the lichenological literature for such globular fungal-algal packages which are not soredia (symbiotic vegetative propagules; see below). Microfilamentous and microglobose thalli are also formed by various lichen-forming basidiomycetes (see Oberwinkler, Chap. 12, this Vol.).

B. Crustose Lichens

The majority of lichen-forming ascomycetes, representatives of all orders which include lichenized taxa (except Peltigerales with no crustose forms; review: Honegger 1997), often form inconspicuous crustose thalli on or within the substratum. Examples include endo- and epiphleodal thalli (Fig. 2e), thalli developing on and partly within other lichens (lichenicolous lichens; Poelt and Doppelbauer 1956; Friedl 1987), on bryophytes (bryophilous lichens; Poelt 1985), or leaves of higher plants (foliicolous lichens; Hawksworth 1988a; Farkas and Sipman 1993), cryptoendolithic crusts in Antarctic cold deserts (Friedmann 1982; Friedmann et al. 1993), thalli developing on and within limestone and other calcareous substrata such as shells of marine molluscs (limpets) or cirripedia (barnacles), on historical buildings and monuments, frescoe paintings included, on granitic rock and on or within anthropogenic substrata such as concrete, glass, plastics etc. (Ascaso and Wierzchos 1994; Piervittori et al. 1994, 1996; Sipman 1994). As can be seen from this rudimentary list, there are many biologically interesting phenomena to be studied in crustose lichens. There are also some conflicts with regard to lichen biodeterioration of historical monuments. However, it is obvious that crustose lichens have caused less damage to the cultural heritage of mankind throughout the centuries and millenia than the combustion of fossil fuels, which has occurred on a large scale in the industrialised era.

Crustose thalli are either non-stratified (no distinct layers recognizable, thus referred to as homoiomerous) or stratified (heteromerous), with a cortical layer at their periphery, an algal layer and a medulla which is in close contact with the substratum. Such taxa are often areolate, effigurate to lobate (review: Büdel and Scheidegger 1996). Some have a distinct non-lichenized prothallus on which areoles develop upon successful capture of a compatible photobiont (Clayden 1998). There are numerous intermediate forms between crustose and stratified foliose taxa. Examples are squamulose lichens (Fig. 4a): small thalli with internal stratification, but often with the lower surface adhering more or less closely to the substratum. Despite their taxonomic diversity and numeric significance, crustose lichens are treated only marginally in this chapter; the focus is on the morphologically complex foliose and fruticose forms, the most advanced vegetative structures in the fungal kingdom, which have been explored in more detail at the ultrastructural level than the crustose taxa.

C. Gelatinous Lichens

The Collemataceae (Lecanorales; approx. 250 spp., 8 genera; photobionts: *Nostoc* spp.) and the Lichinaceae (Lichinales; approx. 260 spp., 37 genera; photobionts: *Calothrix* spp., *Dichothrix* spp., *Stigonema* spp.), both with cyanobacterial photobionts, are referred to as gelatinous lichens. They differ from all other taxa by their relative compactness, i.e., the lack of gas-filled areas within the thallus, and by the strong impact of the photoautotrophic partner on thalline morphology, colouration and water relations. The fungal partner grows within the prominent gelatinous sheaths of the cyanobacterial colonies (Fig. 3a–f) and develops a close contact with the photobiont cells, usually without penetrating their murein sacculus. These gelatinous sheaths absorb large amounts of water (Fig. 3c–d) and, in the fully hydrated state, provide the thallus with a jelly-like consistency. Thus, the homoiomerous thallus of gelatinous lichens receives its overall morphology and consistency from the photobiont, but its shape, lobation (in Collemataceae) or branching (in Lichinaceae) and the formation of isidia (Fig. 3e) or other thalline outgrowths are under the control of the fungal partner. A thin peripheral fungal cortex is formed in only a few genera (e.g., *Leptogium*, Fig. 3f).

III. Foliose and Fruticose Macrolichens with Stratified (Heteromerous) Thallus

A. Peculiarities and Main Building Blocks

Foliose and fruticose lichen thalli, often referred to as macrolichens, are morphologically and anatomically the most complex vegetative structures in the fungal kingdom. They are, in fact, very elegant culturing chambers for a population of minute photobiont cells and thus the product of a remarkable adaptation of the heterotrophic

partner to the requirements of their photoautotrophic symbiont. In contrast to mutualistic or
parasitic fungal symbioses with higher plants, i.e.,
with a multicellular, quantitatively predominant
photoautotrophic host, it is the fungal partner of
macrolichens which rises above the substratum by
differentiating a three-dimensional thallus, com-
petes for space above ground and secures the gas
exchange and adequate illumination of the photobiont cells. These are retained in a gas-filled zone
within the thallus, where they are actively shifted
by the fungal partner via growth processes and
thus brought into an optimal position with regard
to gas exchange and illumination. In dorsiventrally
organized foliose thalli (Fig. 5a,h) as well as in
radially organized, erect or pendulous fruticose
forms (Fig. 4c) the algal layer is kept in a position
comparable to palisade parenchyma in leaves or
the assimilative tissues in the peripheral cortex of
stems of higher plants. The photobiont cell population of macrolichens has no direct access to
water and dissolved mineral nutrients except via
the fungal apoplast (see Honegger 1997). When
put into an upside-down position, the mycobionts
of dorsiventrally organized lichens are capable of
sensing, in a yet unknown manner, this spatial disturbance and are able to correct it by means of
growth processes (Figs. 5d–f; Honegger 1995); thus
the adequate illumination of the photobiont cell
population is secured. Within non-growing areas
of the lichen thallus, the algal cell population
is subjected to some kind of limiting control
whose molecular basis is not understood. A large
proportion of oversized *Trebouxia* cells were
observed in subapical thalline areas of representatives of the Lecanorales and Teloschistales
(Fiechter 1990; Hill 1985, 1989; Honegger 1996d).
These have an arrested cell cycle, i.e., exceed the
size required for autospore formation without
undergoing mitosis.

It is important to note that only one out of four
lichen-forming fungi has the amazing morphogenetic capacity of differentiating a foliose or fruticose, internally stratified symbiotic phenotype.
However, such morphologically advanced thalli are
produced by representatives of various unrelated
orders. Examples are fruticose thalli in the Lecanorales, Teloschistales, Caliciales and Arthoniales,
band-shaped thalli in the Lecanorales (genera
Pseudevernia and *Evernia*), Teloschistales (*Teloschistes*) and Arthoniales (*Roccella*), or umbilicate-
foliose thalli in the Lecanorales (genera *Umbilicaria* and *Lasallia*) and Verrucariales (*Dermatocar-

pon*). On the contrary, a few very interesting cases
are known where the same fungal species differentiates morphologically completely different symbiotic phenotypes in association with either a
cyanobacterial or a green algal photobiont (see
Sect. III.D, Cephalodiate Species and Photomorphs).

The main building blocks of morphologically advanced macrolichens are conglutinate,
hydrophilic pseudoparenchyma, often found as a
peripheral cortical layer or as a central strand
(Figs. 4b,c; 5h; 6a,b,d–g; 7c–f, 8b), and plectenchyma which are built up by loosely interwoven hyphae with hydrophobic wall surfaces; these
create a gas-filled zone either as a medullary layer
in the thalline interior (Figs. 4c, 5h, 7c–e, 8b, 9i), or
at the periphery of a conglutinate central strand
or tube of fruticose forms (Fig. 9m). The algal
layer is situated at the periphery of the gas-filled
zone (review: Honegger 1993). Morphologically
advanced macrolichen thalli are thus the product
of an impressive hyphal polymorphism.

1. Conglutinate Pseudoparenchyma

Conglutinate pseudoparenchyma are built up by
either periclinally or anticlinally arranged fungal
elements which often secrete very large amounts
of glucans, e.g., of the lichenin type (1-3,1-4-β-
glucans) or of the pustulan type (linear 1-6-β-
glucans, first isolated from *Lasallia pustulata*,
Fig. 7a–i; review: Stone and Clarke 1992). The
anatomy and cell shapes in conglutinate pseudoparenchyma are best seen after removal of
the mucilaginous components (Figs. 6d–g, 9e–f)
with the maceration technique as described by
Anglesea et al. (1982). Conglutinate thalline areas
in the form of peripheral cortical layers and/or
central strands provide mechanical stability,
tensile strength and elasticity to the thallus (Figs.
4b,c; 5a,h; 6a–g). They are involved in passive
water uptake and retention. Peripheral cortical
layers, as found in the majority of foliose and fruticose lichens, are translucent and elastic when
wet, but opaque and more or less brittle when dry.

The peripheral cortex plays a key role in
thalline water relations, light absorbance and
transmission, and gas exchange. Many taxa differentiate an upper and lower cortex which may
differ considerably in anatomy, dimensions and
coloration (Figs. 5h, 7a–f). Light transmission to
the algal layer is influenced by the level of hydration of the cortex and by deposits upon and with-

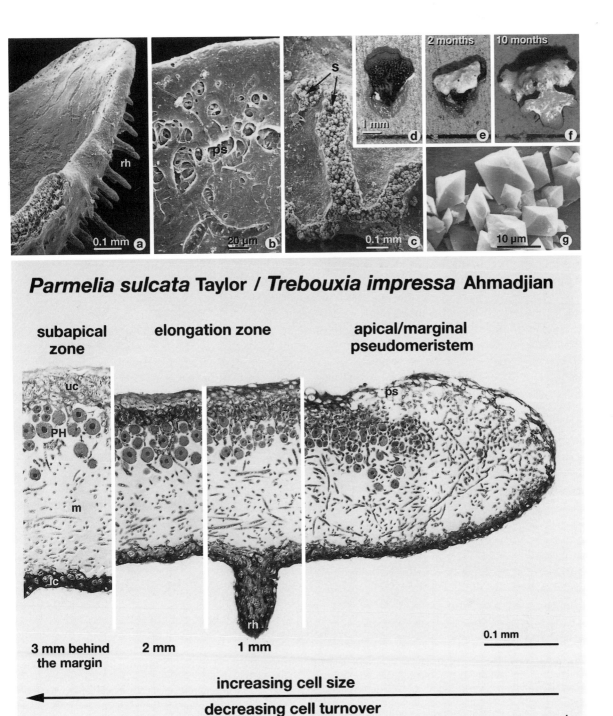

Parmelia sulcata Taylor / *Trebouxia impressa* Ahmadjian

subapical zone	elongation zone	apical/marginal pseudomeristem

3 mm behind the margin   **2 mm**   **1 mm**   0.1 mm

increasing cell size

decreasing cell turnover

h

Fig. 5a–h. *Parmelia sulcata* Taylor (Lecanorales)/*Trebouxia impressa* Ahmadjian, example of a dorsiventrally organised, foliose lichen with marginal/apical growth. **a** Lobe margin with rhizinae developing on the darkly pigmented lower cortex. **b** Detail of the marginal zone with aeration pores (pseudocyphellae) in the conglutinate cortical layer. **c** Marginal and laminal pseudocyphellae of mature thallus areas turn into soralia and release masses of soredia (symbiotic propagules). **d–f** A marginal lobe, fixed to a ceramic substratum in an upside-down position (**d**), incubated outdoors and photographed after 2 (**e**) or 10 months (**f**), respectively. The fungal partner corrected this spatial disturbance by means of growth processes and thus secured the adequate illumination of the photobiont. (After Honegger 1996a). **g** Calcium oxalate crystals (Wedellite) forming a pruina on the thallus surface. **h** Cross-sections along a gradient from the growing margin to the fully elongated, non-growing subapical zone. The photobiont cell population (PH) amounts for less than 20% of thalline biomass (Fiechter 1990); it is actively positioned by the fungal partner underneath the conglutinate upper cortex at the periphery of the gas-filled medullary layer. A large proportion of oversized photobiont cells in the subapical area has exceeded the size required for autospore formation without undergoing mitosis. The dark spot in the centre of *Trebouxia* cells is the pyrenoid of the large, lobate chloroplast (see Fig. 10i); the nucleus is laterally positioned. Abbreviations: *PH* photobiont; *lc* conglutinate lower cortex; *m* medullary layer; *ps* pseudocyphella (aeration pore); *rh* rhizinae; *s* soralia; *uc* conglutinate upper cortex

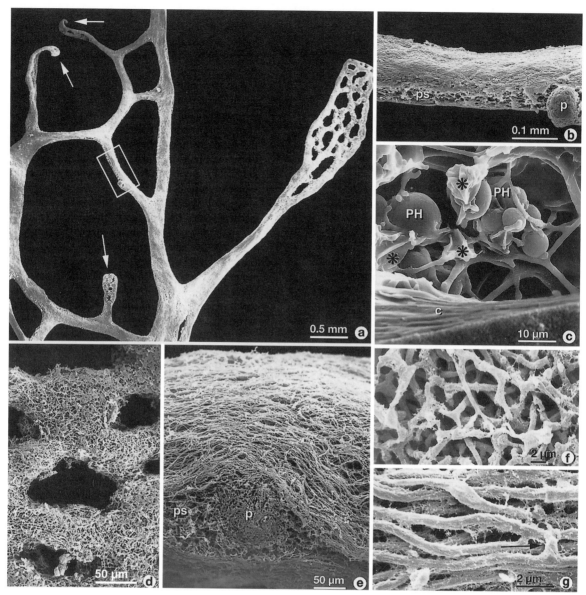

Fig. 6a–g. *Ramalina menziesii* Taylor (Lecanorales)/*Trebouxia* sp., the pendulous, epiphytic lace lichen of the coastal Northwest of North America, fastest-growing lichen due to combined marginal/apical and intercalary growth (Sanders 1992). Each lobe has a usually inrolled apical pseudomeristem and differentiates perforations within the subapical elongation zone. With increasing age the mesh sizes dramatically increase due to continuous elongation, combined with intercalary growth. **a** Detail of the old part of the reticulate thallus with large mesh sizes. *Arrows* point to juvenile lobes differentiating on old meshes. **b** Detail of the boxed area in **a**, with a primordial stage of a new lobe. It differentiates within a pseudocyphella which extends laterally along the flattened mesh. The whole thallus derives its remarkable tensile strength from the prominent conglutinate cortex. **c** LTSEM of the gas-filled interior in a freeze-fractured, old mesh, showing the photobiont layer with freshly divided and mature *Trebouxia* cells. *Asterisks* refer to old, degenerate algal cells. **d–g** Hyphal growth patterns in the conglutinate cortical layer after removal of the gelatinous extracellular material by the maceration technique as described by Anglesea et al. (1982). **d,f** Reticulate hyphal growth pattern near the apical pseudomeristem. **e,g** Parallel-periclinal arrangement of elongated hyphae in an old mesh. A different growth pattern is observed in the lateral pseudocyphella and in the primordium of a young lobe. The *arrow* in **g** points to a hyphal anastomosis. Abbreviations: *PH* photobiont; *c* conglutinate cortex; *p* primordial stage of a new lobe; *ps* pseudocyphella: aeration pore, built up by aerial hyphae with a water-repellent wall surface

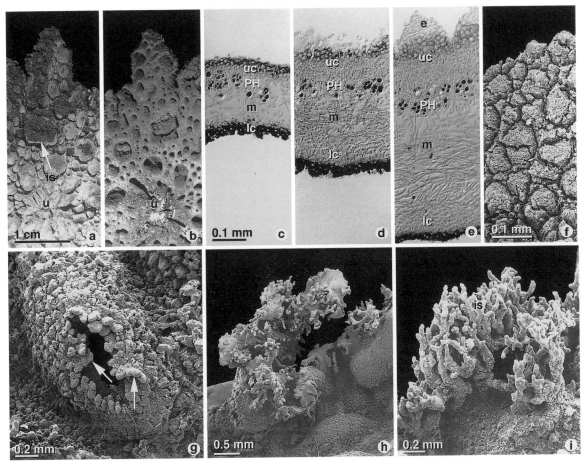

Fig. 7a–i. *Lasallia pustulata* (L.) Mérat (Lecanorales)/*Trebouxia* sp., example of a foliose, umbilicate lichen with irregular-patchy intercalary growth, regenerative growth along wound margins, and isidium formation. **a,b** Laminal view of the upper (**a**) and lower thallus surface (**b**) with central holdfast (umbilicus). Black, coralloid isidia cover the pustules and thallus surface near the fuzzy, eroded margin. **c–d** Cross-sections of thalli of different size (expressed as Ø) and age (same magnification): **c** approx. 5 mm Ø, **d** approx. 2 cm Ø, **e** approx. 5.5 mm Ø, showing increase in thallus thickness with increasing age and the formation of an epinecral layer, composed mainly of walls of walls of dead cells, on the upper cortex. **f** SEM of the epinecral layer. **g** Regeneration of new lobules (*arrows*) along wound margins of an eroded pustule (wind abrasion). **h–i** Differentiation of coralloid isidia on regenerative lobules along wound margins. Abbreviations: *PH* photobiont; *e* epinecral layer; *is* isidium; *lc* lower cortex; *m* medullary layer; *u* umbilicus; *uc* upper cortex

in it. Mycobiont-derived secondary metabolites, many of them brightly coloured, may crystallise within the gelatinous matrix of the cortex or on its surface (reviews: Fahselt 1994; Figs. 9g–h) and thus provide the thallus with its species- or genus-specific coloration. Examples are the bright yellow to orange anthraquinones of *Xanthoria* spp. (Teloschistales), nitrophilous saxicolous and epiphytic species from coastal to alpine habitats whose conspicuous golden yellow to orange tinge often refers to sites which are regularly visited by wild or domestic animals, or vulpinic acid which gives *Letharia vulpina* (L.) Hue, the epiphytic wolf lichen of pine and larch forests, an almost fluorescent sulphur-yellow coloration. Some of these secondary compounds have been shown to absorb, transform or reflect light, others reveal antibiotic or antifeedant properties or form insoluble complexes with metals (review: Fahselt 1994). Epicortical pruina, as formed either by accumulation of dead cortical and rarely algal cells (Valladares 1994; Fig. 7e–f), or, especially under arid conditions, by oxalate crystals (Fig. 5g), or a combination of both, influence light absorbance, water relations and palatibility (review: Büdel and Scheidegger 1996).

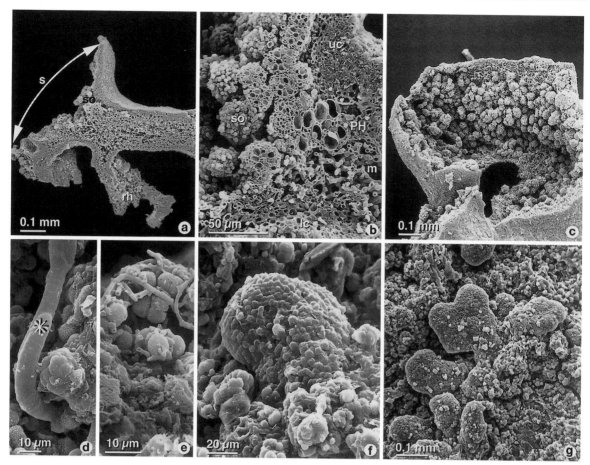

Fig. 8a–h. Reproductive strategies in lichen-forming ascomycetes. **a–c** Formation of soredia (symbiotic propagules) in *Xanthoria fallax* (Hepp) Arnold (Teloschistales)/*Trebouxia* sp. **a** Lateral view of a cross-sectioned thallus lobe which has terminated its growth and differentiates soredia in a marginal soralium. **b** Detail of the transition zone between the soralium and the thallus. Photobiont cells of the algal layer become enveloped by fungal hyphae and undergo autospore formation; thus each soredium contains numerous freshly divided algal cells. The upper cortex is bent backwards due to desiccation (→ hygroscopic movement). **c** Frontal view of the soralium. Upon landing on a suitable substratum, each soredium may form a new thallus; juvenile stages are indistinguishable from those of *X. parietina* (see **e,f**). **d,e** Relichenisation and thallus ontogeny in the sexually reproducing *X. parietina* (L.) Th.Fr. (Teloschistales)/*Trebouxia arboricola* Puymaly. **d** Aerophilic algal community on bark next to adult thalli and primordial stages of *X. parietina*. The most common aerophilic green algae are not compatible partners of lichen-forming fungi, and lichen photobionts are rare outside lichen thalli. The *asterisk* refers to a *Scytonema* colony. **f** Compatible algal cells are contacted by germ tubes of *X. parietina*. **g** Thallus primordium with a peripheral fungal cortex, a growing apical/marginal rim and a non-growing basal zone. **h** Fusion of vicinal, possibly genetically heterogenous thallus primordia. Together, they will form a thallus rosette of astonishing symmetry. Abbreviations: *PH* photobiont; *lc* lower cortex; *m* medullary layer; *rh* rhizina; *s* soralium; *so* soredium; *uc* upper cortex

A massive, uniformly thick peripheral cortex, as observed in a wide range of species, impedes gas exchange. Thus aeration pores such as pseudocyphellae (Figs. 5b,6b) or anatomically more complex cyphellae (examples in Chap. 11, Vol. VA, this Series) are differentiated. Both are built up by hyphae with hydrophobic wall surfaces which, in contrast to the surrounding cortical hyphae, are unwettable and thus facilitate the gas exchange in water-saturated thalli. In mature thalline areas of sorediate species the pseudocyphellae often develop into soralia (Fig. 5b,c). Single cortical cells may switch to filamentous growth and thus contribute to a tomentum; others are involved in the capture of nitrogen-fixing cyanobacteria (Fig. 9a–d), prerequisite for the differentiation of cephalodia (see below). Groups of cortical cells may differentiate into simple or branched rhizinae

(Fig. 5a,h) which serve as holdfasts and facilitate the water uptake from the substratum.

By far not all foliose and fruticose lichen thalli are delimited by a massive cortex, and thus may have fine pores in their discontinuously thick cortical layer. A very interesting situation occurs in the window lichens (Fensterflechten sensu Vogel 1955) whose juvenile, squamulose thalli have a more or less continuous upper cortex and algal layer. With increasing age and thus thallus thickness, conglutinate, anticlinally growing hyphae form massive cones between which the colonies of algal cells are anticlinally arranged (Lumbsch and Kothe 1993). In the hydrated state the conglutinate fungal cones are highly translucent and thus act as light guides or microwindows. This phenomenon was first observed in the terrestrial, xerophilic *Eremastrella tobleri* Vogel in South Africa. The functional analogy of *Eremastrella* spp. to the window plants among the Aizoaceae (genera *Lithops*, *Fenestraria*) and Liliaceae (*Haworthia*) with succulent, truncate leaves and a translucent water storing tissue as a light guide is obvious (Vogel 1955). The same functional principle, i.e., cone-like or band-shaped, conglutinate pseudoparenchyma as microwindows between anticlinally arranged groups of photobiont cells occurs also in lichens of temperate regions. Examples are the crustose, epiphytic *Labyrintha implexa* Malcolm, Elix & Owe-Larsson (Malcolm 1995; Malcolm et al. 1995), or the foliose, terrestrial *Solorina crocea* (L.) Ach. (Fig. 9i–j).

Correlations were found between cortex structure and microclimatic preferences. Granitic boulders are often colonized by *Umbilicaria* and/or *Lasallia* spp., the so-called rock tripes: foliose, umbilicate lichens with a central holdfast and with a peculiar growth pattern (see below; Fig. 7a–f). The top of the boulder may be colonized by species other than those found on the flanks. In eastern Canadian populations this niche segregation was shown to correlate with anatomical differences and thus with different modes of passive water uptake. *Umbilicaria papulosa* (Ach.) Nyl. is wetted most rapidly by large raindrops, which are readily absorbed by the pruinose epicortical layer. *Umbilicaria mühlenbergii* (Ach.) Tuck. absorbs water most readily from runoff and mist-like precipitation via the massive lower cortex, its upper cortex revealing an astonishingly high level of water repellency. *U. papulosa* colonizes the top of boulders and reaches highest productivity in late spring to fall, i.e., the months with rains and thunderstorms, whilst *U. mühlenbergii* settles on the flanks and reveals high productivity from March to December (Larson 1984).

Hygroscopic movements of marginal lobes due to desiccation and rehydration of the cortex during periodic de- and rehydration processes are commonly observed in foliose lichens. In thalli which are attached to the substratum only the ends of peripheral lobes bend backwards during desiccation. Marginal soralia of sorediate species bend backwards upon drying and thus expose their soredia to the wind or to invertebrate vectors (Jahns et al. 1976; Fig. 8a–c). A wide range of taxonomically diverse foliose or fruticose terrestrial lichens with no firm connection to the ground are completely inrolling during drought stress episodes. In this form they can be transported by wind over considerable distances. While rolling over the ground, the brittle thalli are partially fragmented; thus the symbiotic phenotype is successfully dispersed. Such vagrant or erratic lichens (Wanderflechten) are found on all continents in various ecosystems from prairie, steppe and deserts to windswept arctic and alpine sites, with the highest species diversity being recorded in South Africa (review: Rosentreter 1993; Pérez 1997a,b). A large number of morphologically less advanced vagrant lichens do not roll and unroll during drought stress and rehydration events, but instead reveal a more or less globose shape. It was repeatedly suggested that the sweet and nutritious manna of the Israelites might have been *Sphaerothallia esculenta*, a globose lichen which grows attached to rocks but becomes vagrant upon detachment. It is grazed by wild and domestic animals and was consumed by humans in various regions of the Middle East as an addition to flour. However, nowadays *S. esculenta* does not occur on the Sinai Peninsula, and due to its high content in oxalic acid, it is neither a mere delicacy nor suited for human consumption as the sole food over prolonged periods of time. Other potential candidates are *Collema* spp. (gelatinous lichens) which are often overlooked in the desiccated state but very conspicuous upon rehydration by dew (Fig. 3c–e). However, these lichen taxa are not especially sweet-tasting. Therefore ethnobotanists assume sweet gums of trees and shrubs, as commonly found and collected in this area, possibly in combination with lichens, to have been the main components of the biblical manna.

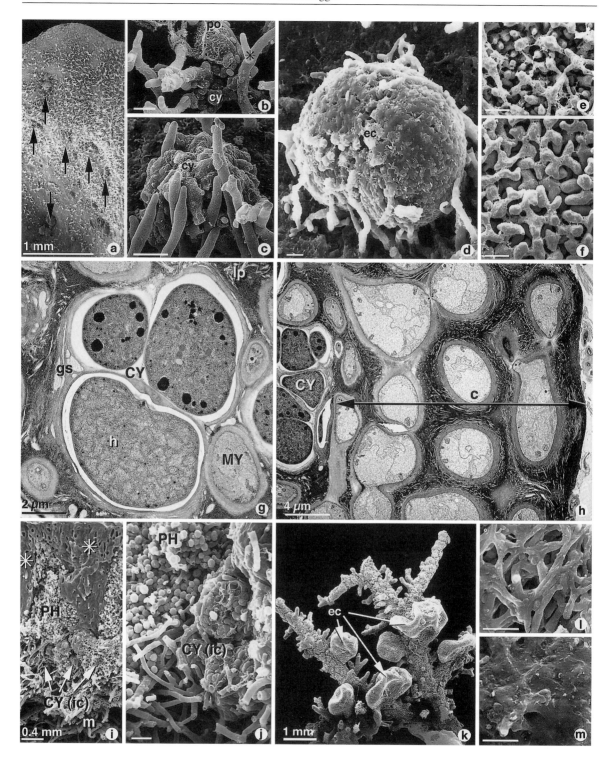

2. Gas-Filled Plectenchyma of the Medullary and Algal Layers

Lichen-forming ascomycetes with morphologically advanced symbiotic phenotype differentiate a system of aerial hyphae with hydrophobic wall surfaces which fulfill functions completely different from the conglutinate, hydrophilic pseudoparenchyma. Their prime function seems to involve the establishment of a functional interface with the cells of the photoautotrophic partner. This includes the differentiation of appressorial or haustorial structures, the short-distance shifting of the photobiont cells into an optimal position within the algal layer by means of growth processes, and the hydrophobic sealing of the apoplastic continuum between the partners. Thus, the routes for passive, apoplastic translocation of solutes from the photobiont to the mycobiont and from the algal layer to the thalline surface and vice versa are established. At the same time the aerial hyphae facilitate the gas exchange of the photoautotrophic partner by creating a gas-filled intrathalline space.

In many lichens the wall surface hydrophobicity of the gas-filled plectenchyma is so prominent that it is very difficult to infiltrate thallus fragments with aqueous solutions such as fixatives

for conventional preparative procedures for electron microscopy. The chemical nature of this wall surface hydrophobicity has been discussed. Mycobiont-derived secondary metabolites were assumed to be prime candidates (Armaleo 1993). These often crystallize in large quantities on medullary hyphae and even on the photobiont cells (Honegger 1986b; Fig. 10h). In contrast to secondary metabolites of non-lichenized fungi, these mainly polyphenolic lichen compounds are almost insoluble in aqueous systems once they are crystallized (Fahselt 1994). However, a large number of lichen-forming ascomycetes produce no medullary secondary metabolites; yet no free water was found on their wall surfaces in LTSEM studies of fully hydrated, freeze-fractured specimens (Fig. 6c; Honegger and Peter 1994; Scheidegger 1994; Honegger 1995; Scheidegger et al. 1995; Honegger et al. 1996b). A mycobiont-derived, proteinaceous wall surface layer covers the medullary hyphae (Honegger 1982) and spreads over the photobiont cells at the immediate contact site (Honegger 1984, 1985, 1991, 1996a). This mycobiont-derived proteinaceous coat appears as a rodlet pattern (Fig. 10d) on walls of young, growing hyphae and photobiont cells, but may achieve, upon coverage by phenolic and lipidic compounds, an irregularly tessellated

---

Fig. 9a–m. Cephalodiate lichens: lichen-forming ascomycetes which associate with green algal taxa (primary photobiont, donor of fixed carbon) and acquire diazotrophic cyanobacteriae as providers of fixed nitrogen. These cyanobacterial partners are housed in external or internal cephalodia. **a–h** *Peltigera aphthosa* (L.) Willd. (Peltigerales)/*Coccomyxa* sp., acquisition of *Nostoc punctiforme* (Kütz.) Hariot and differentiation of corticate, external cephalodia; these can be recognized as *black dots* (*arrows* in **a**) on the surface of the bright green, foliose thalli, **b,c** Hair-like fungal protrusions of the cortical layer near the lobe margin capture cyanobacterial colonies. The compatible cyanobiont, but not pollen grains (as in **b**) or other epibionts, trigger branching of the hairs (*asterisk* in **b**) and subsequent differentiation of a compact, conglutinate cortical layer around the cyanobacterial colony (**d**). The cephalodial cortex (**f**) differs anatomically from the thalline cortex (**e**), as seen after removal of the gelatinous extracellular material. Under the microaerobic conditions within the cephalodium a high percentage of heterocysts differentiates among the cyanobacterial cells (**g**). **h** Mycobiont-derived secondary metabolites crystallise within the gelatinous extracellular material of the cephalodial cortex. **i,j** *Solorina crocea* (L.) Ach. (Peltigerales)/*Coccomyxa* sp., a terrestrial, arctic-alpine lichen growing on late snowbeds, stonefields, windswept ridges etc. Cross-section of the foliose thallus; the green algal photobiont is located in islets between massive cones of the fungal cortex (*asterisks*) which is built up by anticlinally growing hyphae. In the fully hydrated state these cortical cones act as light guides. Colonies of *Nostoc* sp. are kept in a more or less continuous layer underneath the cortex and green algal cells (see text). The medullary layer is bright orange from mycobiont-derived solorinic acid (anthraquinone) which crystallises on the hyphal surfaces. **k–m** *Stereocaulon ramulosum* (Sw.) Räuschel (Lecanorales)/*Pseudochlorella* sp., a successful coloniser of disturbed habitats such as lava fields, forms fruticose thalli with a conglutinate central axis (not visible). The green algal photobiont is kept near the thallus periphery underneath a loosely interwoven plectenchyma which facilitates gas exchange (surface view in **l**). Cyanobacterial colonies (either *Gloeocapsa*, *Scytonema*, *Stigonema* or *Nostoc* sp.) are incorporated in conspicuous, sac-like external cephalodia with conglutinate cortex (surface view in **m**). Abbreviations: *CY* cyanobacterial photobiont; *MY* mycobiont; *PH* green algal photobiont; *c* cephalodial cortex; *cy* free-living cyanobacterial colonies; *ec* external cephalodium; *gs* gelatinous sheath of the cyanobacterial photobiont; *h* heterocyst; *ic* internal cephalodium; *lp* crystalline lichen products (= mycobiont-derived secondary metabolites); *m* medullary layer; *po* pollen grain. *Bars* 10 μm unless otherwise stated

pattern (Fig. 10e,f). The structural and functional similarity of this rodlet layer with hydrophobins of non-lichenized fungi is obvious; these were shown to play essential roles in the establishment of mutualistic and pathogenic interactions of fungi with plant and animal hosts (reviews: Honegger 1997; Wösten and Wessels 1997). A hydrophobin protein from the symbiotic phenotype of *Xanthoria parietina* (L.) Th.Fr., a species with no medullary lichen products but with considerable wall surface hydrophobicity (Honegger and Peter 1994; Honegger et al. 1996b), was successfully isolated and characterized (Scherrer et al. 2000). The isolated protein revealed interfacial self-assembly, a characteristic feature of class 1 hydrophobins (review: Wösten and Wessels 1997), into a rodlet layer with structural similarity to the rodlet layer at the mycobiont-photobiont interface of the lichen (Scherrer et al. 2000).

Proteinaceous, hydrophobic wall surface layers of fungal origin with rodlet or irregularly tessellated pattern were so far found in a wide range of lichen-forming ascomycetes. Medullary secondary metabolites of the fungal partner may crystallize on and within this hydrophobic proteinaceous coat and thus significantly enhance its hydrophobicity (Fig. 10g,h).

B. Ontogeny and Growth

Every stable symbiosis of physically closely connected partners raises questions about the establishment of the symbiotic state, about the coordinated growth of the partners and the regulatory mechanisms of pattern formation in the symbiotic phenotype. These aspects have been rather superficially studied in lichens, and in a descriptive rather than experimental mode. The main problem is the near impossibility of growing synchronous resyntheses under axenic culturing conditions.

1. Acquisition of a Compatible Photobiont

The majority of lichen-forming ascomycetes reproduce sexually and thus have to reestablish the symbiotic state after each reproductive cycle, but very little is known about how and how often this proceeds in nature. Only a few lichenized ascomycetes keep photobiont cells in their ascomata and disperse ascospores together with the compatible symbiont. Examples are hymenial photobionts in the gelatinous perithecial matrix of some Verrucariales, and epithecial photobionts on the ascomata of some foliicolous tropical lichens (review: Honegger 1993).

Morphologically advanced lichen-forming fungi are highly discriminative in their selection of algal or cyanobacterial partners. As the most common photobiont species are not abundant in the aposymbiotic state, finding a compatible symbiont seems to be a major problem for all other lichen-forming fungi which form sexual or asexual, aposymbiotic propagules (Tschermak-Woess 1988; Fig. 8d). Acquisition of compatible photo-

Fig. 10a–i. Mycobiont-photobiont relationships: haustorial types in Lecanorales with green algal photobionts of the genus *Trebouxia* (not identified at the species level). **a** *Buellia punctata* (Hoffm.) Massal., epiphytic crustose species with no internal stratification. The cellulosic cell wall of the photobiont is pierced by finger-shaped, transparietal fungal haustoria. **b** *Cladonia arbuscula* (Wallr.) Flotow, fruticose, erect reindeer lichen with intraparietal haustoria; these do not pierce the algal cell wall. (After Honegger 1996b). **c** *Cladonia caespiticia* (Pers.) Flörke: intraparietal haustoria are ensheathed by the modified algal wall. The mycobiont-derived hydrophobic wall surface layer spreads from the fungal over the algal wall surface (see **f**). **d,f** Freeze-etch replicas of the mycobiont-photobiont interface in *Cladonia macrophylla* (Schaerer) Stenhammar. **d** A thin, semicrystalline rodlet layer covers the wall surface of young, growing hyphae; it spreads over young autospores of the photobiont. **e** The rodlet layer becomes obscured by phenolic and lipidic compounds and thus achieves an irregularly tessellated pattern. **f** The irregularly tessellated wall surface layer of fungal origin seals the apoplastic continuum between the fungal and algal partner as a very hydrophobic coat; see text for explanations. **g–i** Mycobiont-photobiont interface in Parmeliaceae (approx. 2300 spp.). **g** *Cetrelia olivetorum* (Nyl.) Culb. & C. Culb. SEM of a conventionally fixed, dehydrated and critical point-dried specimen. Mycobiont-derived crystalline secondary metabolites were dissolved during the preparative procedure, leaving needle-shaped impressions (*asterisks*) in the mycobiont-derived, proteinaceous wall surface layer on fungal and algal cells; in freeze-etch preparations this proteinaceous coat reveals the same irregularly tessellated pattern as in **e–f**. **h–i** *Parmelia tiliacea* (Hoffm.) Ach. **h** SEM of a chemically untreated, freeze-dried specimen. Fungal and algal wall surfaces in the gas-filled thalline interior are covered by crystals of mycobiont-derived secondary metabolites. **i** A very short, intraparietal fungal haustorium protrudes into the algal cell wall. Abbreviations: *MY* mycobiont; *PH* photobiont; *h* transparietal haustorium; *ip* intraparietal haustorium; *lp* crystalline lichen products (= mycobiont-derived secondary metabolites); *n* (fungal) nucleus; *py* pyrenoid; *sl* mycobiont-derived, hydrophobic wall surface layer

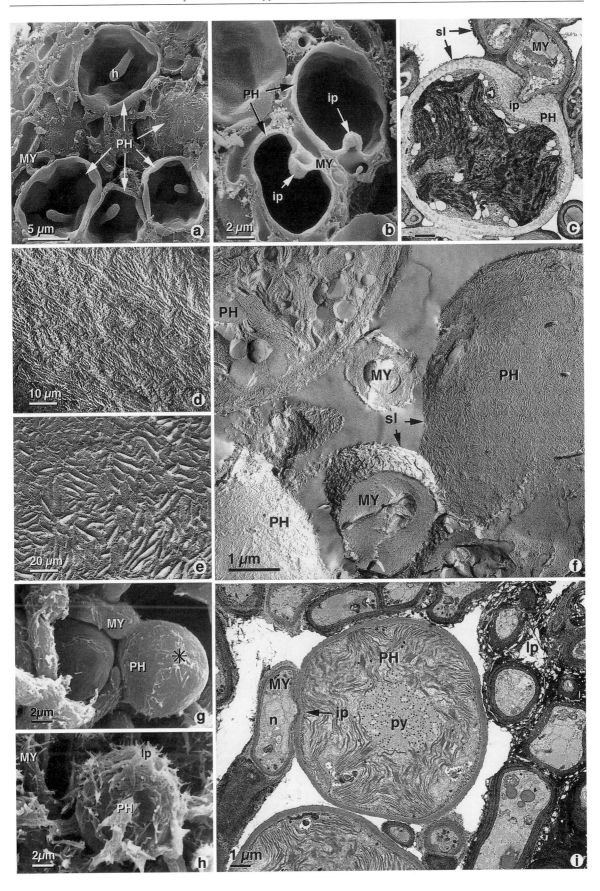

bionts by theft was observed in various, but by far not all, parasitic lichens (also termed lichenicolous lichens) which use other species not only as a suitable substratum, but also as the source of compatible photobiont cells. Germ tubes of these mainly crustose species selectively invade the thalli of other taxa and establish a symbiosis with their photobionts (Poelt and Doppelbauer 1956; Friedl 1987). Reports on such "cleptobiosis" in morphologically advanced taxa such as the foliose *Xanthoria parietina* (L.) Th.Fr. (Teloschistales), a very widespread, sexually reproducing species whose germ tubes were assumed to colonize the photobiont cells (*Trebouxia* sp.) contained in prethalli, thalli or symbiotic propagules of adjacent *Physcia* spp. (Ott 1987a,b) were not supported by molecular analyses (Beck et al. 1998). *X. parietina* was shown to disperse very efficiently by means of thallus fragmentation (Honegger 1996d; Honegger et al. 1996a). Thallus fragmentation is likely to be the most common and successful mode of lichen dispersal; examples are the reindeer lichens of the genus *Cladina*, which are very brittle when dry, and thus are easily fragmented upon trampling. A large number of lichen-forming fungi disperse efficiently by means of symbiotic propagules such as soredia (Figs. 5c, 8a–c) or isidia (Figs. 3e, 7a,h,i). These, together with thallus fragments, enable lichen-forming fungi to invade large areas regardless of the availability of extraneous compatible photobiont cells.

Laboratory experiments and observations in the natural habitat indicate that most lichen-forming ascomycetes with morphologically advanced symbiotic phenotype are moderately specific to specific with regards to their photobiont, i.e., accept either several closely related or only one species, but are highly selective (reviews: Bubrick et al. 1985; Honegger 1996c). A specific algal-binding protein was isolated from homogenates of lichen thalli and from the aposymbiotically cultured mycobiont of *Xanthoria parietina* (Bubrick and Galun 1980; Hersoug 1983) and was immunocytochemically localized on the surfaces of germ tube walls (Bubrick et al. 1981). Lectins with a specific affinity to compounds of the gelatinous sheath of cyanobacterial photobionts were isolated from thallus homogenates of lichens with cyanobacterial photobiont (Lallemant and Savoye 1985; Kardish et al. 1991). Recognition between bionts is likely to be a multistep process. The initial contact of mycobiont hyphae and compatible or incompatible photobiont cells is probably a fairly unspecific event.

2. Prethallus Formation

Freshly resynthesized associations (both natural and artificial) and most symbiotic propagules such as soredia and non-stratified isidia first develop a crustose, non-corticate and internally non-stratified prethallus in which the photobiont cells are more or less densely enveloped by mycobiont hyphae (Fig. 8e; Schuster 1985; Schuster et al. 1985; Honegger 1993, 1996c). One or several germ tubes may be involved in prethallus formation (reviews: Fahselt 1996; Jahns 1988). Prethalli can be formed either by single or groups of soredia or by the fusion of neighbouring prethalli. Free algal cells can be incorporated into the developing prethallus (Schuster 1985; Schuster et al. 1985). The fungal and algal cells of resulting lichen thalli are thus not necessarily genetically homogenous (review: Fahselt 1996). Some lichen-forming ascomycetes produce internally stratified, polarized symbiotic propagules such as phyllidia, blastidia or dorsiventrally organized isidia (Honegger 1987, 1997). These are, in fact, minute thalli which, upon detachment from the mother thallus and landing on an appropriate substratum, may grow into a mature thallus.

3. Differentiation of a Stratified Thallus

Under appropriate microclimatic conditions, the prethallus of compatible associations develops into a mature, stratified thallus with its species-specific morphology, anatomy and metabolism. The first recognizable event in thallus differentiation is the secretion of mucilaginous material by peripheral fungal cells. This mucilaginous secretion, followed by a changing growth pattern of peripheral hyphae, results in the differentiation of the peripheral, pseudoparenchymatous cortex (review: Honegger 1993; Fig. 8f). Internal stratification and polarization of the thalline primordium occur simultaneously with the differentiation of the peripheral cortex. In the first recognizable primordial stages of juvenile thalli the non-growing basal and the growing apical poles are already determined (Fig. 8f). Potentially genetically heterogenous, vicinal primordial stages may fuse, as was observed under natural (reviews: Honegger 1993, 1996c; Fig. 8g) and artificial conditions (Honegger 1996d, 1998). Mature thallus rosettes appear as an operational entity, but may reveal an astonishing level of genetic complexity (review: Fahselt 1996).

C. Growth and Cell Turnover

1. Growth Rates

Lichens are often referred to as extremely slow-growing and long-living microorganisms. Lichens of extreme climates such as desert, arctic-antarctic or alpine ecosystems achieve full metabolic activity only during very short periods per year and survive most of the time in a state of dormancy (reviews: Kappen 1988, 1993). Consequently only very low cell turnover rates and minimal annual size increases (usually <1 mm year,) are recorded. Some epi- and endolithic crustose species are assumed to live for centuries; the most extreme estimates are in the range of 9000 years. However, lichens of temperate, subtropic or tropic climates have annual growth rates in the range of millimetres to centimetres. The highest growth rates are achieved in moist, coastal-influenced regions by species with combined apical-marginal and intercalary growth such as the lace lichen and lungwort of the Pacific Northwest of North America (see below). Various inhabitants of disturbed sites are rather short-lived, their life span being restricted to a few months to a few years (Poelt and Vezda 1990; Jahns and Ott 1997).

Longevity of lichens per se requires critical interpretation. Pseudomeristematic zones of macrolichens persist in time as do meristems of perennial plants. However, in old trees, no metabolically active cell is centuries old; the same applies for old lichen thalli. Their senescent areas are either overgrown, or disintegrating and breaking off, the remaining thallus becomes ring-shaped unless it is able to regenerate new lobes along the wound margins (Honegger et al. 1996a). In arctic climates it may take centuries to achieve a wide ring diameter, but only the youngest portions, i.e., the pseudomeristematic marginal rim and adjacent zones, a few decades old, are retained.

Lichen growth measurements were used for lichenometric dating of rock surfaces in arctic-alpine ecosystems or at archaeological sites (reviews: Innes 1988; Winchester 1988). The growth rates of individual lobes were studied either in situ, or after transplantation and/or other manipulations such as partial painting or removal of vicinal lobes (review: Honegger 1993). Marginal lobe growth was shown to correlate with productive capacity, i.e., contents in photobiont- and mycobiont-derived mobile carbohydrates (polyols; Armstrong and Smith 1993, 1994). The maintenance of symmetry, i.e., circularity in foliose thalli, is a particularly interesting aspect as the lobes of a thalline rosette do not grow equally fast, are metabolically unequally productive and are not necessarily genetically uniform (see above). From their analysis of growth rates and carbohydrate contents of individual lobes of a dorsiventrally organized, foliose *Parmelia* sp. over a 22-month period, Armstrong and Smith (1992) concluded that the maintenance of thallus symmetry is the result of equalizing fluctuations among lobes with unequal growth rates and productive capacities.

2. Growth Patterns

Lichen growth rates have received considerable attention, but only a few detailed analyses of growth patterns have so far been carried out. Based on morphological and physiological studies, three different types of growth can be distinguished in foliose and fruticose lichens (reviews: Honegger 1993, 1996c):

1. *Predominantly marginal/apical growth* (Fig. 5a–h), with a marginal/apical pseudomeristem with relatively small mycobiont and photobiont cells and high cell division rates in both and with an adjacent elongation zone in which both fungal and algal cells undergo a limited cell turnover prior to reaching their mature dimensions. With increasing distance from the marginal/apical pseudomeristem an increasing number of oversized photobiont cells with an arrested cell cycle are observed (Hill 1985, 1989); these exceed the minimum size required for autospore formation without undergoing mitosis. This growth type is very widespread among foliose taxa (especially Parmeliaceae) which fix themselves to the substratum by means of rhizinae (e.g., *Parmelia sulcata* Taylor; Fig. 5a–h) and in various fruticose taxa.

Ascomata or asexual symbiotic propagules such as isidia or soredia are produced either in subapical thalline areas with no impact on the growth pattern of the lobe or branch (Fig. 5c), or in the marginal/apical pseudomeristem of lobes or branches, possibly throughout the entire thallus, which terminate their growth and eventually die, a pattern analogous to semelparous plants upon flowering (Fig. 8a–c).

2. *Combined marginal/apical and intercalary growth*, with a marginal/apical pseudomeristem, an adjacent elongation zone and with considerable intercalary growth. The fastest-growing lichen so far described, the reticulate, pendulous lace lichen or fish net of the coastal northwest of North America (*Ramalina menziesii* Taylor; Fig. 6a–g), with a linear size increase of up to 43 cm year^{-1}, follows this pattern (Sanders 1989, 1992; Sanders and Ascaso 1995), and so does the reticulate-tubular *Cladia retipora* (Labill.) Nyl., which has an impressive growth symmetry due to apical pseudomeristems which divide dichotomously in astonishing regularity (Honegger 1993). Combined apical/marginal and intercalary growth also occurs in the foliose lungworts such as *Lobaria oregana* (Tuck.) Müll. Arg. and *L. pulmonaria* (L.) Hoffm., punctually fixed, fast-growing lichens which may amount to an impressive biomass on branches or boles, e.g., in the deciduous rainforests of the Pacific North West (Honegger 1993).

3. *Irregular, patchy intercalary growth* as observed in the foliose, umbilicate thalli of Umbilicariaceae (Fig. 7a–i). These epilithic taxa fix themselves to the substratum by means of a central umbilicus. With increasing age, their margins, probably the oldest parts of the thallus, become fuzzy and eroded (Hestmark 1997). Considerable increase in thallus thickness with increasing age is another peculiarity of the Umbilicariaceae (Fig. 7c–e).

D. Cephalodiate Species and Photomorphs

Approximately 3–4% of lichen-forming ascomycetes are cephalodiate, i.e., incorporate a green algal primary photobiont in their algal layer as the donor of fixed carbon and acquire one or several species of nitrogen-fixing cyanobacteria as secondary photobionts; these provide the symbiotic system with fixed nitrogen. The taxonomic affiliation of cephalodiate lichens was reviewed by James and Henssen (1976). In most taxa these cyanobacterial partners are housed in gall-like structures either on the thallus surface (external cephalodia; Fig. 9a–f, k–m) or in its interior (internal cephalodia). Each cephalodium results from

the successful capture of a compatible cyanobacterial partner. Recognition and specificity in cephalodiate lichens is poorly understood. A remarkably high specificity was noted with molecular techniques in *Peltigera aphthosa* (L.) Willd. with *Nostoc punctiforme* (Kütz.) Hariot as the cyanobacterial partner in cephalodia (Paulsrud et al. 1998; Fig. 9a–f). The very conspicuous cephalodia of *Stereocaulon ramulosum* (Sw.) Räuschel (Fig. 9k–m) may contain taxonomically diverse cyanobacteria, representatives of the genera *Gloeocapsa*, *Scytonema*, *Stigonema* and *Nostoc* having been identified. Usually only one cyanobacterial partner is harboured per cephalodium, but already early lichenologists described cases where two species were contained in the same cephalodium (Fries 1866; Schwendener 1869). Similarly, the crustose-scaly *Placopsis* spp. harbour either *Scytonema*, *Stigonema* or *Nostoc* spp. in their cephalodia (Ott et al. 1977).

In only a few cephalodiate species has the process of cyanobacterial capture and incorporation been studied in detail; experimental investigations are missing. Hair-like fungal protrusions of either the upper [e.g., *P. aphthosa* (L. Willd.); Babikoff 1878] or the lower thallus surface, here in the form of a tomentum (e.g., *Lobaria* or *Nephroma* spp.; Jordan 1970; Jordan and Rickson 1971) or of loose hyphae [e.g., on the veins in *P. venosa* (L.) Hoffm. or *Solorina saccata* (L.) Ach.; Jahns et al. 1995] sense compatible cyanobacterial partners (Fig. 9a–c). Upon contact, they change their growth pattern, start branching, release mucilaginous material and ensheath the cyanobacterial colony (Fig. 9d,f,h). The process of internalization of cephalodia by means of fungal growth processes, as observed in various Stictaceae (*Lobaria*, *Nephroma*, *Sticta* spp. etc.) includes the ensheathment of compatible cyanobacteria by tomental hyphae, differentiation of a cortex around the cyanobacterial colony, and dissociation of the lower thalline cortex followed by positioning of the developing cephalodium within the medulla underneath the green algal layer (Jordan 1970; Jordan and Rickson 1971). The fungal partners differentiate a massive, conglutinate cortex around external cephalodia (Fig. 9d,g–h,k,m); the same applies for many, but not all internal cephalodia. The cephalodial cortex differs anatomically from the thalline cortex as seen after removal of the conglutinate extracellular material around cortical hyphae (Fig. 9e–f); thus the cyanobacterial partner triggers the expression of a different morphotype in the lichen-forming ascomycete than the green algal photobiont. This applies to a very high degree in morphologically dissimilar photomorphs of the same fungal partner in association with either a green alga or cyanobacterium (see below).

Under the microaerobic conditions within cephalodia the cyanobacterial colonies differentiate a high proportion of heterocysts (Fig. 9g). In a comparative study of young and mature cephalodia of *P. aphthosa*, the increasing heterocyst frequency correlated with an increasing level of nitrogen fixation (Englund 1977; review: Nash 1996). In this species new cyanobacterial colonies are captured in apical to subapical thalline areas; on older zones the hair-like fungal protrusions of the cortical surface break off (Fig. 9a). Young, grey to black cephalodia are thus situated near the lobe margin, mature, brownish ones with a very high level of nitrogen fixation in the subapical zone, and senescent, yellow to brown, less productive ones in the oldest central parts of the thallus.

An interesting situation occurs in the arctic-alpine, peltigeracean *Solorina crocea* (L.) Ach. which harbours a very large number of cyanobacterial colonies without special ensheathment in a somewhat discontinuous layer which resides underneath the green algal photobiont and the cortex (Fig. 9i–j). It remains to be seen whether this cyanobacterial layer results from multiple acquisitions near the growing thalline margin or whether *S. crocea* is capable of simultaneously controlling and maintaining two completely different photobionts within its thallus whose very interesting anatomy is described above (Sect. III.A.1).

Several species of cephalodiate lichen-forming ascomycetes among the Peltigerales are known to express two different symbiotic phenotypes in association with either the green alga as the primary photobiont and the cyanobacterium in cephalodia, or with the cyanobacterium as the only (photoautotrophic and diazotrophic) partner. Such cases have been referred to as lichen chimerae, morphotype pairs, photomorphs, photosymbiodemes, phototypes, etc. The different symbiotic phenotypes may be morphologically more or less identical except for their coloration: vivid green or greyish brown, respectively; such photomorphs were usually classified as different species of the same genus. Only with molecular techniques was it possible to ascertain that in some of these photomorphs the same fungal partner is involved

(Armaleo and Clerc 1990; Goffinet and Bayer 1997; Goffinet and Goward 1998). However, morphologically completely dissimilar phenotypes were classified in different genera. An example is the foliose *Lobaria amplissima* (Scop.) Forss., which has a green algal photobiont and coralloid cephalodia, and the independent coralloid *Dendriscocaulon umhausense* (Auersw.) Degelius, which harbours a cyanobacterial photobiont (review: James and Henssen 1976). The current dilemma of lichen taxonomists was summarized by Jörgensen (1998): the name of the lichen refers to the fungal partner but the species description is based on morphological and chemical features of its symbiotic phenotype; thus, two morphologically distinctly different lichens should carry the same species name. Which photomorph should be given nomenclatural priority? Similar problems had to be solved in non-lichenized fungi (anamorph-teleomorph relationships) and in various other groups of organisms (e.g., polypemedusae pairs).

IV. Mycobiont-Photobiont Interfaces

Correlations were found between the type of mycobiont-photobiont interaction, the taxonomic affiliation of the partners involved, the fine structure and composition of the photobiont cell wall, and the level of morphological complexity of the symbiotic phenotype (Tschermak 1941; Plessl 1963; Honegger 1984, 1986a, 1991, 1992).

Intragelatinous fungal protrusions within the gelatinous sheaths of cyanobacterial colonies are by far the most common interactions of lichen-forming ascomycetes with cyanobacterial photobionts (Fig. 9g; Honegger 1997). Wall-to-wall apposition is commonly found in crustose and foliose taxa with green algal photobionts of the genera *Coccomyxa* or *Elliptochloris*. These contain an enzymatically non-degradable, sporopollenin-like biopolymer located in a very thin, trilaminar outermost wall layer (Brunner and Honegger 1985; Honegger 1991). Resistant trilaminar sheaths of former mother cell walls persist and accumulate in the algal layer of lichen thalli (Honegger and Brunner 1981).

Finger-shaped, transparietal (intracellular) haustoria which penetrate the algal cell wall are typically formed by taxa with morphologically very simple crustose thalli and with green algal photobionts with cellulosic walls such as *Trentepohlia* or *Trebouxia* spp. (Tschermak 1941; Plessl 1963; Honegger 1986a; Matthews et al. 1989; Fig. 10a). However, by far not all lichen-forming fungi with crustose, non-stratified thallus and cellulosic photobiont cell walls differentiate such haustorial structures. Even among morphologically simple taxa intraparietal haustoria (see below) are widespread and quite common.

Two types of intraparietal haustoria were distinguished in ascomycetous lichens. The first is characterised by a short fungal infection peg which becomes ensheathed by the modified wall of the *Trebouxia* photobiont (Fig. 10b,c). This haustorial type was found in a range of Lecanorales with crustose-effigurate to squamulose thalli, especially in Cladoniaceae, reindeer lichens (*Cladina* spp.) included (Honegger 1986a). An even less prominent infection peg, barely recognizable unless sectioned in a median plane (Fig. 10i), is formed by foliose and fruticose Lecanorales (especially Parmeliaceae) and Teloschistales, which all have photobionts of the genus *Trebouxia*. This common and widespread type of intraparietal haustorium is of astonishing simplicity as compared to haustorial structures of other biotrophic fungi such as powdery mildews (Erysiphales; Giese et al. 1997), rusts (Mendgen 1997) or arbuscular mycorrhizal fungi (see Gianinazzi-Pearson et al., Chap. 1 this Vol). No attempts are made by these fungi to establish a large exchange surface in contact with the plasmamembrane of their photoautotrophic partner. Nutrients are mobilized within the apoplast. An analogous situation occurs in ectomycorrhizae (Debaud et al. 1997). With their pokilohydrous water relations lichen-forming fungi and their photoautotrophic partners are subjected to continuously fluctuating levels of hydration ranging between desiccation and water saturation. Thus, leachage products of drought-stressed mycobiont and photobiont cells are released into the apoplast. The hydrophobic sealing of the apoplastic continuum and thus of the routes of passive solute translocation between both partners of the lichen symbiosis by means of a mycobiont-derived, hydrophobic cell wall surface layer (see Sect. III.A.2) is thus of central importance for the functioning of the symbiotic relationship. The main fluxes of nutrients between the partners of the lichen symbiosis are assumed to occur during the periodic desiccation and dehydration cycles (Honegger 1997). Experimental data are missing, but haustorial complexes or intragelatinous fungal protrusions are likely, during periods of optimal thalline hydration, to be involved in active uptake of nutrients from the photoautotrophic partner. These may include soluble carbohydrates and degradation products of algal cell walls or gelatinous sheaths of cyanobacterial partners. Such wall or sheath polymers are available at the immediate contact site and upon autospore formation and degradation of mother cell walls (Honegger 1985).

V. Lichenicolous Fungi and Fungal Endophytes of Lichen Thalli

Approximately 100 species (300 genera) of lichenicolous fungi gain their nutrition from lichens (Hawksworth 1983, 1988b; Rambold and Triebel 1992). As experimental data are lacking, it is often difficult to interpret the biology of these multiple symbioses. Some lichenicolous fungi, probably very oligotrophic species, sporulate abundantly without causing major damage to the fungal or algal partner of the lichen; these are often referred to as parasymbionts or commensalists, while others may have a devastating effect. The majority of lichen-inhabiting fungi seems to be exclusively lichenicolous, but others are also known to be saprobes or pathogens of aerophilic algae or plants. A common and widespread example is the basidiomycete *Athelia arachnoidea* (Berk.) Jülich, a perennial and very destructive parasite of numerous lichen-forming ascomycetes and their photobionts, which was shown with molecular techniques to be the sexual state of *Rhizoctonia carotae* Rader, a postharvest disease of carrots (Adams and Kropp 1996). Some lichenicolous fungi infect a wide range of unrelated taxa, but the majority reveals a remarkable selectivity or even host specificity (Hawksworth 1983; Triebel and Rambold 1988; Rambold and Triebel 1992). Approximately 40 species of lichenicolous fungi are cecidogenous, i.e., induce very interesting gall formations on their host thalli (review: Hawksworth and Honegger 1994).

A wide range of inconspicuous, symptomless inhabitants of lichen thalli were discovered by means of culturing experiments only (Petrini et al. 1990; Girlanda et al. 1997); without immunological methods it is often impossible to recognize foreign hyphae within lichen thalli. Among these fungal endophytes are taxa which are also known as either saprobes, parasites or endophytes of higher plants, but a wide range of species has so far been found exclusively in lichen thalli (Petrini et al. 1990; Girlanda et al. 1997).

VI. Conclusions and Perspectives

With their numerical significance and ecological importance, their manifold adaptations to life under extreme conditions, their wealth in morphologically and functionally fascinating symbiotic phenotypes, lichen-forming ascomycetes and their photoautotrophic partners are intriguing for biologists with an interest in diversity of life styles and life strategies. Current opinions on the functioning of the symbiotic relationship in lichens are largely based on combined structural, physiological and biochemical investigations. However, lichen-forming fungi and their photobionts are definitely not the favourites of modern molecular biologists: they are by far too slow in their growth rate, too difficult to handle experimentally, and economically too insignificant. Accordingly, experimental lichenologists are more than ever a minority group among mycologists (review: Honegger 1998).

In recent years, outstanding contributions have been made by molecular systematicists on the taxonomy and phylogeny of lichen-forming fungi and their photoautotrophic partners, but the biology and genetics of the symbiotic relationship proper remain difficult to approach at the molecular level. Tools are available, but a suitable model system is so far lacking in which the differentiation of defined developmental stages of the symbiotic phenotype could be reproducibly and predictably induced in due time and analysed. Why should such investigations be of scientific interest? "The lichens... remain largely unexplored at the molecular level and yet may turn out to be the Holy Grail for understanding plant-fungus recognition phenomena" (Talbot 1998).

Acknowledgments. My sincere thanks are due to Jean-Jacques Pittet for invaluable help with the artwork, to Marcella Trembley for helpful comments on the manuscript, and to the Swiss National Science Foundation for generous financial support (grant No. 31-52981.97).

References

Adams GC, Kropp BR (1996) *Athelia arachnoidea*, the sexual state of *Rhizoctonia carotae*, a pathogen of carrot in cold storage. Mycologia 88:459–472

Ahmadjian V (1993) The lichen symbiosis. John Wiley, New York

Anglesea D, Veltkamp C, Greenhalgh GH (1982) The upper cortex of *Parmelia saxatilis* and other lichen thalli. Lichenologist 14:29–38

Armaleo D (1993) Why do lichens make secondary products? In: 15th Int Botanical Congress, Yokohama, 1993 Abstr 11 pp

Armaleo D, Clerc P (1990) Lichen chimeras: DNA analysis suggests that one fungus forms two morphotypes. Exp Mycol 5:1–10

Armstrong RA, Smith SN (1992) Lobe growth variation and the maintenance of symmetry in foliose lichen thalli. Symbiosis 12:145–158

Armstrong R, Smith S (1993) Radial growth and carbohydrate levels in the lichen *Parmelia conspersa* on north and south facing rock surfaces. Symbiosis 15:27–49

Armstrong RA, Smith SN (1994) The levels of ribitol, arabitol and mannitol in individual lobes of the lichen *Parmelia conspersa* (Ehrh. ex Ach.) Ach. Environ Exp Bot 34:253–260

Ascaso C, Wierzchos J (1994) Structural aspects of the lichen-rock interface using back-scattered electron imaging. Bot Acta 107:251–256

Babikoff J (1878) Du développement des céphalodies sur le thallus du lichen *Peltigera aphthosa* Hoffm. Bull Acad Imp Sci St Pétersbourg 24:548–559

Beck A, Friedl T, Rambold G (1998) Selectivity of photobiont choice in a defined lichen community: inferences from cultural and molecular studies. New Phytol 139:709–720

Brunner U, Honegger R (1985) Chemical and ultrastructural studies on the distribution of sporopolleninlike biopolymers in six genera of lichen phycobionts. Can J Bot 63:2221–2230

Bubrick P, Galun M (1980) Proteins from the lichen *Xanthoria parietina* which bind to phycobiont cell walls. Correlation between binding patterns and cell wall cytochemistry. Protoplasma 104:167–173

Bubrick p, Galun M, Frensdorff A (1981) Proteins from the lichen *Xanthoria parietina* which bind to phycobiont cell walls. Localization in the intact lichen and cultured mycobiont. Protoplasma 105:207–211

Bubrick P, Frensdorff A, Galun M (1985) Selectivity in the lichen symbiosis. In: Brown DH (ed) Lichen physiology and cell biology. Plenum, New York, pp 319–334

Büdel B, Scheidegger C (1996) Thallus morphology and anatomy. In: Nash TH (ed) Lichen biology. Cambridge University Press, Cambridge, pp 37–64

Clayden SR (1998) Thallus initiation and development in the lichen *Rhizocarpon lecanorinum*. New Phytol 139:685–695

deBary A (1866) Morphologie und Physiologie der Pilze, Flechten und Myxomyceten. W Engelmann, Leipzig

Debaud J-C, Marmeisse R, Gay G (1997) Genetics and molecular biology of the fungal partner in the ectomycorrhizal symbiosis *Hebeloma cylindrosporum* x *Pinus pinaster*. In: Carroll GC, Tudzynski P (eds) The Mycota V, part B (Plant relationships). Springer, Berlin Heidelberg New York pp 95–115

Englund B (1977) The physiology of the lichen *Peltigera aphthosa*, with special reference to the blue-green phycobiont (*Nostoc* sp.). Physiol Plant 41:298–304

Fahselt D (1994) Secondary biochemistry of lichens. Symbiosis 16:117–165

Fahselt D (1996) Individuals, populations and population ecology. In: Nash TH (ed) Lichen biology. Cambridge University Press, Cambridge, pp 181–198

Farkas EE, Sipman HJM (1993) Bibliography and checklist of foliicolous lichenized fungi up to 1992. Trop Bryol 7:93–148

Fiechter E (1990) Thallusdifferenzierung und intrathalline Flechtenstoffverteilung bei Parmeliaceae (Lecanorales, lichenisierte Ascomyceten). Inaugural-Dissertation, Philosophische Fakultät II, Universität Zürich, Zürich

Friedl T (1987) Thallus development and phycobionts of the parasitic lichen *Diploschistes muscorum*. Lichenologist 19:183–191

Friedl T (1995) Inferring taxonomic positions and testing genus level assignments in coccoid lichen algae: a phylogenetic analysis of 18s ribosomal RNA sequences from *Dictyochloropsis reticulata* and from members of the genus *Myrmecia* (Chlorophyta, Trebouxiophyceae cl. nov.). J Phycol 31:632–639

Friedl T, Büdel B (1996) Photobionts. In: Nash TH (ed) Lichen biology. Cambridge University Press, Cambridge, pp 8–23

Friedmann EI (1982) Endolithic microorganisms in the Antarctic cold desert. Science 215:1045–1053

Friedmann EI, Kappen L, Meyer MA, Nienow JA (1993) Long-term productivity in the cryptoendolithic microbial community of the Ross desert, Antarctica. Microb Ecol 25:51–69

Fries TM (1866) Beiträge zur Kenntnis der sogenannten Cephalodien bei den Flechten. Flora 49:17–25

Galloway DJ (1996) Lichen biogeography. In: Nash TH (ed) Lichen biology. University of Cambridge Press, Cambridge, pp 199–216

Giese H, Hippe-Sanwald S, Somervilee S, Weller J (1997) *Erysiphe graminis*. In: Carroll GC, Tudzynski P (eds) The Mycota V, part B (Plant relationships). Springer, Berlin Heidelberg New York, pp 55–78

Girlanda M, Isocrono D, Bianco C, Luppi-Mosca AM (1997) Two foliose lichens as microfungal ecological niches. Mycologia 89:531–536

Goffinet B, Bayer RJ (1997) Characterization of mycobionts of photomorph pairs in the Peltigeraceae (lichenized ascomycetes) based on internal transcribed spacer sequences of the nuclear ribosomal DNA. Fungal Gen Biol 21:228–237

Goffinet B, Goward T (1998) Is *Nephroma silvae-veteris* the cyanomorph of *Lobaria oregana*? Insights from molecular, chemical and morphological characters. In: Glenn MG, Harris RC, Dirig R, Cole MS (eds) Lichenographia Thomsoniana: North American lichenology in honour of John W Thomson. Mycotaxon, Ithaca, NY, pp 41–52

Hawksworth DL (1983) A key to the lichen-forming, parasitic, parasymbiotic and saprophytic fungi occurring on lichens in the British Isles. Lichenologist 15:1–44

Hawksworth DL (1988a) Effects of algae and lichen-forming fungi on tropical crops. In: Agnihotri VP, Sarbhoy AK, Kumar D (eds) Perspectives of mycopathology. Malhorta Publishing House, New Delhi, pp 76–83

Hawksworth DL (1988b) The variety of fungal-algal symbioses, their evolutionary significance, and the nature of lichens. Bot J Linn Soc 96:3–20

Hawksworth DL, Honegger R (1994) The lichen thallus: a symbiotic phenotype of nutritionally specialized fungi and its response to gall producers. In: Williams MAJ (ed) Plant galls: organisms, interactions, populations. Clarendon Press, Oxford, pp 77–98

Hawksworth DL, Kirk PM, Sutton BC, Pegler DN (1995) Ainsworth & Bisby's dictionary of the fungi. CAB International, Cambridge University Press, Cambridge

Hersoug LG (1983) Lichen protein affinity towards walls of cultured and freshly isolated phycobionts and its relationship to cell wall cytochemistry. FEMS Microbiol Lett 20:417–420

Hestmark G (1997) Growth from the centre in an umbilicate lichen. Lichenologist 29:379–383

Hill DJ (1985) Changes in photobiont dimensions and numbers during co-development of lichen symbionts. In: Brown DH (ed) Lichen physiology and cell biology. Plenum, New York, pp 303–317

Hill DJ (1989) The control of the cell cycle in microbial symbionts. New Phytol 112:175–184

Honegger R (1982) Cytological aspects of the triple symbiosis in *Peltigera aphthosa*. J Hattori Bot Lab 52:379–391

Honegger R (1984) Cytological aspects of the mycobiont-phycobiont relationship in lichens. Haustorial types, phycobiont cell wall types, and the ultrastructure of the cell wall surface layers in some cultured and symbiotic myco- and phycobionts. Lichenologist 16:111–127

Honegger R (1985) Fine structure of different types of symbiotic relationships in lichens. In: Brown DH (ed) Lichen physiology and cell biology. Plenum, New York, pp 287–302

Honegger R (1986a) Ultrastructural studies in lichens. I Haustorial types and their frequencies in a range of lichens with trebouxioid phycobionts. New Phytol 103:785–795

Honegger R (1986b) Ultrastructural studies in lichens. II Mycobiont and photobiont cell wall surface layers and adhering crystalline lichen products in four Parmeliaceae. New Phytol 103:797–808

Honegger R (1987) Isidium formation and the development of juvenile thalli in *Parmelia pastillifera* (Lecanorales, lichenized ascomycetes). Bot Helv 97:147–152

Honegger R (1991) Functional aspects of the lichen symbiosis. Annu Rev Plant Physiol Plant Mol Biol 42:553–578

Honegger R (1992) Lichens: mycobiont-photobiont relationships. In: Reisser W (ed) Algae and symbioses. plants, animals, fungi, viruses, interactions explored. Biopress, Bristol, pp 255–275

Honegger R (1993) Developmental biology of lichens. New Phytol 125:659–677

Honegger R (1995) Experimental studies with foliose macrolichens: fungal responses to spatial disturbance at the organismic level and to spatial problems at the cellular level during drought stress events. Can J Bot 73 (Suppl 1):569–578

Honegger R (1996a) Structural and functional aspects of mycobiont-photobiont relationships in lichens compared with mycorrhizae and plant pathogenic interactions. In: Nicole M, Gianinazzi-Pearson V (eds) Histology, ultrastructure and molecular cytology of plant-microorganism interactions. Kluwer, Dordrecht, pp 157–176

Honegger R (1996b) Mycobionts. In: Nash TH (ed) The biology of lichens. Cambridge University Press, Cambridge, pp 25–36

Honegger R (1996c) Morphogenesis. In: Nash TH (ed) The biology of lichens. Cambridge University Press, Cambridge, pp 65–87

Honegger R (1996d) Experimental studies of growth and regenerative capacity in the foliose lichen *Xanthoria parietina*. New Phytol 133:573–581

Honegger R (1997) Metabolic interactions at the mycobiont-photobiont interface in lichens. In: Carroll GC, Tudzynski P (eds) The Mycota V, part A (Plant relationships). Springer, Berlin Heidelberg New York, pp 209–221

Honegger R (1998) The lichen symbiosis – what is so spectacular about it? Lichenologist 30:193–212

Honegger R, Brunner U (1981) Sporopollenin in the cell wall of *Coccomyxa* and *Myrmecia* phycobionts of various lichens: an ultrastructural and chemical investigation. Can J Bot 59:2713–2734

Honegger R, Peter M (1994) Routes of solute translocation and the location of water in heteromerous lichens visualized with cryotechniques in light and electron microscopy. Symbiosis 16:167–186

Honegger R, Conconi S, Kutasi V (1996a) Field studies on growth and regenerative capacity in the foliose macrolichen *Xanthoria parietina* (Teloschistales, Ascomycotina). Bot Acta 109:187–193

Honegger R, Peter M, Scherrer S (1996b) Drought-induced structural alterations at the mycobiont-photobiont interface in a range of foliose macrolichens. Protoplasma 190:221–232

Innes JL (1988) The use of lichens in dating. In: Galun M (ed) CRC Handbook of lichenology, vol 3. CRC Press, Boca Raton, pp 75–91

Jahns HM (1988) The lichen thallus. In: Galun M (ed) CRC Handbook of lichenology, vol 1. CRC Press, Boca Raton, pp 95–143

Jahns HM, Ott S (1997) Life strategies in lichens – some general considerations. Bibl Lichenol 67:49–67

Jahns HM, Tuiz-Dubiel A, Blank L (1976) Hygroskopische Bewegungen der Sorale von *Hypogymnia physodes*. Herzogia 4:15–23

Jahns HM, Klöckner P, Ott S (1995) Development of thalli and ascocarps in *Solorina sponagiosa* (Sm.) Anzi and *Solorina saccata* (L.) Ach. Bibl Lichenol 57: 241–251

James PW, Henssen A (1976) The morphological and taxonomic significance of cephalodia. In: Brown DH, Hawksworth DL, Bailey RH (eds) Lichenology: progress and problems. Academic Press, London, pp 27–77

Jordan WP (1970) The internal cephalodia of the lichen genus *Lobaria*. Bryologist 73:669–681

Jordan WP, Rickson FR (1971) Cyanophyte cephalodia in the lichen genus *Nephroma*. Am J Bot 58:562–568

Jörgensen PM (1998) What shall we do with the blue-green counterparts? Lichenologist 30:351–356

Kappen L (1988) Ecophysiological relationships in different climatic regions. In: Galun M (ed) CRC Handbook of lichenology, vol 2. CRC Press, Boca Raton, pp 37–100

Kappen L (1993) Lichens in the antarctic region. In: Friedman EI (ed) Antarctic microbiology. Wiley-Liss, New York, pp 433–490

Kardish N, Silberstein L, Fleminger G, Galun M (1991) Lectin from the lichen *Nephroma laevigatum* Ach. Localization and function. Symbiosis 11:47–62

Lallemant R, Savoye D (1985) Lectins and morphogenesis: facts and outlooks. In: Brown DH (ed) Lichen physiology and cell biology. Plenum, New York, pp 335–350

Lange OL, Kilian E, Ziegler H (1986) Water vapour uptake and photosynthesis of lichens: performance differences in species with green and blue-green algae as phycobionts. Oecologia 71:104–110

Lange OL, Green TGA, Ziegler H (1988) Water status-related photosynthesis and carbon isotope discrimination in species of the lichen genus *Pseudocyphellaria* with green and blue-green photobionts and in photosymbiodemes. Oecologia 75:394–411

Larson DW (1984) Habitat overlap/niche segregation in two *Umbilicaria* lichens: a possible mechanism. Oecologia 62:118–125

Lumbsch HT, Kothe HW (1993) Development of the thalli of *Eremastrella crystallifera* and their taxonomic significance (Psoraceae, Ascomycotina). Nova Hedwigia 57:19–32

Malcolm WM (1995) Light transmission inside the thallus of *Labyrintha implexa* (Porpidiaceae, lichenized Ascomycetes). In: Farkas EE, Lücking R, Wirth V (eds) Scripta Lichenologica – Lichenological papers dedicated to Antonin Vezda. Bibl Lichenol 58: 275–280

Malcolm WM, Elix JA, Owe-Larsson B (1995) *Labyrintha implexa* (Porpidiaceae), a new genus and species from New Zealand. Lichenologist 27:241–248

Matthews SW, Tucker SC, Chapman RL (1989) Ultrastructural features of mycobionts and trentepohliaceous phycobionts in selected subtropical crustose lichens. Bot Gaz 150:417–438

Meier JL, Chapman RL (1983) Ultrastructure of the lichen *Coenogonium interplexum* Nyl. Am J Bot 70:400–407

Mendgen K (1997) The Uredinales. In: Carroll GC, Tudzynski P (eds) The Mycota V, part B (Plant relationships). Springer, Berlin Heidelberg New York, pp 79–94

Nash TH (1996) Nitrogen, its metabolism and potential contriburion to ecosystems. In: Nash TH (ed) Lichen biology. University of Cambridge Press, Cambridge, pp 121–135

Ott S (1987a) Reproductive strategies in lichens. Bibl Lichenol 25:81–93

Ott S (1987b) Sexual reproduction and developmental adaptations in *Xanthoria parietina*. Nord J Bot 7:219–228

Ott S, Przewosnik R, Sojo F, Jahns HM (1997) The nature of cephalodia in *Placopsis contortuplicatus* and other species of the genus. Bibl Lichenol 67:69–84

Paulsrud P, Rikkinen J, Lindblad P (1998) Cyanobiont specificity in some *Nostoc*-containing lichens and in a *Peltigera aphthosa* photosymbiodeme. New Phytol 139:517–524

Pérez FL (1997a) Geoecology of erratic globular lichens of *Catapyrenium lachneum* in a high Andean Paramo. Flora 192:241–259

Pérez FL (1997b) Geoecology of erratic lichens of *Xanthoparmelia vagans* in an equatorial Andean Paramo. Plant Ecol 129:11–28

Petrini O, Hake U, Dreyfuss MM (1990) An analysis of fungal communities isolated from fruticose lichens. Mycologia 82:444–451

Piervittori R, Salvadori O, Laccisaglia A (1994) Literature on lichens and biodeterioration of stonework. 1. Lichenologist 26:171–192

Piervittori R, Salvadori O, Laccisaglia A (1996) Literature on lichens and biodeterioration of stonework. 2. Lichenologist 28:471–483

Plessl A (1963) Über die Beziehungen von Pilz und Alge im Flechtenthallus. Oesterr Bot Z 110:194–269

Poelt J (1985) Ueber auf Moosen parasitierende Flechten. Sydowia Ann Mycol Ser II 38:241–254

Poelt J, Doppelbauer H (1956) Ueber parasitische Flechten. Planta 46:467–480

Poelt J, Vezda A (1990) Ueber kurzlebige Flechten. Bibl Lichenol 38:377–394

Rambold G, Triebel D (1992) The inter-lecanoralean associations. Bibl Lichenol 48:3–201

Rosentreter R (1993) Vagrant lichens in North America. Bryologist 96:333–338

Sanders W (1989) Growth and development of the reticulate thallus in the lichen *Ramalina menziesii*. Am J Bot 76:666–678

Sanders WB (1992) Comparative in situ studies of thallus net development in morphologically distinct populations of the lichen *Ramalina menziesii*. Bryologist 95:192–204

Sanders WB, Ascaso C (1995) Reiterative production and deformation of cell walls in expanding thallus nets of the lichen *Ramalina menziesii* (Lecanorales, Ascomycetes). Am J Bot 82:1358–1366

Scheidegger C (1994) Low-temperature scanning electron microscopy: the localization of free and perturbed water and its role in the morphology of the lichen symbionts. Cryptogam Bot 4:290–299

Scheidegger C, Schroeter B, Frey B (1995) Structural and functional processes during water vapour uptake and desiccation in selected lichens with green algal photobionts. Planta 197:399–409

Scherrer S, de Vries OMH, Dudler R, Wessels JGH, Honegger R (2000) Interfacial self-assembly of fungal hydrophobins of the lichen-forming ascomycetes *Xanthoria parietina* and *X. ectaneoides*. Fun Gen Biol (in press)

Schuster G (1985) Die Jugendentwicklung von Flechten. Bibl Lichenol 20:3–206

Schuster G, Ott S, Jahns HM (1985) Artificial cultures of lichens in the natural environment. Lichenologist 17:247–253

Schwendener S (1869) Die Algentypen der Flechtengonidien. Universitätsbuchdruckerei Schultze, Basel

Sipman HJM (1994) Foliicolous lichens on plastic tape. Lichenologist 26:311–312

Stone BA, Clarke AE (1992) The chemistry and biology of (1–3)-β-glucans. La Trobe University Press, Melbourne

Talbot NJ (1998) Plants and fungi: friends and enemies. Trends Microbiol 6:250–251

Taylor TN, Remy W, Hass H, Kerp H (1995) Fossil arbuscular mycorrhizae from the Early Devonian. Mycologia 87:560–573

Taylor TN, Hass H, Kerp H (1997) A cyanolichen from the Lower Devonian Rhynie chert. Am J Bot 84:992–1004

Triebel D, Rambold G (1988) *Cecidonia* and *Phacopsis* (Lecanorales): zwei lichenicole Pilzgattungen mit cecidogenen Arten. Nova Hedwigia 47:279–309

Tschermak E (1941) Untersuchungen über die Beziehungen von Pilz und Alge im Flechtenthallus. Oesterr Bot Z 90:233–307

Tschermak-Woess E (1988) The algal partner. In: Galun M (ed) CRC Handbook of lichenology, vol 1. CRC Press, Boca Raton, pp 39–92

Tschermak-Woess E, Bartlett J, Peveling E (1983) *Lichenothrix riddlei* is an ascolichen and also occurs in New Zealand – light and electron microscopical investigations. Plant Syst Evol 143:109–115

Valladares F (1994) Texture and hygroscopic features of the upper surface of the thallus in the lichen family Umbilicariaceae. Ann Bot 73:493–500

Vogel S (1955) Niedere "Fensterpflanzen" in der südafrikanischen Wüste. Beitr Biol Pflanz 31:45–135

Winchester V (1988) An assessment of lichenometry as a method for dating recent stone movements in two stone circles in Cumbria and Oxfordshire. Bot J Linn Soc 96:57–68

Wösten HAB, Wessels JGH (1997) Hydrophobins, from molecular structure to multiple functions in fungal development. Mycoscience 38:363–374

11 Morphology and Phylogeny of Ascomycete Lichens

S. Ott[1] and H.T. Lumbsch[2]

CONTENTS

I. Introduction

A wide range of taxa in the higher ascomycetes (Pezizomycotina sensu Eriksson and Winka 1997) forms some sort of symbiosis with algae and cyanobacteria. These associations include fungal parasites on algae (Hawksworth 1987; Lewin 1995), mycophycobioses (Kohlmeyer 1967; Kohlmeyer and Kohlmeyer 1972), lichenicolous fungi (Hawksworth 1983, 1988a; Clauzade et al. 1989) and lichens (Nash 1996). Naturally, the delimitation of these associations is not clear and intermediate forms exist. However, an attempt to

define these terms is made in Table 1, following Hawksworth (1988a). An exhaustive treatment on fungal-algal associations in the large order Lecanorales can be found in Rambold and Triebel (1992). This chapter concentrates on the morphology, phylogeny and classification of lichen-forming ascomycetes. Only a brief treatment can be given here and the reader is referred to other recent books for more information (e.g., Ahmadjian 1993; Nash 1996). Lichens are symbiotic organisms typically composed of a fungus and a photosynthetic partner, which may be a cyanobacterium, a green alga or both. It is now generally accepted that lichens are a nutritional group of fungi associated with photosynthetic organisms and not a monophyletic group. This knowledge has been recently supported by molecular data as well (Gargas et al. 1995a). The name of a lichen refers to its mycobiont, and approximately 15000–20000 species are known. Most of the fungal partners belong to the Ascomycetes and only few are Basidiomycetes, Chapter 12, this volume by Oberwinkler deals with the latter.

II. Morphology and Anatomy

The symbiotic relationship of fungi and algae or cyanobacteria in lichens is often so well established that a special morphological structure, the lichen thallus, is formed. The lichen thallus is quite different in appearance from the single bionts. Thus it is understandable that the dual nature of these organisms was not discovered before Schwendener (1869). Usually lichens are interpreted as fungi which have aquired a special type of nutrition in form of a symbiosis with algae or cyanobacteria. Although this is true, such a somewhat restrictive view may hinder a full understanding of morphological and physiological adaptations of these dual organisms. The morphogenetic influence of the photobiont is strong, as

[1] Heinrich Heine Universität, Botanisches Institut, Universitätsstraße 1, 40225 Düsseldorf, Germany
[2] Universität Essen, Fachbereich 9/Botanik, 45117 Essen, Germany

The Mycota IX
Fungal Associations
Hock (Ed.)
© Springer-Verlag Berlin Heidelberg 2001

Table 1. Definitions of different associations between higher ascomycetes and algae or/and cyanobacteria

Term	Definition	Example
Fungal parasites on algae	Parasitic association in which the mycobiont is inhabitant	*Epigloea* spp. on *Coccomyxa* (Döbbeler 1984)
Mycophyobioses	Mutualistic association in which the mycobiont is inhabitant	*Mycosphaerella ascophylli* Cotton on *Ascophyllum* or *Pelvetia* (Kohlmeyer and Kohlmeyer 1972)
Lichenicolous fungi	Fungi growing on lichens forming a parasitic, commensalistic or mutualistic association with them	*Phacopsis huuskonenii* Räsänen on *Letharia* spp. (Hawksworth 1982; Triebel et al. 1995)
Lichens	Stable self-supporting association in which the mycobiont is exhabitant	*Peltigera aphthosa* (L.) Willd. containing *Coccomyxa* and *Nostoc* (Honegger 1982b)

shown in the case of photosymbiodemes (see below), and at least some of the bionts rarely occur in nature as single bionts. Species of the genus *Trebouxia*, the most common phycobiont in lichens, are seldom reported free-living (e.g., Tschermak-Woess 1978; Mukhtar et al. 1994; Matthes-Sears et al. 1997), as is also the case with mycobionts (e.g., Hale 1957; Bubrick et al. 1984). However, other photobionts occurring in lichens, such as *Trentepohlia* or *Nostoc*, are very common, free-living (Tschermak-Woess 1988; Nash 1996). Lichens are complex systems of interacting symbiotic partners which have evolved in processes of coevolution. The nature of lichens and the principles of their construction may be more easily understood and interpreted when these symbiotic organisms are understood as an evolution parallel to higher plants (Jahns and Ott 1997). Both groups had to manage similar problems regarding the adaptation to living on land. In morphology and anatomy some striking convergent solutions to these problems can be found in lichens and higher plants. These are, especially, structures which involve photosynthesis, water regime, and stabilization (Jahns and Ott 1997).

The close contact between both partners in the lichen symbiosis goes beyond an attack of a parasite to its host. The fungus produces hyphae which form appressoria or haustoria which penetrate the cell wall but usually not the plasmalemma of the photobiont (Honegger 1985b, 1986a, 1988; Büdel and Rhiel 1987). With the exception of very few poorly adapted lichens, only dead algal cells are digested. It is important to note that the fungus secretes a hydrophobic matrix on the hyphal surface (Honegger 1991). In this way, fungus and alga produce a structural and functional unit in which the algal partner can only

be supplied with water and nutrient salts by the fungus. The fungus, on its part, receives photosynthesis products from the algae and – in cyanobacteria – nitrogen compounds (Feige and Jensen 1992). Further, there is evidence for the exchange of additional substances (e.g., phytohormones) in connection with a largely unexplored regulation of the symbiosis (Ott and Zwoch 1992; Schieleit and Ott 1996). Beside the adaptations in the contact zone between the bionts, there exist numerous morphological and anatomical adaptations to life on land and especially to survival in extreme environments.

A. Thallus Morphology and Growth Forms

The growth forms of lichens primarily reflect the necessity to expose a sufficient area with algae to light for photosynthesis. The simplest growth form in this respect is a thin crust which covers the substrate. However, such crustose lichens are only superficially simple in structure and some anatomical variations can be observed which are discussed below. In crustose lichens a general growth principle of lichens can already be observed which differs from growth in non-lichenised ascomycetes. Since the lichen thallus has a long life span and reaches, especially in crustose forms (e.g., *Rhizocarpon*), an age of at least several decades, a continuous exchange of nutrients and a controlled growth of both bionts over a long period of time must be guaranteed. The mycobiont of most species produces the main part of the biomass. In a few cases, such as filamentous lichens (e.g., *Racodium*) or gelatinous lichens (e.g., *Collema*), the algal partner produces the larger amount of biomass. In filamentous lichens, a filamentous alga is surrounded by fungal hyphae and in gelatinous lichens a gelatinous matrix is produced by the cyanobiont.

Most lichens are stratified and also show a differentiation in their morphology. Crustose lichens are often separated into areoles which may be effigurate and then form some sort of lobes at the margins with apical growth. Such marginally lobate forms may have a foliose appearance, but can be distinguished by being completely adnate to the substrate. Foliose lichens form rhizines or other forms of attachment organs with which they are connected to their substrate. Foliose lichens are more competetive than crustose forms since they usually cannot be overgrown, as crustose lichens often are (Degelius 1940). There are also advantages regarding the water regime in foliose lichens. A water film may occur between the thallus and the substrate which is gradually taken up by the lower surface of the lichen. However, a loosely attached thallus requires anatomical adaptations concerning the stability and gas exchange. Some forms show marginal growth with meristematic margins and a central part which dies off. This is especially conspicuous in *Arctoparmelia centrifuga*. Other lichens which are only connected by a holdfast to their substrate obviously need to be able to live longer in order to prevent them from falling from their substrate. Umbilicate lichens, such as *Dermatiscum* or *Umbilicaria*, are examples for this growth form or fruticose lichens, such as *Ramalina* spp.

Fruticose lichens grow erect or pendulous on their substrate. Supportive and border tissues are of special importance for this group, which have to ensure bending strength in erect forms and tensile strength in pendulous forms. The development of these tissues is correlated with further complex adaptations regarding the water regime and gas exchange. Phyllocladia are small lobules at erect thallus parts (e.g., in *Cladonia*) and represent a convergence to phanerogams and increase the photosynthetically active surface (Jahns 1988; Jahns and Ott 1997).

Water uptake occurs over the whole surface of the lichen. However, some parts of the cortex may be hydrophobic due to the presence of secondary metabolites. The growth form plays an important role for the adaptation of water uptake in different habitats. Deeply divided fruticose lichens are able to use fog, snow or dew very efficiently (e.g., Lange et al. 1990), while foliose lichens can collect rainwater on the their lower surfaces (Ott 1989). In some lichens, such as the reindeer lichen genus *Cladina*, the thallus grows on older and decaying parts which form a water reservoir.

B. Thallus Anatomy

A small number of species consists of a largely undifferentiated layer of fungal hyphae and algae (homoiomerous thallus). However, in several lichens with homoiomerous thallus there is some sort of differentiation. The genus *Leptogium*, for example, has a pseudoparenchymatous cortex. In most cases, the anatomy is more complex and a well-defined algal layer can be distinguished (heteromerous thallus). The algae occur in areas with appropriate insolation. Taxa occurring in environments with high insolation have a thick cortex or reduce high radiation by secondary metabolites localised in the upper cortex. The medulla below the algal layer consists only of fungal hyphae. In some crustose lichens, such as *Placynthiella* spp., the thallus consists of globose corticate structures which enclose the photobiont. These structures are called goniocysts.

The mycobionts are usually quite specific in their choice of a photobiont. Related species may have different photobionts, such as *Peltigera* spp. with green algae or cyanobacteria, but a fungal species usually forms a symbiosis with only one or few algal species (Hawksworth 1988a; Friedl 1989). Some species with green algae as primary photobionts additionally form symbioses with cyanobionts. These secondary photobionts are usually included in special structures which are called cephalodia. Cephalodia may grow on the thallus surface and are visible as small verrucae (external cephalodia; e.g., *Peltigera aphthosa*) or occur in the medulla (internal cephalodia; e.g., *Nephroma arcticum*). The cephalodia improve the supply of nitrogen for the mycobiont by N_2 fixation of the cyanobacteria which show a high percentage of heterocyst frequency (Hitch and Millbank 1975b; Englund 1977). In some Peltigerineae a mycobiont forms different thalli with a phycobiont and cyanobiont. These cases are called photosymbiodemes (Tønsberg and Holtan-Hartwig 1983; Ott 1988; Armaleo and Clerc 1991). In some of these photosymbiodemes the mycobiont even forms morphologically very different thalli with the different photobionts (James and Henssen 1976), such as foliose and fruticose morphs. This indicates a large morphogenetic influence of the photobionts in lichen symbioses.

The thallus of numerous crustose lichens is covered by an ahyphal polysaccharide layer which is called the epinecral layer. In most macrolichens and some crustose lichens a cortex is developed by the mycobiont. Such a cortex may be restricted to the upper side or may occur on the lower side of the thallus as well. The cortex is often covered by a polysaccharide layer, the epicortex. The cortex in lichens plays only a limited role against water loss, in contrast to the epidermis in higher plants. In fact, the poikilohydric thalli are dependent on losing water rapidly under extreme environmental conditions.

The importance of supportive and border tissues for the stability of foliose and fruticose lichens has already been mentioned. In some species this is ensured by a thick and multi-layered cortex (e.g., in *Sphaerophorus*). Different orientation of cortical hyphae is another way of lichens to stabilize their thalli. In some *Cladonia* spp. a cylindrical supportive tissue is formed around a central medullary hole. In other genera, such as *Usnea*, a central cord is formed to ensure stability and may be interpreted as a convergence to lianes. However, central cords are also found in lichens in arid regions, such as *Seirophora* or *Simonyella*, and are there interpreted as water reservoirs (Poelt 1983; Feige et al. 1992).

Within the lichen thallus, water is transported apoplastically through the fungal hyphae (Honegger 1991). The capillary cavities in the medulla in higher-developed lichens are free of water, which is important for the gas exchange. To ensure this, the surface of the fungal hyphae is hydrophobic due to hydrophobins and lichens substances (Honegger 1986b). When soaked, the gelatinous cortices do not have capillary cavities and thus special structures are formed by lichens to ensure gas exchange in these situations with high photosynthetic activity (Lange et al. 1993). Numerous lichens form pores which can be of different structure. Pseudocyphellae are cracks in the cortex, while cyphellae are corticate hollows. Both types of structures are functionally involved to increase gas exchange in the thallus.

C. Vegetative Reproduction

Sexual reproduction in lichen-forming fungi requires a subsequent relichenization. This is probably not as uncommon as usually supposed. An indication is the large number of lichens producing ascomata. Some lichens have developed special strategies as adaptation to this. A few species, such as *Endocarpon* or *Sporopodium*, possess hymenial algae which are simultaneously distributed with the fungal spores. Other lichens are juvenile parasites and may take over the photobiont of their hosts (Ott 1987b; Hawksworth 1988a; Ott et al. 1995). Further, there is evidence that mycobionts can survive for some time as saprophytes or in loose association with algae unsuitable for a close symbiosis (Ott 1987a). Despite this, numerous lichens form vegetative propagules which distribute both partners. The simplest form of vegetative propagation is by fragmentation. Several special vegetative diaspores are formed by numerous lichens.

Small groups of algae surrounded by fungal hyphae separated from the algal layer are called soredia. These soredia may be granular or farinose and occur in cracks of the thallus surface which are called soralia. Usually, several soredia fuse to form a new thallus. In some cases soredia may have a cortex and are then called schizidia (Poelt 1995). However, there are numerous variations of soredia and intermediate forms, and thus it does not seem appropriate to use every special form with separate terms.

Several different types of soralia can be distinguished. Their form is often species-specific. They may be globose, labiate or split-like and may be marginal or laminal. In some lichens the whole surface is sorediate and no definite soralia can be distinguished.

Isidia are the other most common type of amphigenous diaspores occurring in lichens. They are corticate outgrowths of the thallus which break off and function as a single propagule. Ecorticate outgrowths are called pseudoisidia. The morphology of isidia is very diverse and usually a good taxonomic character at the species level. They may be simple or branched, globose or cylindrical. Some isidia resemble small foliose lobules and may be called phyllidia. There are intermediate forms between soredia and isidia, including sorediate isidia. In some cases isidia do not function as propagules, but increase the photosynthetically active surface or almost replace the thallus, as in some terricolous arctic-boreal *Pertusaria* spp. In other cases, isidia may germinate on decaying mother thalli and are available for regeneration, thus forming a new thallus on the mother thallus, e.g., in *Parmelia saxatilis* (Jahns 1984; Ott et al. 1993). The decaying old thallus serves as a water reservoir.

Some groups of lichens have special forms of vegetative propagules, including hormocystangia in the genus *Lempholemma* (Henssen 1968), parasoredia in *Umbilicaria* (Codogno et al. 1989), thlasidia in *Gyalideopsis* (Poelt 1986), or blastidia in different genera (Poelt and Petutschnig 1992; Poelt 1995).

The determination of vegetative structures is much lower in lichens than in higher plants. The number of specialised cells and pseudotissues is low. However, this disadvantage is compensated by a high flexibility of the structures (Jahns and Ott 1994). For example, cortices may grow out and include algae and thus form an apothecial margin. These apothecial margins show a large amount of variation as discussed below. Due to the high flexibility of thallus structures, the taxonomic importance of these characters is limited and usually not suitable for classification at higher ranks.

D. Ascoma Morphology and Anatomy

The ascomata of lichenised ascomycetes do not principally differ from those of non-lichenised ascomycetes. They are usually long-lasting and thus have often well-developed supportive tissues. In some ascomata the photobiont is included in a thalline margin. In general, the fruiting body is a structure formed by the mycobiont, and its induction and development are independent of the presence of photobionts (Jahns et al. 1995a; Jørgensen and Jahns 1987).

The same ascoma types as in non-lichenised ascomycetes can be found in lichens, such as apothecia, perithecia, pseudothecia or hysterothecia. Cleistothecia do not occur in lichen-forming fungi. The morphologically defined fruiting body types occur in separate clades of ascomycetes (Berbee and Taylor 1995), but some intermediate forms exist as well. In some groups, such as *Thelocarpon* (Poelt and Hafellner 1975), the opening of the ascomata varies. Thus, the systematic importance of the ascoma types is limited in some cases. Further, molecular results suggest that discomycetes are at least a paraphyletic group (Gargas and Taylor 1995).

The ontogeny of the ascomata is used in systematics of both non-lichenised and lichenised ascomycetes as a character set at family and higher levels. Several review papers give an overview of the developmental patterns observed in groups of lichen-forming fungi and discuss the taxonomic relevance of this character set (Henssen and Jahns 1974; Henssen 1976; Henssen et al. 1981; Parguey-Leduc and Janex-Favre 1981; Letrouit-Galinou and Bellemere 1989; Döring and Lumbsch 1998). Two different principal types of ascoma ontogeny can be observed; the ascohymenial and the ascolocularous development. In the ascolocularous development type the asci arise in cavities of an initially developed stroma. This stroma will form the monokaryotic parts of the mature ascoma. In the ascohymenial type, the ascomata are formed after the dikaryotisation. The monokaryotic hyphae which surround the ascogonia and later form the fruiting body are called generative tissue (Henssen 1969a). The generative tissue can be distinguished from other parts of the thallus through its thin-walled meristematic hyphae which stain deeper blue in lactophenol cotton-blue. The concept of the generative tissue is controversially disputed (Hammer 1993), since its hyphae do not differ qualitatively from other hyphae of the vegetative thallus. This is generally true, but it seems appropriate to distinguish the generative tissue, since it is the first anatomically observable part of the ascoma primordium (Jahns et al. 1995b). The mode of fertilization is not resolved yet, but there is evidence that it does not differ from the different modes occurring in non-lichenised ascomycetes. The significance of ascolocularous and ascohymenial ascoma development was discovered by Nannfeldt (1932) and used by him for the distinction of supraordinal taxa in the ascomycetes. Fungi with ascolocularous ascoma development are currently placed in the order Dothideales s. lat. and only few lichenised taxa occur in this group. The large majority of lichen fungi belongs to the group of ascohymenial fungi.

There are several types of ascohymenial ascoma development. In the gymnocarpous development the hymenium arises at the upper part of the primordium, while a development with an enclosed hymenium in the centre of a primordium which only opens with an ostiolum is called angiocarpous development. In the hemiangiocarpous development, the initially closed hymenium bursts open at a later stage. Within these three types, numerous variants of the development can be found. The type of ascohymenial can be of systematic importance; e.g., the members of the Peltigerineae are characterised by a hemiangiocarpous ontogeny. However, in other groups, taxa with different ascoma development can be found, as in the Agyriales, which contain the gymnocarpous Anamylopsoraceae and hemiangiocarpous Agyriaceae (Lumbsch 1997a). As with other characters, ascoma ontogeny is useful as a taxonomic discriminator only when it correlates with other characters.

Different types of ascomatal margins can be distinguished. The margin may consist either of purely fungal hyphae and then it does not differ from ascomatal margins in other ascomycetes. As in these, the margin is called either exciple in discomycetes, peridium in pyrenomycetes or pseudoperidium in loculascomycetes. In numerous lichens, the ascomatal margin may contain algae. This margin may be an addition to the exciple and the ascoma is then called zeorine. In several lichens the thalline margin almost completely replaces the true exciple, especially in apotheciate forms. Such apothecia are called lecanorine, while apothecia with a true exciple only are called biatorine, or lecideine, when carbonaceous. The true exciple may be present at the margin (annular) or completely surround the hymenium (cupulate). The different apothecial margin types are descriptive and often have low taxonomic significance, since numerous intermediate forms exist. An algae-containing apothecial margin may be simply a continuation of the vegetative thallus which is raised with the growing fruiting body. In other cases, the generative tissue raises algae upwards or incorporates them secondarily during the ascoma development. The number of possible convergencies and special types is very high and it is often impossible to conclude from the mature ascoma as to its development. Thus, the taxonomic importance of characters of the mature ascomatal margin is limited, if the development of these structures is not examined.

Below the hymenium, which consists of asci and sterile hyphae, a thin layer with ascogenous hyphae can be observed which is called the subhymenium. Below the subhymenium, a further part can be distinguished, the hypothecium, which may or may not be pigmented.

The hymenium consists of asci and vegetative hyphae which are called hamathecium (Eriksson 1981). The hamathecium can consist of paraphyses

which are apically free hyphae. Paraphysoids are reticulately branched hyphae which are primarily not free apically. They may occur in mature ascomata (e.g., *Pertusaria*) or be superceded by paraphyses (e.g., *Trapelia*). In other groups no paraphysoids are formed, but true paraphyses grow into the primordium (e.g., *Coccotrema*). In ascolocularous fungi, pseudoparaphyses may be present or an interthecial tissue is formed. Beside the sterile hyphae in the hymenium, there are further sterile hyphae in the ascomatal centre which also belong to the hamathecium. In perithecial forms often sterile hyphae occur near the ostiolum and are called periphyses. A special form of sterile hyphae are the lateral paraphyses (periphysoids sensu Henssen and Jahns 1974) occurring in the Gyalectales and Ostropales which grow horizontally into the ascomatal centre at the hymenial level (Henssen 1995c). The upper part of the hymenium often differs in pigmentation, granulation or other characters from the rest of the hymenium and is called epihymenium. In some taxa, such as the Pertusariales, a compact epithecium is formed (Henssen 1976).

Beside the ascoma ontogeny, the type of meiosporangium, the ascus, is used as a primary character to distinguish different groups among ascomycetes. There are three main types of asci, i.e., prototunicate, unitunicate and bitunicate, which all occur in lichen-forming fungi. In prototunicate asci there is no mechanism for spore ejaculation and the spores are released by decay of the ascus wall. In lichenised ascomycetes prototunicate asci are secondarily reduced, as molecular studies support (Wedin and Tibell 1997; Wedin et al. 1998). Unitunicate asci may either open with an operculum (operculate) or are inoperculate. Operculate asci do not occur in lichenised fungi. Unitunicate asci in the Lecanorales have two walls which do not separate, or only slightly, when opening (Honegger 1978, 1980, 1982c). In bitunicate asci the two walls are separated in the spore release. The outer wall or exoascus is not flexible and bursts open, the inner wall or endoascus prolongs until or above the upper surface of the hymenium and releases the ascospores. Since most unitunicate asci in lichenized ascomycetes also have two walls, it was proposed to use the term fissitunicate and non-fissitunicate instead of bitunicate and unitunicate. In unitunicate asci different opening mechanisms were observed. The most common is the rostrate type of most Lecanorales. The ascus prolongs above the hymenium surface with an apical apparatus. The apical tips of the asci have different structures which show different amyloid reactions. With the help of these amyloid reactions, different ascus types were distinguished and used in the classification (e.g., Hafellner 1984). Although these types are generally of much importance, a considerable amount of variation in ascus types within natural groups has also been demonstrated (Rambold et al. 1994). Further, it was emphasized (Baral 1992) that structures may alter in herbarium material and that there is a danger of misinterpretations due to artefacts.

As in non-lichenized ascomycetes, the most common number of ascospores per ascus is eight. The number, however, can be reduced to one, two or four. Polyspored asci occur in numerous groups and recent studies show that this character has been largely overestimated in previous studies (Hafellner and Casares-Porcel 1992; Hafellner 1995). The form of ascospores is usually species-specific. Beside simple spores, one-septate, transversally septate and muriform spores occur. The form of the cell lumen also differs. while most taxa have spores with rectangular cells, lenticular cells are characteristic for the ascospores in such groups as Pyrenulaceae or Thelotremataceae. A special type of ascospore is bipolarous to plurilocularous spores in the Teloschistaceae. The ascospores may be surrounded by a gelatinous perispore and are then called halonate, as in *Rhizocarpon* spp. They may or may not show amyloid reactions, such as in members of the Thelotremataceae.

E. Anamorph

Almost 60% of the known lichen-forming fungi are known to produce conidial anamorphs (Vobis and Hawksworth 1981; Hawksworth 1988b). In most cases these propagules complement the ascomycetous teleomorphs, but some species and also genera are not known to produce teleomorphs, such as *Siphula* (Mathey 1971; Kantvilas 1987, 1994). The conidia are usually produced in conidiomata, which are in most cases globose to pyriform in shape opening with a single pore, the ostiolum. These structures are called pycnidia (Vobis 1980). However, other forms of conidiomata occur in lichen-forming fungi as well, including helmet-shaped campylidia (Serusiaux 1986; Vezda 1986; Aptroot and Sipman 1993b; Malcolm and Vezda 1994), erect hyphophores (Vezda 1979), and cushion-like sporodochia

(Tibell 1991). In some cases, conidia are produced directly from single hyphae rather than aggregated hyphae. Such anamorphs occur for example in the Coniocybaceae (Honegger 1985a; Tibell 1997). In the cases mentioned above, conidia are produced by conidiogenous cells. The conidiogenous cells can be simple or branched and anastomosing. Different types of conidiophores, according to the ramification of the conidiogenous cells, were distinguished by Vobis (1980). Vobis (1980) also studied the ontogeny of the pycnidia and distinguished several types.

There exist further types of asexual spores in lichenised ascomycetes which do not arise from conidiophores. These are thallospores (or thalloconidia) which occur in the Umbilicariaceae (Hestmark 1990, 1991a,b,c) and a variety of crustose lichens (Poelt and Obermayer 1990).

In certain groups of lichenised fungi more than one type of conidia is known in a single species. This is well documented in the genus *Micarea* where three different conidial types are known in some species (e.g., *M. denigrata*) and have been termed micro-, meso- and macroconidia according to their size (Coppins 1983). Different conidia are known from several groups of lichens, including the Gyalectaceae, Lecanoraceae and Strigulaceae (Letrouit-Galinou 1973; Etayo and Diederich 1993; Roux and Bricaud 1993).

The role of conidia in lichens has been disputed for many years and there is some confusion concerning the terminology for conidiomata and conidia – some authors prefer to use the term spermatia and spermogonia (Johnson 1954; Poelt 1980; Thell 1995). Although there are indications that at least some of the conidia may function as spermatia (Honegger 1984), there is no definite proof for this. Some conidia were shown to grow out and thus obviously function as vegetative propagules (Vobis 1977). Both roles cannot be assumed to be mutually exclusive, and thus it seems unfortunate to link terms with functions in these cases. The neutral terms conidia and conidiomata, which are commonly used in mycology, should be applied here.

Conidiomata and conidia have also been used in the taxonomy of lichens. Conidial differences provided help in distinguishing two morphologically similar *Punctelia* spp. (Culberson and Culberson 1980). Also the size and shape of conidia was used as a character for the circumscription of genera, for example in the Parmeliaceae (Elix 1993) or Physciaceae (Moberg 1977).

III. Chemistry

The investigation of secondary metabolites plays an important part in taxonomic and biological studies in lichenised ascomycetes. This method has had a long tradition in lichenology since differences in thallus coloration due to lichen substances have always been used to distinguish species (e.g., *Parmeliopsis ambigua, P. hyperopta*). The first chemical tests were introduced by Nylander (1866), who employed spot tests with reagents to produce characteristic colour changes in the lichen tissue. Later, more reagents and other methods such as microcrystallisation or paper chromatography were used. The investigation of lichen substances was vastly improved by the introduction of thin-layer chromatography, for which Culberson and colleagues developed standardised methodologies (Culberson 1972; Culberson and Johnson 1976, 1982; Culberson et al. 1981), and – more recently – high performance liquid chromatography, for which Feige et al. (1993) developed a standardised method for the identification of phenolic lichen substances in an elution gradient. Data on chromatographic behaviour of lichen substances and additional information are now available on computer (Elix et al. 1988; Mietzsch et al. 1993). The different methodologies for the examination of lichen substances are summarized by Culberson and Elix (1989) and Huneck and Yoshimura (1995).

Almost 750 secondary products have been found in lichenised ascomycetes and further substances are continuously being found. The lichen substances belong to different substance classes (C. Culberson 1969, 1970; C. Culberson et al. 1977; Elix 1996) and are produced via the acetate-polymalonate, mevalonic acid and shikimic acid pathways (Mosbach 1969). These metabolites accumulate on the outside walls of the mycobiont hyphae. The majority of secondary metabolites in lichens is synthethised by the acetate-polymalonate pathway and belongs to substance groups, such as depsides and depsidones. Substances unique to lichens usually belong to these groups, while substances in other groups – or closely related substances – might occur in other groups of either plants or fungi as well. A list of the most important substance classes of secondary metabolites occurring in lichens is given in Table 2.

Table 2. Major classes of lichen substances

I. Acetate-polymalonate pathway
  A. Fatty acids and related compounds
  B. Phenolic compounds
    1. Anthraquinones
    2. Benzyl esters
    3. Chromones
    4. Depsides
    5. Depsidones
    6. Depsones
    7. Dibenzofuranes, usnic acids and related compounds
    8. Diphenyl ethers
    9. Monocyclic compounds
    10. Naphthoquinones
    11. Napthopyran
    12. Tridepsides
    13. Xanthones

II. Mevalonic acid pathway
  A. Steroids
  B. Terpenoids

III. Shikimic acid pathway
  A. Pulvinic acid derivatives
  B. Terphenylquinones

Chemical data can be used as taxonomic characters at different levels, and their taxonomic importance has been one of the most controversial issues in lichen systematics, as discussed by numerous authors (e.g., W. L. Culberson 1969, 1986; Hawksworth 1976; Brodo 1978, 1986; Rogers 1989; Lumbsch 1998). Secondary metabolites have been widely used at species and subspecific level, but this cannot be discussed here and the reader is referred to the above-cited references for further information. At generic level, chemical information if often used as a correlating character. Moberg (1977) excluded the genus *Phaeophyscia* from *Physcia* based on absence of atranorin and presence of different conidia. Metabolic data are also used in Parmeliaceae for generic delimitation, as summarized by Elix (1993). Among crustose lichens examples are less obvious; with the current trend of defining smaller natural units in crustose groups, however, some genera are chemically well defined. *Rhizoplaca* for example, is characterised by placodiolic or pseudoplacodiolic acids (Leuckert et al. 1977; McCune 1987), and chemistry played an important role in the distinction of subgenera in *Pertusaria* by Archer (1993).

Some metabolic data may have significance at family level. In the genus *Haematomma* two groups were recognised based mainly on their different ascus type. These groups are now regarded as belonging to different families, the Haematom-maceae and Ophioparmaceae (Hafellner 1984; Rogers and Hafellner 1988). Both groups are morphologically similar due to the presence of red pigments in the apothecia, but the structure of the quinones present in both families was shown to be different (Bruun and Lamvik 1971; Huneck et al. 1991), thus supporting the separation of the two groups.

At order level metabolic data are less useful, but the occurrence of secondary metabolites nonetheless reveals certain patterns at this level. While some orders, such as the Arthoniales, Lecanorales, or Pertusariales, contain a number of substances of various classes, others, such as the Gyalectales, lack any lichen compounds. The presence of the substance classes among the orders with lichenised fungi differs as well. While depsones are restricted to the Pertusariales, depsides or quinones are present in numerous groups.

IV. Molecular Investigations

Molecular data provide a powerful tool to study the different aspects in the biology of organisms, including population genetics and phylogeny. The introduction of molecular methods into lichenology has already demonstrated the potential power of molecular techniques (cf. Gargas et al. 1995a; Bridge and Hawksworth 1998) in this field as well. Specific DNA isolation protocols for lichen fungi have been developed (Armaleo and Clerc 1995; Grube et al. 1995). The introduction of fungal-specific primers (Gargas and Taylor 1992; Gardes and Bruns 1993; Gargas and DePriest 1996; Crespo et al. 1997) has considerably enhanced the application of molecular techniques in lichenology. The majority of the recent molecular studies focuses on systematic questions at different taxonomic levels (e.g., DePriest 1993a; Gargas and Taylor 1995; Gargas et al. 1995a; Lutzoni and Vilgalys 1995a,b; Tehler 1995b; Wedin and Tibell 1997; Arup and Grube 1998; Crespo and Cubero 1998; Mattsson and Wedin 1998; Stenroos and DePriest 1998a; Thell 1998; Thell et al. 1998; Wedin et al. 1998; Winka et al. 1998). Their significance for the classification at ordinal and family level is discussed in section V (Classification). Further, molecular data were employed to study species concepts in lichenised fungi, with regard to derived sterile populations (Kroken and Taylor

1998; Lohtander et al. 1998a,b), photosym-biodemes (Armaleo and Clerc 1991) or chemical variation (Groner and LaGreca 1997). The potential power of small insertions and group I introns for resolving evolutionary relationships among genera and species of lichen-forming fungi was recently demonstrated by Stenroos and DePriest (1998b). However, molecular studies in lichenology are not restricted to phylogenetic and taxonomic questions, but address numerous other issues, such as presence and variation of group I introns (DePriest 1993b; Gargas et al. 1995b; Grube et al. 1996) or genetic diversity and specificity of symbiotic cyanobacteria (Miao et al. 1997; Paulsrud and Lindblad 1998), as well as questions of molecular evolution with regard to mutualism (Lutzoni and Pagel 1997). A recent workshop on progress in molecular studies in lichens (Graz, August 1998) demonstrated that molecular methods will be increasingly used in a variety of different issues in lichenology.

V. Classification

The modern classification which includes the lichenised ascomycetes in a system with non-lichenised fungi, based on mycological characters such as ascus structure and ascoma development, started with Santesson (1952, 1953), although the main ideas had already been developed by Nannfeldt (1932). The systems proposed by Hale (1967) and Hale and Culberson (1970) followed their concepts. Comprehensive systems were proposed by Henssen and Jahns (1974) and Poelt (1973). In the meantime, numerous changes have been proposed, especially since Hafellner's (1984) critical reevaluation of the variation in ascus structure within the Lecanoraceae and Lecideaceae s. ampl. and the creation of numerous new families based mainly on ascus types. An overview of his revisions is given in Hafellner (1988). From 1982 to 1993, Eriksson, and Eriksson and Hawksworth published an *Outline of the Ascomycetes* including lichenised fungi (Hawksworth and Eriksson 1994) each year (Eriksson and Hawksworth 1993). Their classification scheme forms the basis of our proposed classification of lichenised ascomycetes. All orders which include lichenised members are mentioned and the orders are briefly characterised by a standardised description. All families including lichenised species are mentioned and relevant literature is given. An overview of the orders arranged into supraordinal taxa is presented by Eriksson and Winka (1997).

Recent years have seen a remarkable increase in interest in molecular data for systematic purposes in the systematics of the Ascomycota. Particular attention has been paid to the small subunit of the ribosomal DNA. This trend has, with some delay, also appeared in lichenology, as discussed above. Thus, some differences to the system proposed by Eriksson and Hawksworth (1993) resulting from current studies are briefly discussed. Another systematic arrangement was presented by Tehler (1996), whose treatment was also based on Eriksson and Hawksworth (1993). Poorly known families, such as the Moriolaceae (Bachmann 1926), are not further treated here.

A. Agyriales

Ascomata apothecial; ascoma development ascohymenial, hemiangiocarpous or gymnocarpous; exciple cupulate or annular, zeorine, lecideine or biatorine; hamathecium with true paraphyses replacing paraphysoids, weakly or non-amyloid; ascus unitunicate, *Trapelia*-type, *Schaereria*-type or similar forms; ascospores hyaline, non- to transversely septate; pycnidia with conidiophores of type II, V, VII (Vobis 1980), conidia bacilliform, filiform; thallus crustose, placodioid or squamulose, primary photobiont a green alga or absent; chemistry: benzyl esters and derivatives, orcinol depsides and β-orcinol depsidones. Most taxa lichenised.

Families: Agyriaceae s.l. (incl. Rimulariaceae, Trapeliaceae), Anamylopsoraceae, Elixiaceae Schaereriaceae (Hertel 1970; Hafellner 1984; Hertel and Rambold 1990; Rambold and Triebel 1990; Lumbsch et al. 1995b, 2000; Lunke et al. 1995).

This order has hitherto been recognised as a suborder Agyriineae in the Lecanorales. It differs from that order, however, in having non-lecanoralean asci, which open by eversion (Bellemère and Letrouit-Galinou 1987; Lumbsch 1997b) and weakly amyloid hymenial gel. The order is similar to the Leotiales, but differs in having thicker asci. Recent molecular studies (Lumbsch et al. 2000) support the recognition as a separate order.

B. Arthoniales

Ascomata apothecial or hysterothecial; ascoma development ascohymenial, hemiangiocarpous or gymnocarpous; exciple absent or annular, zeorine, or lecideine; hamathecium with paraphysoids, weakly or non-amyloid; ascus bitunicate, fissitunicate, amyloid or non-amyloid; ascospores hyaline or brown, transversely septate to muriform; pycnidia with conidiophores of type I, II, III, IV (Vobis 1980); conidia bacilliform, filiform or ovoid; thallus crustose, placodioid, foliose or fruticose, photobiont a green alga or absent; chemistry: aliphatic acids, chromones, dibenzofuranes, naphthopranes, orcinol depsides, β-orcinol depsides, β-orcinol depsidones, quinones, terpenes and xanthones. Most taxa lichenised.

Families: Arthoniaceae, Chrysotrichaceae, Roccellaceae (incl. Opegraphaceae) (Henssen and Jahns 1974; Barr 1987; Tehler 1990, 1995a; Henssen and Thor 1994; Letrouit-Galinou et al. 1994; Grube 1998; Myllys et al. 1998).

The Arthoniales represent a group of ascomycetes with bitunicate asci and ascohymenial ascoma development. This shows that the classification of Ascohymeniales and Ascoloculares by Nannfeldt (1932) and the distinction on the basis of ascus structure by Luttrell (1951, 1973) is not congruent. In our opinion the Loculoascomycetes should be restricted to those ascomycetes which have both ascolocular ascoma development and bitunicate asci. However, bitunicate asci also occur in the probably paraphyletic group of ascohymenial fungi (cf. Tehler in Hawksworth 1994, p. 416). The term Zwischengruppe was used by Henssen and Jahns (1974) and Henssen and Thor (1994) to circumscribe this phenomenon. We prefer to interpret the Arthoniales as bitunicate members of the ascohymenial fungi, in the same way that the Pyrenulales and Verrucariales are considered ascohymenial pyrenomycetes with bitunicate asci (Parguey-Leduc and Janex-Favre 1981), thus not being extraordinary.

C. Dothideales s. lat.

Ascomata pseudothecial; ascoma development ascolocularous; with or without pseudoperidium; hamathecium with pseudoparaphyses or interascal pseudoparenchyma, weakly or non-amyloid; ascus bitunicate, fissitunicate; ascospores hyaline or brown, non- to transversely septate or muriform; pycnidia with conidiophores of different types, also hyphomycetous anamorphs known, conidiophores and conidia of different types; thallus crustose, photobiont a green alga, cyanobacteria or absent; chemistry: no lichen substances. Most taxa non-lichenised.

Families with lichenised taxa: Arthopyreniaceae, Dacampiaceae, Pyrenotrichaceae, Xanthopyreniaceae (Luttrell 1951, 1973; Janex-Favre 1970; von Arx and Müller 1975; Eriksson 1981; Barr 1987; Harris 1995; Henssen 1995b; Berbee 1996).

The Dothideales as accepted here are clearly a collective and heterogeneous group. We have followed the broad concepts of von Arx and Müller (1975) and Eriksson and Hawksworth (1993) rather than those of Luttrell (1951) and Barr (1987), who divided the order into several groups, simply because only very few lichenised taxa are involved and we are not familiar with this large group of mostly non-lichenised ascomycetes. Furthermore, Henssen (1987) suggested that the hamathecial structure, which was used as an important character for the acceptance of orders among the Loculoascomycetes, may vary within some groups of these fungi. We therefore follow a more conservative view here for practical reasons only. However, recent preliminary molecular data suggest that at least two major clades, i.e., the Dothideales sensu stricto and the Pleosporales, can be distinguished among the ascolocularous fungi (Berbee 1996). The Xanthopyreniaceae are only provisionally kept separate from Arthopyreniaceae following Harris (1995) and the Mycoporaceae are interpreted as non-lichenised following Lumbsch (1999).

D. Gyalectales

Ascomata apothecial (± perithecioid in early stages of development); ascoma development ascohymenial, hemiangiocarpous; exciple cupulate, zeorine or biatorine; hamathecium with true paraphyses growing into a cavity, weakly or non-amyloid; with or without lateral paraphyses; ascus unitunicate, thin-walled, amyloid; ascospores hyaline, non- to transversely septate to muriform; pycnidia with conidiophores of type I (Vobis 1980), conidia bacilliform to ovoid; thallus crustose or filamentous; photobiont a green alga (*Trentepohlia*)

or cyanobacteria; chemistry: no lichen substances. All taxa lichenised.

Family: Gyalectaceae (Henssen and Jahns 1974; Henssen 1976).

The ascoma development of the Gyalectales and Ostropales is very similar and might suggest a close relationship. However, the two orders can be distinguished by their different asci and the amyloidity of the hymenial gelatine. The differences in ascus type and hymenial characters indicate that the similarity in ascoma ontogeny is due to convergence. The presence or absence of lateral paraphyses, which is often mentioned, cannot be regarded as important in distinguishing the Gyalectales and Ostropales, since this character varies within the Ostropales (e.g., within the Thelotremataceae, cf. Hale 1980).

E. Icmadophilales ad int.

Ascomata apothecial; ascoma development ascohymenial, gymnocarpous or hemiangiocarpous; exciple annular, biatorine; hamathecium with true paraphyses replacing paraphysoids, weakly or non-amyloid; ascus unitunicate, helotial, amyloid or only outer ascus wall faintly amyloid; ascospores hyaline, non- to transversely septate; pycnidia of lichenised members with conidiophores of type VI (Vobis 1980), conidia of various shape; thallus crustose, squamulose or foliose, photobiont a green alga or absent; chemistry: orcinol depsides and β-orcinol depsides and β-orcinol depsidones. Most taxa lichenised.

Families with lichenised taxa: Icmadophilaceae, Siphulaceae (Honegger 1983; Gierl and Kalb 1993; Rambold et al. 1993; Eriksson and Hawksworth 1996; Stenroos and DePriest 1998a).

While in earlier systems (Poelt 1973; Henssen and Jahns 1974) the Icmadophilaceae were placed in the Lecanorales, they were subsequently regarded as members of the Leotiales, due to their ascus type (Chadefaud 1960; Honegger 1983). However, recent 18S rDNA data suggest that *Icmadophila ericetorum* might not belong here and that similarities in ascus structure may be explained as convergence (Eriksson and Hawksworth 1996). This assumption is confirmed by Stenroos and DePriest (1998a).

F. Lecanorales s. lat. (incl. Caliciales, Lichinales, Peltigerales, Teloschistales)

Ascomata apothecial; ascoma development asco-hymenial, hemiangiocarpous or gymnocarpous; exciple cupulate or annular, zeorine, lecanorine, lecideine or biatorine; hamathecium with true paraphyses, amyloid; ascus lecanoralean (= unitunicate, apically thickened), amyloid, or prototunicate; ascospores hyaline or brown, non- to transversely septate to muriform; pycnidia with conidiophores of different types, conidia of various shape; thallus of different growth forms; photobiont a green alga or/and cyanobacteria; chemistry: aliphatic acids, benzyl ester, chromones, depsides, depsidones, dibenzofuranes, diphenyl ether, pulvinic acid derivatives, quinones, steroids, terpenes, and xanthones. Most taxa lichenised.

Families: Acarosporaceae, Arctomiaceae, Bacidiaceae, Baeomycetaceae, Biatorellaceae, Brigantiaeaceae, Caliciaceae, Calycidiaceae, Catillariaceae, Cladiaceae, Cladoniaceae, Coccocarpiaceae, Collemataceae, Coniocybaceae, Crocyniaceae, Ectolechiaceae, Fuscideaceae, Gloeoheppiaceae, Gypsoplacaceae, Haematommaceae, Heppiaceae, Heterodeaceae, Hymeneliaceae (incl. Eigleraceae), Lecanoraceae (incl. Candelariaceae), Lecideaceae, Letrouitiaceae, Lichinaceae, Lobariaceae, Loxosporaceae, Megalosporaceae, Micareaceae, Miltideaceae, Mycoblastaceae, Nephromataceae, Ophioparmaceae, Pannariaceae, Parmeliaceae s. lat. (incl. Alectoriaceae, Anziaceae, Hypogymniaceae, Usneaceae), Peltigeraceae, Peltulaceae, Phlyctidaceae, Physciaceae, Pilocarpaceae, Placynthiaceae, Porpidiaceae, Psoraceae, Ramalinaceae, Rhizocarpaceae, Sphaerophoraceae, Stereocaulaceae, Teloschistaceae, Umbilicariaceae, Vezdaeaceae (Hakulinen 1954; Henssen 1963, 1965, 1969a,b, 1970, 1979, 1994, 1995a; Jahns 1970; Schmidt 1970; Poelt 1973; Keuck 1977, 1979; Filson 1978; Hafellner et al. 1979; Poelt and Hafellner 1980; Bellemère and Letrouit-Galinou 1981; Hafellner and Bellemère 1981a,b,c; Henssen et al. 1981; Ahti 1982; Krog 1982; Honegger 1983; Sipman 1983; Hafellner 1984, 1995; Tibell 1984, 1987; Timdal 1984, 1990, 1992; Hertel and Rambold 1995; Eriksson and Hawksworth 1986, 1996; Vezda 1986; Henssen and Dobelmann 1987; Rogers and Hafellner 1988; Rambold 1989; Hafellner and Casares-Porcel 1992; Kärnefelt and Thell 1992; Elix 1993; Gierl and Kalb 1993; Rambold et al. 1993, 1994; Wedin 1993; Jörgensen 1994; Lücking et al. 1994; Eriksson and Strand 1995; Gargas and Taylor 1995; Gargas et al. 1995a; Lumbsch et al. 1995a; Lutzoni and Brodo 1995; Staiger and Kalb 1995; Thell et al. 1995; Lumbsch 1997a,b; Wedin and Tibell 1997; Döring and Lumbsch 1998; Mattsson and Wedin 1998; Stenroos and DePriest 1998a; Wedin et al. 1998).

The Lecanorales is the largest order among lichenised ascomycetes and is probably polyphyletic, an assumption which is supported by molecular data (Stenroos and DePriest 1998a). In our delimitation more than 50 families are included in the Lecanorales. The circumscription of this order varies between authors. While Hafellner (1994) has a rather broad concept of this order and includes the Pertusariales as well as the Peltigerales and Teloschistales (but excludes the Lichinales), Eriksson and Hawksworth (1993) have a more restricted concept of the Lecanorales. On the basis of differences in ascus structure and ascoma development they excluded the Lichinales, Peltigerales, Pertusariales and Teloschistales as separate orders. We have followed a middle course in this matter for practical reasons. The Pertusariales are well separated from the Lecanorales in ascus structure and ascoma development and no intermediates are known to us. Further, molecular data support a distinction of this group. Therefore we have accepted this group as a distinct order. The Lichinales, Peltigerales and Teloschistales are included here in the Lecanorales, since intermediate families occur which cannot be related to one or the other group at present.

The Peltulaceae are somewhat intermediate between the Lecanorales and Lichinales, since the ascus in this group is lecanoralean, but the ontogeny of the fruiting bodies is similar to that of the Lichinaceae. Furthermore, *Pyrenopsis*, a member of the Lichinaceae, and related taxa have unitunicate asci, rather than the prototunicate asci typical of Lichinaceae. The Pannariaceae might belong either to the Lecanorales or Peltigerales, and the Placynthiaceae seem to be somewhat isolated in the Peltigerineae. Moreover, analysis of 18S rDNA showed that the Peltigerineae clustered with the Lecanorales s.str. (Eriksson and Strand 1995). Therefore, we prefer to accept the Peltigerineae as a suborder in the Lecanorales rather than as an independent order. Regarding Teloschistales, the Fuscideaceae show features characteristic of this order, but might also belong to Lecanorales s.str. Since the circumscription of the group is still in flux, we are recognizing the group as a suborder for the time being. Molecular data will surely help to distinguish natural groups among the Lecanorales s.lat. much better than it has been possible so far.

Some families with prototunicate asci (e.g., Caliciaceae, Sphaerophoraceae) were formerly placed in the Caliciales, which was defined mainly by the presence of a mazaedium and passive spore dispersal. Most taxa have more or less stalked ascomata. While Poelt (1973), following Zahlbruckner (1926), regarded the order as "very uniform", Henssen and Jahns (1974) only tentatively placed the non-lichenised, unitunicate Mycocaliciaceae here. Tibell (1984, in Hawksworth 1994, p. 393), as did von Höhnel (1919), believed the Caliciales to be a polyphyletic group that includes different lineages of lichenised fungi which have lost active spore dispersal, and suggested that most families in this order should be placed incertae sedis. Tehler (1996) in contrast, placed the Caliciales and Lichinales (which we included in Lecanorales) as sister groups to the unitunicate and bitunicate ascomycetes. In the case of Caliciales and Lichinales, we cannot see any supporting characters other than the prototunicate asci to suggest any close relationship of the two groups. However, it seems obvious to us that the prototunicate asci of Lichinales (or Lichinineae) are secondary and must be interpreted in connection with the degradation of asci and spores in this group (Henssen and Jahns 1974). Moreover, some members of this group have unitunicate asci. Molecular data support that the prototunicate asci in some Caliciales are also secondary and that the group is polyphyletic (Gargas et al. 1995a; Wedin and Tibell 1997; Wedin et al. 1998). Consequently, the lichen-forming families are here placed within the Lecanorales in which the taxa cluster in the phylogenetic hypothesis inferred from rDNA data. Other fungi formerly placed in the Caliciales, such as the Sphinctrinaceae, do not belong here. The Baeomycetaceae are tentatively placed in the Lecanorales and may belong to a separate order or represent a cladoniiform offshot of the Pertusariales. Further molecular data are necessary to decide this question.

Rambold and Triebel (1992) proposed a new subordinal classification of the Lecanorales which was largely adopted by Hafellner et al. (1993). They distinguish the Acarosporineae, Agyriineae, Cladoniineae, Lecanorineae, Lichineae (separate order in Hafellner et al. 1993), Peltigerineae, Pertusariineae, Teloschistineae and Umbilicarineae. We treat the Pertusariineae as a separate order, as discussed above. The distinction of the Teloschistineae and Umbilicarineae has to be restudied. However, molecular data suggest that the Umbilicarineae should be separated at ordinal level (DePriest et al. 1997; Stenroos and DePriest 1998a). We accept the Lichineae and Peltigerineae as suborders as discussed above. The Acarosporineae seem to be a heterogeneous assemblage whose members do not have many characters in common. Examinations of the ascus structure and ascoma ontogeny are urgently needed, as well as more molecular data of other taxa and genes.

It was proposed to divide the core group of the Lecanorales into two suborders, the Cladoniineae and Lecanorineae, based on different staining abilities of the ascal tholus with iodine (e.g., Rambold and Triebel 1992). However, we are not aware of any other correlating characters. A new classification of the Lecanorineae s.lat. should be postponed until careful studies have been carried out to determine whether any partition is necessary. Since some anatomical characters are difficult to interpret, molecular data might prove helpful in this case as well. Interestingly, the character used by Rambold and Triebel (1992) and Hafellner (1994) to distinguish Cladoniineae and Lecanorineae is questioned by Hafellner (1994, p. 317) himself. He refers to *Sporopodium*, which is tentatively placed in the Cladoniineae because it produces campylidia, although its ascus type does not fit in this suborder. These strange types of conidiomata, however, were shown by Lücking et al. (1994) to be weak taxonomic characters, since they occur in different groups of tropical lichens. Further, Printzen (1995) regarded *Biatora*, a genus of the Lecanorineae s.str., as closely related to *Micarea*, a genus with an ascus type typical for "Cladoniineae". Recent molecular studies confirm that the Cladoniineae do not form a monophyletic group (Stenroos and DePriest 1998a; Wedin et al. 1998). A strict distinction of Cladoniineae and Lecanorineae would produce unnatural groups, and leave a large number of genera and families not placed in either of both suborders.

G. Ostropales (incl. Graphidales)

Ascomata apothecial or hysterothecial (± perithecioid in early stages of development); ascoma development ascohymenial, hemiangiocarpous; exciple cupulate, biatorine, lecideine or zeorine; hamathecium with true paraphyses, non-amyloid, with or without lateral paraphyses; ascus unitunicate, non-amyloid; ascospores hyaline or brown, transversely septate or muriform; pycnidia with conidiophores of different types; conidia of various shape, filiform; thallus crustose; photobiont a green alga or absent; chemistry: orcinol depsides, β-orcinol depsidones, quinones, terpenes and xanthones.

Families with lichenised taxa: Gomphillaceae, Graphidaceae, Stictidaceae, Solorinellaceae, Thelotremataceae (Santesson 1952; Gilenstam 1969; Sherwood 1977, 1987; Hale 1980, 1981; Hafellner 1984; Vezda 1987; Vezda and Poelt 1990; Winka et al. 1998).

Sherwood (1977) proposed to maintain a distinction between the Graphidales with mostly lichenised members and the Ostropales, which includes mostly non-lichenised fungi, based on different spore septation types. The fundamental difference between the different spore septation types was questioned by Lumbsch et al. (1997). Recent molecular data (Winka et al. 1998) support a close relationship of the two groups and hence the Graphidales are here included in the Ostropales.

The family concept in this order is rather obscure and needs restudying. The Graphidaceae and Thelotremataceae are generally distinguished by their fruiting bodies being hysterothecia in the former and apothecia in the latter. Some authors distinguish both families based on presence or absence of periphysoids (Sherwood 1977) or slight differences in the ascomatal development (Henssen and Jahns 1974). However, overlaps occur in all cases and it may be necessary to unite these two families. The Stictidaceae are said to have typically long-filiform ascospores, while the Odontotremataceae – a family with non-lichenised fungi – is assumed to differ in having ovoid spores (Sherwood 1987). However, some of the genera with lichenised taxa currently placed in the Stictidaceae have ovoid spores (e.g., *Absconditella*). The placement of the Gomphillaceae in the Ostropales is tentative, since, according to Hafellner (1984), the asci in this family are bitunicate. If this is indeed the case, the family should be separated as an independent order. The relation of Gyalectales and Ostropales is discussed under the former.

H. Patellariales

Ascomata apothecial or hysterothecial; ascoma development ascolocular; with a well-developed peridium; hamathecium with paraphysoids, non- or weakly amyloid; ascus bitunicate; ascospores more or less hyaline, transversely septate to muriform; pycnidia not known in lichenised taxa; thallus crustose, photobiont a green alga or absent; chemistry: orcinol depsides, β-orcinol depsides, β-orcinol depsidones and pulvinic acid derivatives. Most taxa non-lichenised.

Family with lichenised taxa: Arthrorhaphidaceae (Bellemère 1967; Poelt and Hafellner 1976; Barr 1987; Obermayer 1994; Hafellner and Obermayer 1995).

According to Obermayer (1994), the Arthrorhaphidaceae show similarities to the Lahmiales and may belong there; alternatively, the Lahmiales and Patellariales might be synonymous. The whole group is poorly known and needs critical examination.

I. Pertusariales

Ascomata apothecial; ascoma development ascohymenial, hemiangiocarpous; exciple cupulate, zeorine; hamathecium with paraphysoids, amyloid, hemiamyloid or non-amyloid; ascus unitunicate, *Megaspora* or *Pertusaria* type; ascospores hyaline, thick-walled, non- to transversely septate; pycnidia with conidiophores of type III, IV (Vobis 1980), conidia bacilliform, filiform; thallus crustose or squamulose, photobiont a green alga; chemistry: aliphatic acids, depsides, depsidones, depsones, and xanthones. All taxa lichenised.

Families: Megasporaceae, Pertusariaceae (incl. Ochrolechiaceae sensu Harris 1990) (Letrouit-Galinou 1960; Henssen 1976; Honegger 1982a; Lumbsch et al. 1994, 1995a).

This order was included in the Lecanorales (as Pertusariineae) by Henssen and Jahns (1974) and Hafellner et al. (1993). We follow Eriksson and Hawksworth (1993) here, since the group is well separated from the Lecanorales by the ascoma development and the different ascus type and no intermediate families exist. Moreover, molecular data support such a treatment (DePriest et al. 1997; Stenroos and DePriest 1998a). The group contains two families. While the Megasporaceae is monotypic, the Pertusariaceae includes four genera. The core group of this family, the genus *Pertusaria*, shows a remarkable variability in secondary metabolites. The formerly associated

Coccotremataceae differ in ascoma development (Henssen 1976) and are probably closely related to pyrenocarpous lichens rather than to the Pertusariales; they are therefore treated here as incertae sedis.

J. Pyrenulales

Ascomata perithecial; ascoma development ascohymenial, peridium pigmented or not; hamathecium with true paraphyses, weakly or non-amyloid; ascus bitunicate, non-amyloid; ascospores hyaline or brown, transversely septate to muriform, with or without internal wall thickenings; pycnidia with different conidiophore types; conidia of different shape; thallus crustose; photobiont a cyanobacteria, green alga or absent; chemistry: quinones and xanthones. Lichenised and non-lichenised members present.

Families: Monoblastiaceae, Pyrenulaceae, Thelenellaceae (incl. Aspidotheliaceae), Trypetheliaceae (Janex-Favre 1970; Eriksson 1981; Parguey-Leduc and Janex-Favre 1981; Harris 1986, 1989, 1995; Aptroot 1991a,b; Hafellner and Kalb 1995).

Since we have little experience with this group of mainly tropical and subtropical pyrenocarpous crustose lichens, we have largely followed the concepts of Eriksson and Hawksworth (1993). Harris (1995) and others have proposed to include the Pyrenulales in the Melanommatales. This order was included in Dothideales s.lat. by Eriksson and Hawksworth (1993). If the Melanommatales have an ascolocular ascoma development, as a member of the Dothideales should have, they cannot be regarded as closely related to Pyrenulales, which is characterized by ascohymenial development (Parguey-Leduc and Janex-Favre 1981).

K. Trichotheliales

Ascomata perithecial; ascoma development ascohymenial; peridium encrusted with certain pigments; hamathecium with true paraphyses and periphyses, non-amyloid; ascus unitunicate, non-amyloid, with external chitinoid ring structure, rarely reduced; ascospores hyaline, transversely septate to muriform; thallus crustose, photobiont a green alga (Trentepohliales); chemistry: only quinones known. All known taxa lichenised.

Familiy: Trichotheliaceae (Janex-Favre 1970; Aptroot and Sipman 1993a; Hafellner and Kalb 1995; Harris 1995).

This order was recently separated from the Pyrenulales on the basis of different ascal, hamathecial and peridial structure (Hafellner and Kalb 1995). It is a rather small group containing four or five genera with a distribution centre from the tropics to the oceanic extratropics. Members of this order grow on different substrata, but are especially common on bark, leaves and acidic rock. The generic distinction in the Trichotheliaceae is disputed and three concepts currently exist, which all differ considerably (McCarthy 1993, 1995; Hafellner and Kalb 1995; Harris 1995; McCarthy and Malcolm 1997).

L. Verrucariales

Ascomata perithecial; ascoma development ascohymenial; peridium pigmented or not; hamathecium with periphyses, paraphyses mostly absent, non-amyloid; ascus bitunicate, non-amyloid; ascospores hyaline or brown, non- to transversely septate or muriform; thallus crustose, filamentous, placodioid, squamulose or foliose, photobiont belonging to different groups or absent; chemistry: no lichen substances known. Most taxa lichenised.

Family with lichenised taxa: Verrucariaceae (Doppelbaur 1960; Janex-Favre 1970; Henssen 1976, 1995a; Parguey-Leduc and Janex-Favre 1981).

This order is a very uniform group of lichenised and lichenicolous fungi. While the distinction of the group from other orders of pyrenomycetes does not seem to be controversial, the acceptance of monophyletic entities within the Verrucariales is difficult and most genera are artificially distinguished.

M. Incertae sedis

The following families cannot be included in one of the previously mentioned orders and are therefore placed incertae sedis. They cannot be discussed further here; the reader is referred to the cited literature.

Coccotremataceae, Protothelenellaceae, Strigulaceae, Thrombiaceae (Santesson 1952; Poelt 1973; Henssen 1976; Mayrhofer and Poelt 1985; Printzen and Rambold 1995).

Acknowledgments. We wish to thank Prof. Dr. G.B. Feige, R. Guderley and I. Schmitt (all Essen), Prof Dr. H.M. Jahns (Düsseldorf), Dr. W. Sanders (Recife), and M.I. Messuti (Bariloche) for constructive critisism on an earlier draft of this chapter.

References

Ahmadjian V (1993) The lichen symbiosis. Wiley, New York

Ahti T (1982) The morphological interpretation of cladoniform thalli in lichens. Lichenologist 14:105–113

Aptroot A (1991a) A monograph of the Pyrenulaceae (Excluding *Anthracothecium* and *Pyrenula*) and the Requienellaceae, with notes on the Pleomassariaceae, the Trypetheliaceae and *Mycomicrothelia* (lichenised and non-lichenised ascomycetes). Bibl Lichenol 44:1–178

Aptroot A (1991b) Tropical pyrenocarpous lichens. A phylogenetic approach. In: Galloway DJ (ed) Tropical lichens: their systematics, conservation, and ecology. Systematics Association Spec Vol No 43. Clarendon Press, Oxford, pp 253–273

Aptroot A, Sipman HJM (1993a) Trichotheliaceae. Flora Guianas Ser E, 2:1–57

Aptroot A, Sipman H (1993b) *Musaespora*, a genus of pyrenocarpous lichens with campylidia, and other additions to the foliicolous lichen flora of New Guinea. Lichenologist 25:121–135

Archer AW (1993) A chemical and morphological arrangement of the lichen genus *Pertusaria*. Bibl Lichenol 53:1–17

Armaleo D, Clerc P (1991) Lichen chimeras: DNA analysis suggests that one fungus forms two morphotypes. Exp Mycol 15:1–10

Armaleo D, Clerc P (1995) A rapid and inexpensive method for the purification of DNA from lichens and their symbionts. Lichenologist 27:207–213

Arup U, Grube M (1998) Molecular systematics of *Lecanora* subgenus *Placodium*. Lichenologist 30:415–425

Bachmann E (1926) Die Moriolaceen. Nytt Mag Naturvidensk 64:170–228

Baral HO (1992) Vital versus herbarium taxonomy: morphological differences between living and dead cells of ascomycetes, and their taxonomic implications. Mycotaxon 44:333–390

Barr ME (1987) Prodromus to Class Loculoascomycetes. Univ of Massachussets, Amherst

Bellemère A (1967) Contribution a l'étude du développement de l'apothécie chez les discomycètes inoperculés. Bull Soc Mycol Fr 83:395–931

Bellemère A, Letrouit-Galinou MA (1981) The lecanoralean ascus: an ultrastructural preliminary study. In: Reynolds DR (ed) Ascomycete systematics. The Luttrellian Concept. Springer, Berlin Heidelbarg New York, pp 54–70

Bellemère A, Letrouit-Galinou MA (1987) Differentiation of lichen asci including dehiscence and sporogenesis: an ultrastructural survey. Bibl Lichenol 25: 137–161

Berbee ML (1996) Loculoascomycete origins and evolution of filamentous ascomycete morphology based on 18S rRNA gene sequence data. Mol Biol Evol 13:462–470

Berbee ML, Taylor JW (1995) From 18S ribosomal sequence data to evolution of morphology among the fungi. Can J Bot 73 (Suppl 1):S677–S683

Bridge PD, Hawksworth DL (1998) What molecular biology has to tell us at the species level in lichenised fungi? Lichenologist 30:307–320

Brodo IM (1978) Changing concepts regarding chemical diversity in lichens. Lichenologist 10:1–11

Brodo IM (1986) Interpreting chemical variation in lichens for systematic purposes. Bryologist 89:132–138

Bruun T, Lamvik A (1971) Haemaventosin. Acta Chem Scand 25:483–486

Bubrick P, Galun M, Frensdorff A (1984) Observations on free-living *Trebouxia* DePuymaly and *Pseudotrebouxia* Archibald, and evidence that both symbionts from *Xanthoria parietina* (L.) Th. Fr. can be found free-living in nature. New Phytol 97:455–462

Büdel B, Rhiel E (1987) Studies on the ultrastructure of some cyanolichen haustoria. Protoplasma 139:145–152

Chadefaud M (1960) Traité de Botanique Systématique I. Les Végétaux non Vasculaires Cryptogamie. Masson, Paris

Clauzade G, Diederich P, Roux C (1989) Nelikenigintaj fungoj likenlogaj – Ilustrita determinlibro. Bull Soc Linn Prov, Num Spec 1:1–142

Codogno M, Poelt J, Puntillo D (1989) *Umbilicaria freyi* spec. nov. und der Formenkreis von *Umbilicaria hirsuta* in Europa (Lichens, Umbilicariaceae). Plant Syst Evol 165:55–69

Coppins BJ (1983) A taxonomic study of the lichen genus *Micarea* in Europe. Bull Br Mus (Nat Hist) Bot Ser 22:17–214

Crespo A, Cubero OF (1998) A molecular approach to the circumscription and evaluation of some genera segregated from *Parmelia* s. lat. Lichenologist 30:369–380

Crespo A, Bridge PD, Hawksworth DL (1997) Amplification of fungal rDNA-ITS regions from non-fertile specimens of the lichen-forming genus *Parmelia*. Lichenologist 29:275–282

Culberson CF (1969) Chemical and Botanical Guide to Lichen Products. University of North Carolina Press, Chapel Hill

Culberson CF (1970) Supplement to chemical and botanical guide to lichen products. Bryologist 73:177–377

Culberson CF (1972) Improved conditions and new data for the identification of lichen products by a standardized thin-layer chromatographic method. J Chromatogr 72:113–125

Culberson CF, Elix JA (1989) Lichen substances. In: Harborne JB (ed) Methods in plant biochemistry vol 1. Plant Phenolics. Academic Press, London, pp 509–535

Culberson CF, Johnson A (1976) A standardized two-dimensional thin-layer chromatographic method for lichen products. J Chromatogr 128:253–259

Culberson CF, Johnson A (1982) Substitution of methyl tert.-butyl ether for diethyl ether in the standardized thin-layer chromatographic method for lichen products. J Chromatogr 238:483–487

Culberson CF, Culberson WL, Johnson A (1977) Second supplement to chemical and botaincal guide to lichen products. Missouri Botanical Garden, St Louis

Culberson CF, Culberson WL, Johnson A (1981) A standardized TLC analysis of β-orcinol depsidones. Bryologist 84:16–29

Culberson WL (1969) The use of chemistry in the systematics of the lichens. Taxon 18:152–166

Culberson WL (1986) Chemistry and sibling speciation in the lichen-forming fungi: ecological and biological considerations. Bryologist 89:123–131

Culberson WL, Culberson CF (1980) Microconidial dimorphism in the lichen genus *Parmelia*. Mycologia 72:127–135

Degelius G (1940) Studien über die Konkurrenzverhältnisse der Laubflechten auf nacktem Fels. Medd Göteborgs Bot Trädg 14:195–219

DePriest PT (1993a) Molecular innovations in lichen systematics: the use of ribosomal and intron nucleotide sequences in the *Cladonia chlorophaea* complex. Bryologist 96:314–325

DePriest PT (1993b) Small subunit rDNA variation in a population of lichen fungi due to optional group-I introns. Gene 134:67–74

DePriest P, Steenroos S, Ivanova N, Gargas A (1997) Polyphyletic origins of the lichen association in the fungi: phylogenetic analyses of nuclear small subunit ribosomal DNA sequences. Am J Bot 84, Suppl: 9

Döbbeler P (1984) Symbiosen zwischen Gallertalgen und Gallertpilzen der Gattung *Epigloea* (Ascomycetes). Beih Nova Hedwigia 79:203–239

Doppelbaur H (1960) Ein Beitrag zur Anatomie und Entwicklungsgeschichte von *Dermatocarpon miniatum* (L.) Mann. Nova Hedwigia 2:279–286

Döring H, Lumbsch HT (1998) Ascoma ontogeny: is this a character set of any use in the systematics of lichenised Ascomycetes? Lichenologist 30:489–500

Elix JA (1993) Progress in the generic delimitation of *Parmelia* sensu lato lichens (Ascomycotina: Parmeliaceae) and a synoptic key to the Parmeliaceae. Bryologist 96:359–383

Elix JA (1996) Biochemistry and secondary metabolites. In: Nash TH (ed) Lichen biology. Cambridge University Press, Cambridge, pp 154–180

Elix JA, Johnston J, Parker JL (1988) A computer program for the rapid identification of lichen substances. Mycotaxon 31:89–99

Englund B (1977) The physiology of the lichen *Peltigera aphthosa*, with special reference to the blue-green phycobiont (Nostoc. sp.). Physiol Plant 41:298–304

Eriksson O (1981) The families of bitunicate ascomycetes. Opera Bot 60:1–209

Eriksson O, Hawksworth DL (1986) Outline of the ascomycetes – 1986. Syst Ascomycetum 5:185–324

Eriksson OE, Hawksworth DL (1993) Outline of the ascomycetes – 1993. Syst Ascomycetum 12:51–257

Eriksson OE, Hawksworth DL (1996) Notes on ascomycete systematics – Nos. 2024–2139. Syst Ascomycetum 14:101–133

Eriksson OE, Strand A (1995) Relationships of the genera *Nephroma*, *Peltigera* and *Solorina* (Peltigerales, Ascomycota) inferred from 18S rDNA. Syst Ascomycetum 14:33–39

Eriksson OE, Winka K (1997) Supraordinal taxa of Ascomycota. Myconet 1:1–16

Etayo J, Diederich P (1993) *Lecanora schistina* (Nyl.) Arnold, a lichen with dimorphic conidia. Lichenologist 25:365–368

Feige GB, Jensen M (1992) Basic carbon and nitrogen metabolism of lichens. In: Reisser W (ed) Algae and symbioses: plants, animals, fungi, viruses, interactions explored. Biopress, Bristol, pp 277–299

Feige GB, Lumbsch HT, Schmitz KE (1992) Die Ausbildung eines Zentralstranges in der Flechtenfamilie Rocellaceae (Opegraphales; Ascomycotina): Anatomische Untersuchungen an *Simonyella variegata*. Flora 187:159–167

Feige GB, Lumbsch HT, Huneck S, Elix JA (1993) A method for the identification of aromatic lichen substances by gradient high performance liquid chromatography. J Chromatogr 646:417–427

Filson RB (1978) A revision of the genus *Heterodea* Nyl. Lichenologist 10:13–25

Friedl T (1989) Systematik und Biologie von *Trebouxia* (Microthamniales, Chlorophyta) als Phycobiont der Parmeliaceae (lichenisierte Ascomyceten). Ph Dissertation, Bayreuth

Gardes M, Bruns TD (1993) ITS primers with enhanced specifity for basidiomycetes – application to the identification of mycorrhizae and rusts. Mol Ecol 2:113–118

Gargas A, DePriest PT (1996) A nomenclature for fungal PCR primers with examples from intron-containing SSU rDNA. Mycologia 88:745–748

Gargas A, Taylor JW (1992) Polymerase chain reaction (PCR) primers for amplifying and sequencing nuclear 18S rDNA from lichenised fungi. Mycologia 84:589–592

Gargas A, Taylor JW (1995) Phylogeny of discomycetes and early radiation of the apothecial Ascomycotina inferred from SSU rDNA sequence data. Exp Mycol 19:7–15

Gargas A, DePriest PT, Grube M, Tehler A (1995a) Multiple origins of lichen symbioses in fungi suggested by SSU rDNA phylogeny. Science 268:1492–1495

Gargas A, DePriest PT, Taylor JW (1995b) Positions of multiple insertions in SSU rDNA of lichen-forming fungi. Mol Biol Evol 12:208–218

Gierl C, Kalb K (1993) Die Flechtengattung *Dibaeis*. Eine Übersicht über die rosafrüchtigen Arten von *Baeomyces* sens. lat. nebst Anmerkungen zu *Phyllobaeis* gen. nov. Herzogia 9:593–645

Gilenstam G (1969) Studies in the lichen genus *Conotrema*. Ark Bot Ser 2 7:149–179

Groner U, LaGreca S (1997) The "Mediterranean" *Ramalina panizzei* north of the Alps: morphological, chemical and rDNA sequence. Lichenologist 29:441–454

Grube M (1998) Classification and phylogeny in the Arthoniales (lichenised Ascomycetes). Bryologist 101:377–391

Grube M, DePriest PT, Gargas A, Hafellner J (1995) DNA isolation from lichen ascomata. Mycol Res 99:1321–1324

Grube M, Gargas A, DePriest PT (1996) A small insertion in the SSU rDNA of the lichen fungus *Arthonia lapidicola* is a degenerate group-I intron. Curr Gen 29:582–586

Hafellner J (1984) Studien in Richtung einer natürlicheren Gliederung der Sammelfamilien Lecanoraceae und Lecideaceae. Beih Nova Hedwigia 79:241–371

Hafellner J (1988) Principles of classification and main taxonomic groups. In: Galun M (ed) CRC Handbook of lichenology, vol 3. CRC Press, Boca Raton, pp 41–52

Hafellner J (1994) Problems in Lecanorales systematics. In: Hawksworth DL (ed) Ascomycete systematics. Problems and perspectives in the nineties. Plenum Press, New York, pp 315–320

Hafellner J (1995) Towards a better circumscription of the Acarosporaceae (lichenized Ascomycotina, Lecanorales). Cryptogam Bot 5:99–104

Hafellner J, Bellemere A (1981a) Elektronenoptische Untersuchungen an Arten der Flechtengattung *Bombyliospora* und die taxonomischen Konsequenzen. Nova Hedwigia 35:207–235

Hafellner J, Bellemere A (1981b) Elektronenoptische Untersuchungen an Arten der Flechtengattung *Brigantiaea*. Nova Hedwigia 35:237–261

Hafellner J, Bellemere A (1981c) Elektronenoptische Untersuchungen an Arten der Flechtengattung *Letrouitia* gen. nov. Nova Hedwigia 35:263–312

Hafellner J, Casares-Porcel M (1992) Untersuchungen an den Typusarten der lichenisierten Ascomycetengattungen *Acarospora* und *Biatorella* und die daraus entstehenden Konsequenzen. Nova Hedwigia 55:309–323

Hafellner J, Kalb K (1995) Studies in Trichotheliales *ordo novus*. Bibl Lichenol 57:161–186

Hafellner J, Obermayer W (1995) *Cercidospora trypetheliza* und einige weitere lichenicole Ascomyceten auf *Arthrhorhaphis*. Cryptogam Bryol Lichenol 16:177–190

Hafellner J, Mayrhofer H, Poelt J (1979) Die Gattungen der Flechtenfamilie Physciaceae. Herzogia 5:39–79

Hafellner J, Hertel H, Rambold G, Timdal E (1993) A new outline of the Lecanorales. Unpubl manuscript distributed to participants of The 1st Int Workshop on Ascomycete Systematics, Paris, May 11–14, 1993

Hakulinen R (1954) Die Flechtengattung *Candelariella* Müll. Argoviensis, mit besonderer Berücksichtigung ihres Auftretens und ihrer Verbreitung in Fennoskandien. Ann Bot Soc Zool Bot Fenn "Vanamo" 27:1–127

Hale ME Jr (1957) Conidial state of the lichen fungus *Buellia stilligiana* and its relation to *Sporidesmium folliculatum*. Mycologia 49:417–419

Hale ME Jr (1967) The biology of lichens. Arnold, London

Hale ME Jr (1980) Generic delimitation in the lichen family Thelotremataceae. Mycotaxon 11:130–138

Hale ME Jr (1981) A revision of the lichen family Thelotremataceae in Sri Lanka. Bull Br Mus (Nat Hist) Bot Ser 8:227–332

Hale ME Jr, Culberson WL (1970) A fourth checklist of the lichens of the continental United States and Canada. Bryologist 73:499–543

Hammer S (1993) Development in *Cladonia ochrochlora*. Mycologia 85:84–92

Harris R (1986) The family Trypetheliaceae (Loculoascomycetes: lichenised Melanommatales) in Amazonian Brazil. Suppl Acta Amazonica 14:55–80

Harris RC (1989) A sketch of the family Pyrenulaceae (Melanommatales) in eastern North America. Mem NY Bot Gard 49:74–107

Harris RC (1990) Some Florida lichens. Published by the author, New York

Harris RC (1995) More Florida lichens including the 10c tour of the pyrenolichens. Published by the author, New York

Hawksworth DL (1976) Lichen chemotaxonomy. In: Brown DH, Hawksworth DL, Bailey RH (eds) Lichenology. Progress and problems. Academic Press, London, pp 139–184

Hawksworth DL (1982) Secondary fungi in lichen symbioses: parasites, saprophytes and parasymbionts. J Hattori Bot Lab 52:357–366

Hawksworth DL (1983) A key to the lichen-forming, parasitic, parasymbiotic and saprophytic fungi occurring on lichens in the British Isles. Lichenologist 15:1–44

Hawksworth DL (1987) Observations on three algicolous microfungi. Notes R Bot Gard Edinb 44:549–560

Hawksworth DL (1988a) The variety of fungal-algal symbioses, their evolutionary significance, and the nature of lichens. Bot J Linn Soc 96:3–20

Hawksworth DL (1988b) Conidiomata, conidiogenesis, and conidia. In: Galun M (ed) CRC Handbook of lichenology vol I. CRC Press, Boca Raton, pp 181–193

Hawksworth DL (1994) Ascomycete systematics. Problems and perspectives in the nineties. Plenum Press, New York

Hawksworth DL, Eriksson OE (1994) Systema Ascomycetum: the concept. In: Hawksworth DL (ed) Ascomycete systematics. Problems and perspectives in the nineties. Plenum Press, New York, pp 349–355

Henssen A (1963) Eine Revision der Flechtenfamilien Lichinaceae und Ephebaceae. Symb Bot Ups 18:1–123

Henssen A (1965) A review of the genera of the Collemataceae with simple spores (excluding *Physma*). Lichenologist 3:29–41

Henssen A (1968) *Thyrea radiata*, eine *Lempholemma*-Art mit Hormocystangien. Ber Dtsch Bot Ges 81:176–182

Henssen A (1969a) Die Entstehung des Thallusrandes bei den Pannariaceen (Lichens) mit einer generellen Diskussion über die Entwicklung lecanoriner und biatoriner Flechtenapothecien. Ber Dtsch Bot Ges 82:235–248

Henssen A (1969b) Eine Studie über die Gattung *Arctomia*. Sven Bot Tidskr 63:126–138

Henssen A (1970) Die Apothecienentwicklung bei *Umbilicaria* Hoffm. emend Frey. Vortr Gesamtgeb Bot NF 4:103–126

Henssen A (1976) Studies in the developmental morphology of lichenised ascomycetes. In: Brown DH, Hawksworth DL, Bailey RH (eds) Lichenology: Progress and Problems. Academic Press, London, pp 107–138

Henssen A (1979) Problematik der Gattungsbegrenzung bei den Lichinaceen. Ber Dtsch Bot Ges 92:483–506

Henssen A (1987) *Lichenothelia*, a genus of microfungi on rocks. Bibl Lichenol 25:257–293

Henssen A (1994) Contribution to the morphology and species delimitation in *Heppia* sensu stricto (lichenized Ascomycotina). Acta Bot Fenn 150:57–73

Henssen A (1995a) *Psoroglaena costaricensis*, a new lichen from Costa Rica, and remarks on other taxa of the genus *Psoroglaena* (Verrucariaceae). Bibl Lichenol 57:199–210

Henssen A (1995b) Studies on the biology and structure of *Dacampia* (Dothideales), a genus with lichenised and lichenicolous species. Cryptogam Bot 5:149–158

Henssen A (1995c) *Sagiolechia atlantica*, eine neue Flechte von den Atlantischen Inseln (Ascomycotina, Ostropales). Bibl Lichenol 58:123–136

Henssen A, Dobelmann A (1987) Development of the spongiostratum in *Anzia* and *Pannoparmelia*. Bibl Lichenol 25:103–108

Henssen A, Jahns H (1974) Lichens. Georg Thieme, Stuttgart

Henssen A, Thor G (1994) Developmental morphology of the "Zwischengruppe" between Ascohymeniales and Ascoloculares. In: Hawksworth DL (ed) Ascomycete systematics. Problems and perspectives in the nineties. Plenum Press, New York, pp 43–61

Henssen A, Keuck G, Renner B, Vobis G (1981) The Lecanoralean centrum. In: Reynolds DR (ed) Ascomycete systematics. The luttrellian concept. Springer, Berlin Heidelberg New York, pp 138–234

Hertel H (1970) Trapeliaceae – eine neue Flechtenfamilie. Vortr Gesamtgeb Bot NF 4:171–185

Hertel H, Rambold G (1990) Zur Kenntnis der Familie Rimulariaceae (Lecanorales). Bibl Lichenol 38:145–189

Hertel H, Rambold G (1995) On the genus Adelolecia (lichenised Ascomycotina, Lecanorales). Bibl Lichenol 57:211–230

Hestmark G (1990) Thalloconidia in the genus Umbilicaria. Nord J Bot 9:547–574

Hestmark G (1991a) Teleomorph-anamorph relationships in Umbilicaria I. Making the connections. Lichenologist 23:343–359

Hestmark G (1991b) Teleomorph-anamorph relationships in Umbilicaria II. Patterns in propagative morph production. Lichenologist 23:361–380

Hestmark G (1991c) To sex, or not to sex . . . structures and strategies of reproduction in the family Umbilicariaceae (Lecanorales, Ascomycetes). Sommerfeltia, Suppl 3:1–47

Hitch CJB, Millbank JW (1975a) Nitrogen metabolism in lichens. VI. The blue-green phycobiont content, heterocyst frequency and nitrogenase activity in Peltigera species. New Phytol 74:473–476

Hitch CJB, Millbank JW (1975b) Nitrogen metabolism in lichens. VII. Nitrogenase activity and heterocyst frequency in lichens with blue-green phycobionts. New Phytol 75:239–244

Honegger R (1978) The ascus apex in lichenized fungi. I. The Lecanora-, Peltigera- and Teloschistes- types. Lichenologist 10:47–67

Honegger R (1980) The ascus apex in lichenized fungi II. The Rhizocarpon-type. Lichenologist 12:157–172

Honegger R (1982a) The ascus apex in lichenised fungi III. The Pertusaria-type. Lichenologist 14:205–217

Honegger R (1982b) Cytological aspects of the triple symbiosis in Peltigera aphthosa. J Hattori Bot Lab 52:379–391

Honegger R (1982c) Ascus structure and function, ascospore delimitation, and phycobiont cell wall types associated with the Lecanorales (lichenized ascomycetes). J Hattori Bot Lab 52:417–429

Honegger R (1983) The ascus apex in lichenised fungi IV. Baeomyces and Icmadophila in comparison with Cladonia (Lecanorales) and the non-lichenised Leotia (Helotiales). Lichenologist 15:57–71

Honegger R (1984) Scanning electron microscopy of the contact site of conidia and trichogynes in Cladonia furcata. Lichenologist 16:11–19

Honegger R (1985a) The hyphomycetous anamorph of Coniocybe furfuracea. Lichenologist 17:273–279

Honegger R (1985b) Fine structure of different types of symbiotic relationships in lichens. In: Brown DH (ed) Lichen physiology and cell biology. Plenum Press, New York, pp 287–302

Honegger R (1986a) Ultrastructural studies in lichens. I. Haustorial types and their frequencies in a range of lichens with trebouxioid photobionts. New Phytol 103:785–795

Honegger R (1986b) Ultrastructural studies in lichens. II. Mycobiont and photobiont cell wall surface layers and adhering crystalline lichen products in four Parmeliaceae. New Phytol 103:797–808

Honegger R (1988) The functional morphology of cell-to-cell interactions in lichens. In: Scannerini S, Smith D, Bonfante-Fasolo P, Gianinazzi-Pearson V (eds) Cell to cell signals in plant, animal and microbial symbiosis. Springer, Berlin Heidelberg New York, pp 39–53

Honegger R (1991) Functional aspects of the lichen symbiosis. Annu Rev Plant Physiol Plant Mol Biol 42:553–578

Huneck S, Yoshimura I (1995) Identification of lichen substances. Springer, Berlin Heidelberg New York

Huneck S, Culberson CF, Culberson WL, Elix JA (1991) Haematommone, a red pigment from apothecia of Haematomma puniceum. Phytochemistry 30:706–707

Jahns H (1970) Untersuchungen zur Entwicklungsgeschichte der Cladoniaceen unter besonderer Berücksichtigung des Podetien-Problems. Nova Hedwigia 20:1-vi, 1–177

Jahns HM (1984) Morphology, reproduction and water relations – a system of morphogenetic interactions in Parmelia saxatilis. Beih Nova Hedwigia 79:715–737

Jahns HM (1988) The establishment, individuality and growth of lichen thalli. Bot J Linn Soc 96:21–29

Jahns HM, Ott S (1994) Thallic mycelial and cytological characters in ascomycete systematics. In: Hawksworth DL (ed) Ascomycete systematics. Problems and perspectives in the nineties. Plenum Press, New York, pp 57–62

Jahns HM, Ott S (1997) Life strategies in lichens – some general considerations. Bibl Lichenol 67:49–67

Jahns HM, Klöckner P, Ott S (1995a) Development of thalli and ascocarps in Solorina spongiosa (Sm.) Anzi and Solorina saccata (L.) Ach. Bibl Lichenol 57:241–251

Jahns HM, Sensen M, Ott S (1995b) Significance of developmental structures in lichens, especially in the genus Cladonia. Ann Bot Fenn 32:35–48

James PW, Henssen A (1976) The morphological and taxonomic significance of cephalodia. In: Brown DH, Hawksworth DL, Bailey RH (eds) Lichenology: progress and problems. Academic Press, London, pp 27–77

Janex-Favre MC (1970) Recherches sur l'ontogénie, l'organisation et les asques de quelques pyrénolichens. Rev Bryol Lichenol 37:421–650

Johnson G (1954) Ascogonia and spermatia of Stereocaulon. Mycologia 46:337–345

Jørgensen PM (1994) Studies in the lichen family Pannariaceae VI: the taxonomy and phytogeography of Pannaria Del. s. lat. J Hattori Bot Lab 76:197–206

Jørgensen PM, Jahns HM (1987) Muhria, a remarkable new lichen genus from Scandinavia. Notes R Bot Gard Edinb 44:581–599

Kärnefelt I, Thell A (1992) The evaluation of characters in lichenized families, exemplified with the alectorioid and some parmelioid genera. Pl Syst Evol 180:181–204

Kantvilas G (1987) Siphula jamesii, a new lichen from south-western Tasmania. Nord J Bot 7:585–588

Kantvilas G (1994) Siphula elixii, a new lichen from Tasmania and New Zealand. NZ J Bot 32:17–20

Keuck G (1977) Ontogenetisch-systematische Untersuchungen über Erioderma im Vergleich mit anderen cyanophilen Flechtengattungen. Bibl Lichenol 6:1–236

Keuck G (1979) Die systematische Stellung der Ramalinaceae. Ber Dtsch Bot Ges 92:507–518

Kohlmeyer J (1967) Intertidal and phycophilous fungi from Tenerife (Canary Islands). Trans Br Mycol Soc 50:137–147

Kohlmeyer J, Kohlmeyer E (1972) Is *Ascophyllum nodosum* lichenised? Bot Mar 15:109–112

Krog H (1982) Evolutionary trends in foliose and fruticose lichens of the Parmeliaceae. J Hattori Bot Lab 52:303–311

Kroken S, Taylor JW (1998) A multi-locus approach to reveal speciation and life history changes in the genus *Letharia*. Inoculum 49:29

Lange OL, Meyer A, Zellner H, Ullmann I, Wessels DCJ (1990) Eight days in the life of a desert lichen: water relations and photosynthesis of *Teloschistes capensis* in the coastal fog zone of the Namib Desert. Madoqua 17:17–30

Lange OL, Büdel B, Meyer A, Kilian E (1993) Further evidence that activation of net photosynthesis by dry cyanobacterial lichens requires liquid water. Lichenologist 25:175–189

Letrouit-Galinou M (1960) Sur le developpement des apothecies du discolichen *Pertusaria pertusa* Tuck. (=*P. communis* D.C.). CR Seanc Acad Sci Paris 250:3701–3703

Letrouit-Galinou MA (1973) Les pycnospores et les pycnides du *Gyalecta carneolutea* (Turn.) Oliv. Bull Soc Bot Fr 120:373–384

Letrouit-Galinou MA, Bellemère A (1989) Ascomatal development in lichens: a review. Cryptogam Bryol Lichenol 10:189–233

Letrouit-Galinou M-A, Bellemère A, Torrente P (1994) Études ultrastructurales d'asques et d'ascospores chez quelques Opegraphaceae, Roccellaceae et Arthoniaceae. Bull Soc Linn Prov 45:389–416

Leuckert C, Poelt J, Hähnel G (1977) Zur Chemotaxonomie der eurasischen Arten der Flechtengattung *Rhizoplaca*. Nova Hedwigia 28:71–129

Lewin RA (1995) Symbiotic algae: definitions, quantification and evolution. Symbiosis 19:31–51

Lohtander K, Källersjö M, Tehler A (1998a) Dispersal strategies in *Roccellina capensis* (Arthoniales). Lichenologist 30:341–350

Lohtander K, Myllys L, Sundin R, Källersjö M, Tehler A (1998b) The species pair concept in the lichen *Dendrographa leucophaea* (Arthoniales): analyses based on ITS sequences. Bryologist 101:404–411

Lücking R, Lumbsch HT, Elix JA (1994) Chemistry, anatomy and morphology of foliicolous species of the genera *Fellhanera* and *Badimia* (Ascomycetes, Lecanorales). Bot Acta 107:393–401

Lumbsch HT (1997a) A comparison of ascoma ontogeny supports the inclusion of the Eigleraceae in the Hymeneliaceae (Lecanorales). Bryologist 100:180–192

Lumbsch HT (1997b) Systematic studies in the suborder Agyriineae (Lecanorales). J Hattori Bot Lab 83:1–73

Lumbsch HT (1998) The taxonomic use of metabolic data in lichen-forming fungi. In: Frisvad JC, Bridge PD, Arora DK (eds) Chemical fungal taxonomy. M Dekker, New York, pp 345–387

Lumbsch HT (1999) The ascoma development in *Mycoporum elabens* (Mycoporaceae, Dothideales). Plant Biol 1:321–326

Lumbsch HT, Feige GB, Schmitz KE (1994) Systematic studies in the Pertusariales I. Megasporaceae, a new family of lichenised Ascomycetes. J Hattori Bot Lab 75:295–304

Lumbsch HT, Dickhäuser A, Feige GB (1995a) Systematic studies in the Pertusariales III. Taxonomic position of

Thamnochrolechia (lichenised Ascomycetes). Bibl Lichenol 57:355–361

Lumbsch HT, Lunke T, Feige GB, Huneck S (1995b) Anamylopsoraceae – a new family of lichenised ascomycetes with stipitate apothecia (Agyriineae; Lecanorales). Plant Syst Evol 198:275–286

Lumbsch HT, Guderley R, Feige GB (1997) Ascospore sepatation in *Diploschistes* and the taxonomic significance of macro- and microcephalic spore types. Plant Syst Evol 205:179–184

Lumbsch HT, Schmitt I, Döring H, Wedin M (1999) Molecular Systematics of the Agyriales. Mycol Res (in press)

Lunke T, Lumbsch HT, Feige GB (1995) Anatomical and ontogenetical studies on the Schaereriaceae (Agyriineae, Lecanorales). Bryologist 99:53–63

Luttrell ES (1951) Taxonomy of the pyrenomycetes. Univ Mo Stud 24:1–120

Luttrell ES (1973) Loculoascomycetes. In: Ainsworth GC, Sparrow FK, Sussman AS (eds) The Fungi, vol IVA. Academic Press, New York, pp 135–219

Lutzoni FM, Brodo IM (1995) A generic redelimitation of the *Ionaspis-Hymenelia* complex (lichenized Ascomycotina). Syst Bot 20:224–258

Lutzoni FM, Pagel M (1997) Accelerated evolution as a consequence of transitions to mutualism. Proc Natl Acad Sci, USA 94:11422–1427

Lutzoni F, Vilgalys R (1995a) Integration of morphological and molecular data sets in estimating fungal phylogenies. Can J Bot 73 (Suppl 1):S649–S659

Lutzoni F, Vilgalys R (1995b) *Omphalina* (Basidiomycota, Agaricales) as a model system for the study of coevolution in lichens. Cryptogam Bot 5:71–81

Malcolm WM, Vezda A (1994) *Badimiella serusiauxii*, a new genus and species of foliicolous lichens from New Zealand (Ectolechiaceae). Nova Hedwigia 59:517–523

Mathey A (1971) Contribution a l'étude du genre *Siphula* (lichens) en Afrique. Nova Hedwigia 22:795–878

Matthes-Sears U, Gerrath JA, Larson DW (1997) Abundance, biomass, and productivity of endolithic and epilithic lower plants on the temperate-zone cliffs of the Niagara Escarpment, Canada. Int J Plant Sci 158:451–460

Mattsson JE, Wedin M (1998) Phylogeny of the Parmeliaceae – DNA data versus morphological data. Lichenologist 30:463–472

Mayrhofer H, Poelt J (1985) Die Flechtengattung *Microglaena* sensu Zahlbruckner in Europa. Herzogia 7:13–79

McCarthy PM (1993) Saxicolous species of *Porina* Müll. Arg. (Trichotheliaceae) in the Southern Hemisphere. Bibl Lichenol 52:1–134

McCarthy PM (1995) A reappraisal of *Clathroporina* Müll. Arg. (Trichotheliaceae). Lichenologist 27:321–350

McCarthy PM, Malcolm WM (1997) The genera of Trichotheliaceae. Lichenologist 29:1–8

McCune B (1987) Distribution of chemotypes of *Rhizoplaca* in North America. Bryologist 90:6–14

Miao VPW, Rabenau A, Lee A (1997) Cultural and molecular characterization of photobionts of *Peltigera membranacea*. Lichenologist 29:571–586

Mietzsch E, Lumbsch HT, Elix JA (1993) A new computer program for the identification of lichen substances. Mycotaxon 47:475–479

Moberg R (1977) The lichen genus *Physcia* and allied genera in Fennoscandia. Symb Bot Ups 22:1–108

Mosbach K (1969) Zur Biosynthese von Flechtenstoffen, Produkte einer symbiotischen Lebensgemeinschaft. Angew Chem 81:233–244.

Mukhtar A, Garty J, Galun M (1994) Does the lichen alga *Trebouxia* occur free-living in nature: further immunological evidence. Symbiosis 17:247–253

Myllys L, Källersjö M, Tehler A (1998) A comparison of SSU rDNA data and morphological data in Arthoniales (Euascomycetes) phylogeny. Bryologist 101:70–85

Nannfeldt JA (1932) Studien über die Morphologie und Systematik der nicht-lichenisierten inoperculaten Discomyceten. Nova Act Reg Soc Sci Ups Ser 4, 8(2):1–368

Nash TH III (ed) (1996) Lichen biology. Cambridge University Press, Cambridge

Nylander W (1866) Circa novum in studio Lichenum criterium chemicum. Flora 49:198–201

Obermayer W (1994) Die Flechtengattung *Arthrorhaphis* (Arthrorhaphidaceae, Ascomycotina) in Europa und Grönland. Nova Hedwigia 58:275–333

Ott S (1987a) Sexual reproduction and developmental adaptations in *Xanthoria parietina*. Nord J Bot 7:219–228

Ott S (1987b) Reproductive strategies in lichens. Bibl Lichenol 25:81–93

Ott S (1988) Photosymbiodemes and their development in *Peltigera venosa*. Lichenologist 20:361–368

Ott S (1989) Standorte epiphytischer Flechten in einem Dünengebiet. Herzogia 8:149–175

Ott S, Zwoch I (1992) Ethylene production by lichens. Lichenologist 24:73–80

Ott S, Treiber K, Jahns HM (1993) The development of regenerative thallus structures in lichens. Bot J Linn Soc 113:61–76

Ott S, Meier T, Jahns HM (1995) Development, regeneration and parasitic interactions between the lichens *Fulgensia bracteata* and *Toninia caeruleonigricans*. Can J Bot 73 (Suppl 1):S595–S602

Parguey-Leduc A, Janex-Favre MC (1981) The ascocarps of ascohymenial pyrenomycetes. In: Reynolds DR (ed) Ascomycete systematics. The luttrellian concept. Springer, Berlin Heidelberg New York, pp 103–123

Paulsrud P, Lindblad P (1998) Sequence variation of the tRNA[Leu] Intron as a marker for genetic diversity and specificity of symbiotic cyanobacteria in some lichens. Appl Environ Microbiol 64:310–315

Poelt J (1973) Classification. In: Ahmadjian V, Hale ME (eds) The lichens. Academic Press, New York, pp 599–632

Poelt J (1980) Eine diözische Flechte. Plant Syst Evol 135:81–87

Poelt J (1983) Musterbeispiele analoger Lagerdifferenzierung bei Flechten: *Almbornia, Speerschneidera, Seirophora* gen. nov. Flora 174:439–445

Poelt J (1986) Morphologie der Flechten – Fortschritte und Probleme. Ber Dtsch Bot Ges 99:3–29

Poelt J (1995) On lichenized asexual diaspores in foliose lichens – a contribution towards a more differentiated nomenclature (lichens, Lecanorales). Cryptogam Bot 5:159–162

Poelt J, Hafellner J (1975) Schlauchpforten bei der Flechtengattung *Thelocarpon*. Phyton (Austria) 17:67–77

Poelt J, Hafellner J (1976) Die Flechte *Neonorrlina trypetheliza* und die Familie Arthrorhaphidaceae. Phyton (Austria) 17:213–220

Poelt J, Hafellner J (1980) *Apatoplaca* – genus novum Teloschistacearum (Lichens). Mitt Bot Staatssamml Münch 16:503–528

Poelt J, Obermayer W (1990) Über Thallosporen bei einigen Krustenflechten. Herzogia 8:273–288

Poelt J, Petutschnig W (1992) *Xanthoria candelaria* und ähnliche Arten in Europa. Herzogia 9:103–114

Printzen C (1995) Die Flechtengattung *Biatora* in Europa. Bibl Lichenol 60:1–275

Printzen C, Rambold G (1995) Aphanopsidaceae – a new family of lichenised ascomycetes. Lichenologist 27:99–103

Rambold G (1989) A monograph of the saxicolous lecideoid lichens of Australia (Excl. Tasmania). Bibl Lichenol 34:1–345

Rambold G, Triebel D (1990) *Gelatinopsis, Geltingia* and *Phaeopyxis*: three helotialean genera with lichenicolous species. Notes R Bot Gard Edinb 46:375–389

Rambold G, Triebel D (1992) The inter-lecanoralean associations. Bibl Lichenol 48:1–201

Rambold G, Triebel D, Hertel H (1993) Icmadophilaceae, a new family in the Leotiales. Bibl Lichenol 53:217–240

Rambold G, Mayrhofer H, Matzer M (1994) On the ascus types in the Physciaceae (Lecanorales). Plant Syst Evol 192:31–40

Rogers RW (1989) Chemical variation and the species concept in lichenised ascomycetes. Bot J Linn Soc 101:229–239

Rogers RW, Hafellner J (1988) *Haematomma* and *Ophioparma*: two superficially similar genera of lichenised fungi. Lichenologist 20:167–174

Roux C, Bricaud O (1993) Studo de la genro *Strigula* (Lichens, Strigulaceae) en S-Francio Graveco de la makrokonidioj. Bull Soc Linn Prov 44:117–134

Santesson R (1952) Foliicolous lichens I. A revision of the taxonomy of the obligately foliicolous, lichenised fungi. Symb Bot Ups 12:1–590

Santesson R (1953) The new systematics of lichenised fungi. Proc 7th Int Bot Congr Stockholm 1950:809–810

Schieleit P, Ott S (1996) Ethylene production and 1-aminocyclopropane-1-carboxylic acid content of lichen bionts. Symbiosis 21:223–231

Schmidt A (1970) Ascustypen in der Familie Caliciaceae (Ordnung Caliciales). Vortr Gesamtgeb Bot NF 4:127–137

Schwendener S (1869) Die Algentypen der Flechtengonidien. Schultze, Basel

Serusiaux E (1986) The nature and origin of campylidia in lichenised fungi. Lichenologist 18:1–35

Sherwood MA (1977) The ostropalean fungi. Mycotaxon 5:1–277

Sherwood MA (1987) The ostropalean fungi III. The Odontotremataceae. Mycotaxon 28:137–177

Sipman HJM (1983) A monograph of the lichen family Megalosporaceae. Bibl Lichenol 18:1–241

Staiger B, Kalb K (1995): *Haematomma*-Studien. I. Die Flechtengattung *Haematomma*. Bibl Lichenol 59:1–198

Stenroos SK, DePriest PT (1998a) SSU rDNA phylogeny of cladoniiform lichens. Am J Bot 85:1548–1559

Stenroos SK, DePriest PT (1998b) Small insertions at a shared position in the SSU rDNA of Lecanorales (lichen-forming Ascomycetes). Curr Genet 33:124–130

Tehler A (1990) A new approach to the phylogeny of euascomycetes with a cladistic outline of Arthoniales focussing on Roccellaceae. Can J Bot 68:2458–2492

Tehler A (1995a) Arthoniales phylogeny as indicated by morphological and rDNA sequence data. Cryptogam Bot 5:82–97

Tehler A (1995b) Morphological data, molecular data, and total evidence in phylogenetic analysis. Can J Bot 73 (Suppl 1):S667–S676

Tehler A (1996) Systematics, phylogeny and classification. In: Nash TH (ed) Lichen biology. Cambridge University Press, Cambridge, pp 217–239

Thell A (1995) Pycnoconidial types and their presence in cetrarioid lichens (Ascomycotina, Parmeliaceae). Crypt Bryol Lichenol 16:247–256

Thell A (1998) Phylogenetic relationships of some cetrarioid species in British Columbia with notes on *Tuckermannopsis*. Folia Crypt Estonica 32:113–122

Thell A, Mattsson J-E, Kärnefelt I (1995) Lecanoralean ascus types in the lichenized families Alectoriaceae and Parmeliaceae. Cryptogam Bot 5:120–127

Thell A, Berbee M, Miao V (1998) Phylogeny within the genus *Platismatia* based on rDNA ITS sequences (lichenised Ascomycota). Cryptogam Bryol Lichenol 19:307–319

Tibell L (1984) A reappraisal of the taxonomy of Caliciales. Beih Nova Hedwigia 79:597–713

Tibell L (1987) Australasian Caliciales. Symb Bot Ups 27:1–279

Tibell L (1991) The anamorph of *Tylophoron moderatum*. Mycol Res 95:290–294

Tibell L (1997) Anamorphs in mazaediate lichenised fungi and the Mycocaliciaceae ("Caliciales s. lat."). Symb Bot Ups 32, 1:291–322

Timdal E (1984) The delimitation of *Psora* (Lecideaceae) and related genera, with notes on some species. Nord J Bot 4:525–540

Timdal E (1990) Gypsoplacaceae and *Gypsoplaca*, a new family and genus of squamiform lichens. Bibl Lichenol 38:419–427

Timdal E (1992) A monograph of the genus *Toninia* (Lecideaceae, Ascomycetes). Op Bot 110:1–137

Tønsberg T, Holtan-Hartwig J (1983) Phycotype pairs in *Nephroma*, *Peltigera* and *Lobaria* in Norway. Nord J Bot 3:681–688

Triebel D, Rambold G, Elix JA (1995) A conspectus of the genus *Phacopsis* (Lecanorales). Bryologist 98:71–83

Tschermak-Woess E (1978) *Myrmecia reticulata* as a phycobiont and free-living *Trebouxia* – the problem of *Stenocybe septata*. Lichenologist 10:69–79

Tschermak-Woess E (1988) The algal partner. In: Galun M (ed) CRC Handbook of lichenology vol I. CRC Press, Boca Raton, pp 39–92

Vezda A (1979) Flechtensystematische Studien XI. Beitrage zur Kenntnis der Familie Asterothyriaceae (Discolichens). Folia Geobot Phytotaxon 14:43–94

Vezda A (1986) Neue Gattungen der Familie Lecideaceae s. lat. (Lichens). Folia Geobot Phytotaxon 21:199–219

Vezda A (1987) Flechtensystematische Studien XII. Die Familie Gomphillaceae und ihre Gliederung. Folia Geobot Phytotaxon 22:179–198

Vezda A, Poelt J (1990) Solorinellaceae, eine neue Familie der lichenisierten Ascomyceten. Phyton (Austria) 30:47–55

Vobis G (1977) Studies on the germination of lichen conidia. Lichenologist 9:131–136

Vobis G (1980) Bau und Entwicklung der Flechten-Pycnidien und ihrer Conidien. Bibl Lichenol 14:1–141

Vobis G, Hawksworth DL (1981) Conidial lichen-forming fungi. In: Cole GT, Kendrick B (ed) Biology of conidial fungi, vol 1. Academic Press, New York, pp 245–273

von Arx JA, Müller E (1975) A re-evaluation of the bitunicate ascomycetes with keys to families and genera. Stud Mycol 9:1–159

von Höhnel F (1919) Fragmente zur Mykologie. Nr 1188. Sitzungsber Akad Wiss Wien, Math Naturwiss Kl Abt 1, 128:616–621

Wedin M (1993) A phylogenetic analysis of the lichen family Sphaerophoraceae (*Caliciales*); a new generic classification and notes on character evolution. Plant Syst Evol 187:213–241

Wedin M, Tibell L (1997) Phylogeny and evolution of Caliciaceae, Mycocaliciaceae, and Sphinctrinaceae (Ascomycota), with notes on the evolution of the prototunicate ascus. Can J Bot 75:1236–1242

Wedin M, Tehler A, Gargas A (1998) Phylogenetic relationships of Sphaerophoraceae (Ascomycetes) inferred from SSU rDNA sequences. Plant Syst Evol 209:75–83

Winka K, Ahlberg C, Eriksson O (1998) Are there lichenised Ostropales? Lichenologist 30:455–462

Zahlbruckner A (1926) Lichens. B. Spezieller Teil. In: Engler A (ed) Die natürlichen Pflanzenfamilien, 2nd edn, vol 8. Bornträger, Leipzig, pp 61–270

12 Basidiolichens

F. Oberwinkler

CONTENTS

I. Introduction

Originally, basidiolichens were considered to be exclusively tropical lichens, best represented by the genera *Cora* and *Dictyonema*. In his monograph on *Clavaria* and allied genera, Corner (1950) treated several phycophilous clavarioid fungi from temperate zones in the genus *Clavulinopsis* and *Lentaria*. Mycofloristic studies initiated a breakthrough in further recognitions of basidiolichens. *Clavaria mucida* was considered to be an extratropical basidiolichen by Geitler (1955). Poelt reported *Clavulinopsis septentrion-alis* from the Alps (1959) and *Lentaria (Clavaria) mucida* from Bavaria (1962). Finally, Gams (1962)

recognized the association of *Botrydina* and *Coriscium* thalli with *Omphalina* basidiocarps. Poelt and Oberwinkler (1964) analyzed anatomical structures of *Botrydina* and *Coriscium* thalli and basidiocarps of associated *Omphalina* species. A comparative morphology of all known basidiolichen genera was published by Oberwinkler (1970). Based on transmission electron microscopic studies, Roskin (1970) reported on the ultrastructure of the host-parasite interaction in *Dictyonema glabratum (Cora pavonia)*. Detailed ultrastructural analyses were carried out by Oberwinkler (1980, 1984) for representatives of most basidiolichen genera. Three main types of fungus-alga interactions in basidiolichens were found and documented. Recently, a new genus, *Acantholichen*, has been proposed (Jørgensen 1998).

Dictyonema glabratum was and is the favoured basidiolichen for ecophysiological studies (Coxson 1987a,b,c; Feige 1969; Lange 1965; Lange et al. 1994; Larcher and Vareschi 1988).

The axenic culture of mycobionts of basidiolichens is very difficult. Experiments of Langenstein (1994) were successful to separate myco- and phycobionts, to synthesize *Botrydina*, and to characterize essential steps in the ontogeny of the thallus globules.

Lichenized *Omphalina* species and related non-lichenized ones were used as a model for understanding coevolution and the influence of lichenization on evolutionary rates (Lutzoni 1997; Lutzoni and Pagel 1997; Lutzoni and Vilgalys 1995a,b).

Figure 1 is used as a guideline for this chapter. The scheme is considered to facilitate various comparative interpretations in the following text. It comprises and compares structural components (cellular interactions, thalli, basidiocarps) with taxonomic units on the generic level (*Athelia, Athelopsis, Dictyonema, Cora, Cyphellostereum, Lepidostroma, Multiclavula,* and *Omphalina*).

Fakultät für Biologie, Lehrstuhl Spezielle Botanik und Mykologie, Universität Tübingen, Auf der Morgenstelle 1, 72076 Tübingen, Germany

The Mycota IX
Fungal Associations
Hock (Ed.)
© Springer-Verlag Berlin Heidelberg 2001

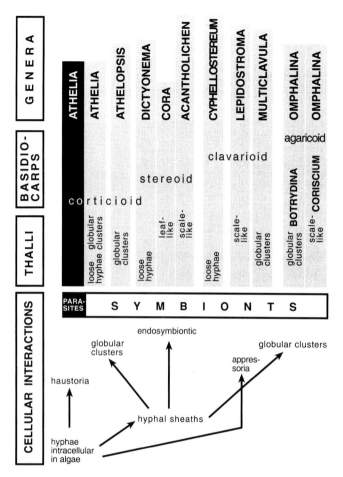

Fig. 1. Scheme of basidiolichen genera and their most important structural features, modified after Oberwinkler (1984). The arrangement of genera reflects increasing complexity of basidiocarps, i.e. corticioid-stereoid-clavarioid-agaricoid, thus corresponding with the general evolution in homobasidiomycetes. The distribution of types of lichen thalli is intermixed. Loose hyphae, globular clusters, and scaly thalli occur convergently. The cellular interactions show an informative distribution pattern: *Athelia* can be parasitized with haustoria but also surrounded by hyphae in globular clusters; hyphal sheaths in the *Dictyonema* group are always associated with endosymbiontic hyphae in the *Scytonema* symbionts; appressoria occur only in *Lepidostroma*; comparatively simple globular clusters are formed by *Multiclavula* species, complex ones in *Botrydina*, and a scale-like association of globules is present in *Coriscium*

II. Morphology of the Lichen Thalli

In basidiolichens exclusively homobasidiomycetes are lichenized. They are obligatorily associated with cyanobacteria and/or green algae. The morphological expression of this symbiontic association is the lichen thallus (Figs. 1, 2f, 3c, 4g,k; 5a,b,e,f).

The most conspicuous basidiolichen thalli are those of *Dictyonema glabratum*, better known as *Cora pavonia* (Fig. 2f). Some morphological and anatomical observations were reported by Tomaselli and Caretta (1969). The cellular composition of the thallus and the fungus-alga compartmentation was studied and illustrated by Oberwinkler (1970). He also compared *Cora* with *Dictyonema* species and discussed the considerable differences in thallus architecture of both taxa. In this comparative morphological study also *Corella* and *Vainiocora* were included. The thallus of these taxa is identical and determined by the

fungus, forming leaf-like structures with upper and lower pseudoparenchymatous layers and central compartments in which *Scytonema* trichomes are densely packed as photosynthetic units. In *Dictyonema* the *Scytonema* trichomes are long filaments ensheathed with mantles of densely longitudinally oriented hyphae which are shortly bent in a wave-like manner (Fig. 2d,e). The same structural characters are present in *Rhipidonema*, a taxon which cannot be separated from *Dictyonema*.

Also some *Athelia* species are associated with *Scytonema* and/or coccomyxoid green algae (Oberwinkler 1970). There are no macroscopically distinct thalli. The fungus-alga interaction occurs in basal layers of the basidiocarp (Fig. 2c). As shown by Oberwinkler (1970), in one fungal fructification both cyanobacteria and green algae can serve as photobionts. In contrast, the closely related *Athelia arachnoidea* and *A. epiphylla* are strong parasites of algae and lichens (Arvidsson 1979; Gilbert 1988; Oberwinkler 1970; Parmasto

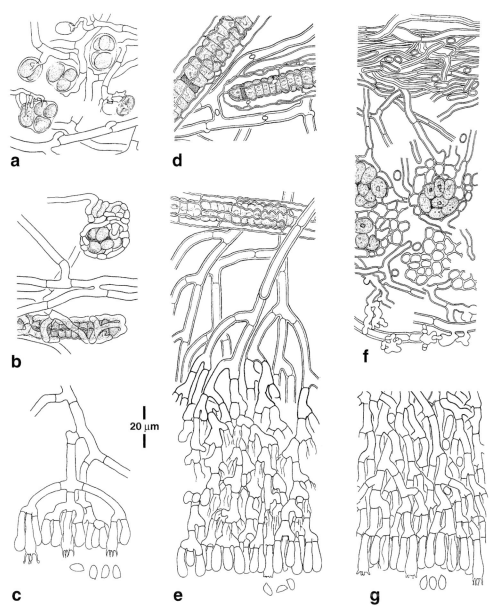

Fig. 2a–g. Fungus-alga interactions, thalli, and basidiocarps of *Athelia* (**a–c**) and *Dictyonema* (**d–g**), rearranged after Oberwinkler (1970); *scale* refers to all figures. **a** Hyphae and haustoria of *Athelia epiphylla* parasitic on unicellular green algae. **b** Hyphae of *Athelia andina* partly ensheathing *Scytonema* (*below*; note central hypha in *Scytonema*) and unicellular green algae (*above*). **c** Section of basidiocarp of *Athelia andina* with loose subhymenial hyphae and non-thickening hymenium with basidia of different developmental stages, and basidiospores. **d** Fungus-alga interaction in *Dictyonema irpicinum*; note clamped hyphae, dense hyphal mantles, and faint structures of central hyphae in *Scytonema* trichomes. **e** Fungus-alga interaction and basidiocarp of *Dictyonema sericeum*; note clampless hyphae (in comparison to *Athelia*), loose subhymenium with thick-walled hyphae, and thickening hymenium, one young basidium and basidiospores. **f** Section through thallus of *Cora pavonia (Dictyonema glabratum)*; note leaf-like architecture of the thallus with upper and lower hyphal layers and central compartments of coiled *Scytonema* trichomes, densely surrounded by hyphae; parts of central hyphae in *Scytonema* are indicated by *darker areas*. **g** Thickening hymenium of *Cora pavonia* with basidia of different age and basidiospores; compare the thickening hymenium with that of *Dictyonema sericeum* and the non-thickening one of *Athelia andina*

1998; Poelt and Jülich 1969b), penetrating the hosts with haustoria (Fig. 2a).

Already Elias Fries (1820) recognized green material at the basis of *Agaricus ericetorum* which in fact is the assemblage of tiny lichenized globules (Fig. 5a,b), later named as *Botrydina vulgaris* by de Brébisson (1844). Because of simple globose thallus structures, *Botrydina vulgaris* has been considered a primitive lichen (Acton 1909). Observations in the field led Gams (1962) to propose specific connections between the sterile lichen thalli *Botrydina* and *Coriscium* with basidiomycetes of the agaricoid genus *Omphalina* (Fig. 5a,e). Poelt and Oberwinkler (1964) and Oberwinkler (1970) studied the micromorphology of the thalli and of the associated *Omphalina* basidiocarps and were able to provide structural evidence for lichenization of *Omphalina* species. Dolipores with perforated parenthesomes in hyphae of *Coriscium* were found by Henssen and Kowallik (1976) and Oberwinkler (1980, 1984). They were independently documented also in *Botrydina* by Boissière (1980) and Oberwinkler (1980, 1984). The globose *Botrydina* thalli have a most complex cell architecture in which algal compartments are inserted in a compact, pseudoparenchymatous hyphal system (Fig. 5b). This was analyzed with the light microscope by Poelt and Oberwinkler (1964) and Oberwinkler (1970) and with the transmission electron microscope by Oberwinkler (1980, 1984). Haustoria are lacking, but there is a dense cell to cell contact of myco- and phycobionts. The *Botrydina* globules are dynamic structures which undergo a strongly regulated ontogeny. During this process, photobionts are able to reproduce mitotically to build up more and more algal compartments, but still within the solid pseudoparenchymatous hyphal structure of the enlarging globules.

Globular clusters of fungus-alga interactions are also present in *Athelopsis* (Coste and Royaud 1994). The globules are formed in the basal layer of the thin corticioid basidiocarps (Fig. 3a,b), thus not representing macroscopically visible thalli.

Rather conspicuous thallus layers are found in several *Multiclavula* species. In *M. mucida* irregular globose thalli (Oberwinkler 1970, 1984; Fig. 4g) are embedded in a gelatinous matrix on wet and decaying wood. Soil-inhabiting *Multiclavula* species have various types of thalli, some appear as uniform crustous layers, others are more separated in granular units. These species have not been studied comparatively.

Also the scale- and leaf-like types of basidiolichen thalli are rather diverse in architecture and do not occur in closely related taxa. The best known of these is *Coriscium viride* (Fig. 5e), described as a sterile lichen by Vainio (1890). The thalli are morphologically distinct and anatomically highly structured (Fig. 5f): there are upper and lower pseudoparenchymatous epidermis-like fungal layers and central loose hyphae with globular chambers for the phytobionts, surrounded by hyphal sheats. The gross morphology varies specifically.

Also the thallus of *Lepidostroma calocerum* (Mägdefrau and Winkler 1967) is scaly with distinct and dense hyphal pseudoparenchyms and loose inner hyphae, but the overall cellular composition is totally different from the *Coriscium* type and any other known basidiolichen thallus (Fig. 4k). This is another example for the multiple convergent evolution of structurally similar and functionally identical photosynthetic organs which are not homologous.

A new genus and species, *Acantholichen pannarioides*, has recently been proposed by Jørgensen (1998). It is based on sterile thalli composed of clampless hyphae with acanthohyphidia and *Scytonema* trichomes (Fig. 3c,d). The lichen grows *Pannaria*-like on debris, mosses, and other lichens.

III. Associations of Mycobionts and Phycobionts and Ultrastructure of Fungus-Alga Interactions

Three main types of fungus-alga interactions occur in basidiolichens (Fig. 1): (1) hyphal sheaths from which haustoria-like endosymbiontic hyphae originate, (2) appressoria, and (3) solid globular clusters formed by densely arranged hyphae surrounding green algae without haustoria.

1. *Scytonema* trichomes are enveloped by hyphae and penetrated by haustoria, originating from the mantle hyphae in *Dictyonema* and *Cora* (Fig. 2d,e,f). This interactive system, which has been misinterpreted as a dual fungal symbiosis with the algal host (Slocum and Floyd 1977), is different from all others known in basidio- and ascolichens. Because of the vitality of the phycobiont and the lack of any signs of significant structural changes, the mycobiont-phycobiont relationship

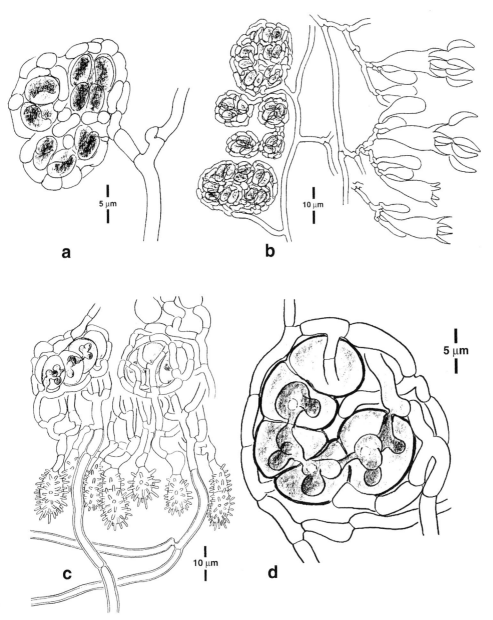

Fig. 3. Fungus-alga interactions and thalli of *Athelopsis* (**a,b**) and *Acantholichen pannarioides* (**c,d**). **a** *Athelopsis* hyphae ensheathing *Coccomyxa* algae; note clamped hypha. **b** Section through *Athelopsis* with lichenized lower part and whole basidiocarp showing the total cellular construction of the basidiolichen in the sporulating stage with basidia in different developmental stages, and basidiospores. **c** Lower part of the thallus of *Acantholichen pannarioides* with hyphae connecting the thallus to the substrate (*below*), acanthohyphidia of the lower surface of the thallus, and two algal chambers with short *Scytonema* trichomes surrounded by hyphae; note the central hypha in *Scytonema* in the left trichome. **d** Detail of the *Scytonema-Acantholichen* interactive system; note the intracellular hypha in *Scytonema* and its origin from an external hypha.

has been interpreted as an endosymbiontic interaction (Oberwinkler 1980; Slocum 1980). Tschermak-Woess (1982) records strong compression of the lichenized trichomes and the lack of cyanophycin in *Dictyonema moorei* studied with the light microscope. Yet there is no proof that these changes are due to mutualistic interactions. The myco-phycobiont interaction in *Acantholichen pannarioides* has also been studied by the author to update this chapter (Fig. 3c,d). It was found that it is essentially the same as in *Cora* and *Dictyonema*.

2. In *Lepidostroma calocerum* hyphae become attached to the algal cells and develop into morphologically distinct appressoria (Oberwinkler 1980, 1984). The attached algae degenerate and collapse (Fig. 4k,l), non-attached ones in the thallus are vital and obviously capable of cell division.

3. Globose hyphal-algal associations are characteristic of *Multiclavula* species (Oberwinkler 1980, 1984; Fig. 4g) and of *Botrydina vulgaris* (Boissière 1980; Oberwinkler 1980, 1984; Fig. 5a,b). Comparable globular clusters of fungal and algal cells develop inside the thalli of *Coriscium viride* (Oberwinkler 1980, 1984; Fig. 5f). Though densely packed, the algal cells are only surrounded by hyphae, and are not penetrated by haustoria. The globular thallus with clustered green algae has also been found in *Athelopsis* (Coste and Royaud 1994; Fig. 3a,b), and in otherwise differently lichenized species of *Athelia* (Oberwinkler 1970; Fig. 2b), *Cyphellosterum* (Oberwinkler 1970 as *Clavaria* sp.; Fig. 4d), and *Lepidostroma* (Oberwinkler 1970). In *Sistotrema brinkmannii* (Oberwinkler 1970) the myco-phycobiont association certainly is facultative because the fungus normally grows on wood without algae. The common *Resinicium bicolor* also is not lichenized but quite often grows together with green algae (Oberwinkler 1970; Poelt and Jülich 1969b).

IV. Fungi

All fungal partners of basidiolichens are homobasidiomycetes which share holobasidia, spores germinating with hyphae, and dolipores with perforated parenthesomes. No lichenized heterobasidiomycetes are known. Thus, basidiolichens belong to one relationship with a high number of non-lichenized species and few lichenized ones.

A. *Athelia, Athelopsis*

Athelia species are corticioid homobasidiomycetes with loose subhymenial hyphae and non-thickening, pellicular hymenia (Fig. 2c). Basidia are short-cylindrical to slightly clavate with two to four sterigmata. As in all basidiolichens, spores are smooth, thin-walled, hyaline and inamyloid. The species delimitation in *Athelia* is difficult

and needs careful revisions, like that for non-lichenized ones of Northern Europe by Eriksson and Ryvarden (1973).

Athelopsis differs from *Athelia* by clavate basidia (Fig. 3b). Often basidiocarps are very thin with a dense subhymenium, but loosely interwoven hyphae also can occur, as in the example illustrated here.

B. *Acantholichen*

The genus is monotypic and based on sterile thalli with clampless hyphae terminating with acanthophyphidia (Fig. 3c). Acanthohyphidia are known from corticioid, stereoid, and cyphelloid basidiomycetes, e.g., *Aleurodiscus* s.1. and *Stereum* s.1. However, they are diverse and not always homologous. *Acantholichen* is unique in hyphal morphology, but has the *Dictyonema* type of fungus-alga interaction. The fungal relationship cannot be derived from the hyphal system of the thallus. In *Cora* and *Dictyonema* basidiocarps develop on the underside of the lichen thalli. This can be expected also for *Acantholichen*. However, it is technically very difficult to examine the underside of the tiny scales in the search for basidiocarps.

C. *Dictyonema, Cyphellostereum*

Cora and *Dictyonema* basidiocarps have been studied by Oberwinkler (1970) and Parmasto (1978). They agreed that the fungi are very similar, representing one genus which appears closely related to *Byssomerulius* (*Meruliopsis*) with saprobic species. Characteristic morphological features for mature basidiocarps are thickening hymenia and loose, thick-walled subhymenial hyphae (Fig. 2e,g). Thus, Parmasto (1978) proposed to consider *Cora* as a synonym of *Dictyonema*, a taxonomy which is meaningful from a mycological point of view.

Cyphellostereum has small and stout, upright growing, simple or split basidiocarps, a monomitic hyphal system, and clavate-cylindrical, slightly bent holobasidia (Fig. 4a,b). Generic delimitation and taxonomic position are not clear.

D. *Lepidostroma, Multiclavula*

The fungal partner of *Lepidostroma calocerum* has been studied in detail by Oberwinkler (1970;

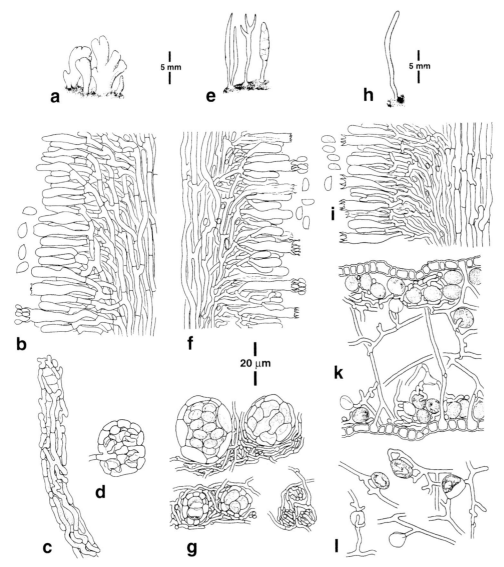

Fig. 4a–k. Basidiocarps, hymenia, thalli, and fungus-alga interactions in *Cyphellostereum* (**a–d**), *Multiclavula mucida* (**e–g**), and *Lepidostroma calocerum* (**h–l**), rearranged after Oberwinkler (1970). **a** Basidiocarps of *Cyphellostereum* with basal lichen thalli. **b** Section through the subhymenium and hymenium with basidia of different ages and basidiospores; note unclamped hyphae. **c** *Scytonema* trichome surrounded by *Cyphellostereum* hyphae. **d** Cluster of green algae enveloped by *Cyphellostereum* hyphae. **e** Basidiocarps of *Multiclavula mucida* with basal lichen thalli. **f** Section through the subhymenium and hymenium with basidia of different ages and basidiospores; note clamped hyphae. **g** Interactive globular units of hyphae and *Coccomyxa* in different developmental stages. **h** Basidiocarp of *Lepidostroma calocerum* with basal lichen thalli. **i** Section through the subhymenium and hymenium with basidia of different ages and basidiospores; note clamped hyphae. **k** Section of thallus with upper and lower pseudoparenchymatic hyphae, neighbouring algal layers, and loose interconnecting hyphae. **l** Fungus-alga interactions by appressoria; note different numbers of appressoria per algal cell, and different developmental stages of algae, incl. dead ones without cytoplasm

Fig. 4h,i). Macroscopically the clavarioid basidiocarps resemble *Calocera* species. The micromorphology is similar to that of *Multiclavula*, i.e., there is a simple, monomitic hyphal system, septa with clamps, a dense subhymenium, and a thickening hymenium. In contrast to *Multiclavula*, the basidia are not suburniform and 4-6-(8)-spored, but clavate-cylindrical and 4-spored.

Multiclavula (Petersen 1967) is a segregate of *Clavaria* s.l., *Lentaria*, and *Clavulinopsis* sensu Corner (1950, 1956) with hyphal morphology as a unifying characteristic, and, except of one species, a mutualistic association with algae or mosses. Also *Clavaria (Lepidostroma) calocera* was included in *Multiclavula* by Petersen (1967).

E. *Omphalina*

Omphalina species are small agarics with a characteristic macromorphology (Fig. 5a,e), i.e., with central stipes, convex to umbilicate pilei, and decurrent lamellae. The micromorphology is also very distinct, but prominent features are lacking (Fig. 5c,e); the hyphae are simple, comparatively loosely interwoven and therefore rather distinct, not gelatinous, with parietal or intrahyphal bright pigments, and without clamps. Trama hyphae are not swollen or only slightly, there are no cystidia in the hymenium and on the pileus. Also subhymenial hyphae are normally loosely interwoven, the hymenium is not much thickening, and basidia are basally slightly tapering and apically inconspicuously swollen, normally with four, but also with two sterimgata. Basidiospores lack any special characters, i.e., they are short cylindrical to subglobous, thin-walled, smooth, and inamyloid. Karyological studies by Lamoure (1968) and Poelt and Jülich (1969a) demonstrated apomictic ontogenies in *Omphalina*.

Redhead and Kuyper (1988) considered fully lichenized omphaloid agarics distinct from their non-lichenized allies in *Omphalina*. The differentiating characters focused on lichen thalli and noticeably thickened basal mycelium in mutualistic species. These distinctions have not been accepted as generically valuable by later workers.

Lutzoni and Vilgalys (1995a) used 17 morphological characters and nuclear ribosomal DNA large subunit sequences to analyze four lichenized *Omphalina* and three non-lichenized ones, and *Arrhenia lobata*. The dendrogram, based on morphological characters as well as one based on combined datasets, clustered the lichenized species in one group. Also in a study with four molecular datasets from 30 species of *Omphalina* and related taxa, Lutzoni (1997) demonstrated that the lichen-forming *Omphalina* species represent a monophyletic group, thus confirming the taxonomic conclusions of Redhead and Kuyper (1987) and Norvell et al. (1994). However, the author (Lutzoni l.c.) concluded: "With the present knowledge of relationships among *Omphalina* species and for nomenclatural stability, I recommend that

the lichenized clade and its sister group be maintained in the same genus."

V. Algae

As in ascolichens, also in basidiolichens the two major lichen-forming algal groups, cyanobacteria and green algae, serve as phycobionts.

Unanimously, the long trichomes of the phycobiont in *Dictyonema* species have been identified as *Scytonema* (e.g., Bornet 1873; Oberwinkler 1970, 1980, 1984; Parmasto 1978; Tschermak-Woess 1982). However, the cyanobacteria in *Dictyonema glabratum (Cora pavonia)* were earlier assigned to *Chroococcus* (Ahmadjian 1958; Bornet 1873; Mattirolo 1881). In reality, they are short coiled trichomes of the *Scytonema* type (Oberwinkler 1970, 1980, 1984; Räsänen 1943; Roskin 1970; Tschermak-Woess 1982; Vainio 1890).

The phycobiont of *Acantholichen pannarioides* was reported as *Scytonema* by Jørgensen (1998). A restudy showed that it is very similar to the *Cora* phycobiont inclusive of the fungus-alga interaction type (Fig. 3c,d).

The association of coccomyxoid green algae with *Lentaria (Multiclavula) mucida* has been described by Geitler (1955). He also found that the algae of the lichenized globules are larger than free-living ones. This was interpreted as a possible retardation of the algal division frequency caused by the mycobiont (Geitler 1956).

Langenstein (1994) isolated *Coccomyxa* from *Botrydina* and *Coriscium* and characterized algal cells in axenic culture. They reproduce asexually by autospores. A comparison of strains from *Omphalina* thalli with *Coccomyxa* isolated from the asolichen *Icmadophila ericetorum* as well as authentic substrains of *Coccomyxa pringsheimii* and *C. subellipsoidea* showed high identity of morphological and reproductive characters. Also the growth of colonies of the various strains was remarkably similar.

VI. Physiology

Because of its large and more or less flat thalli, *Cora pavonia (Dictyonema glabratum)* is a suitable lichen for ecophysiological experiments.

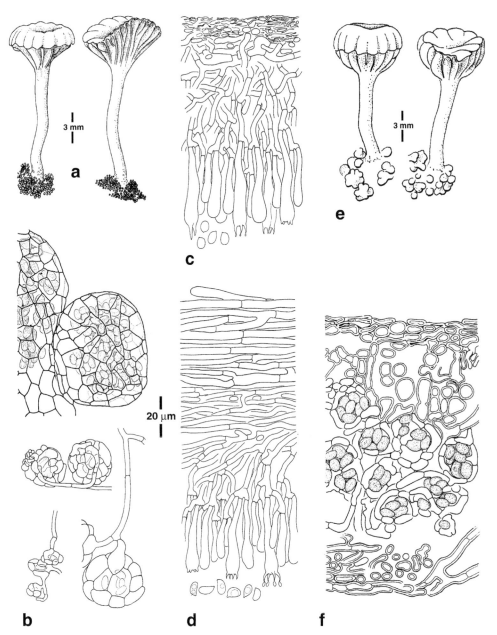

Fig. 5a–f. Basidiocarps, hymenia and thalli in lichenized *Omphalina* species of the *Botrydina* (**a–c**) and *Coriscium* (**d–f**) type, rearranged after Oberwinkler (1970). **a** Basidiocarps of *Omphalina umbellifera* (*O. ericetorum*) with basal lichen thalli. **b** Ontogenetic stages of the *Botrydina* thalli. **c** Transverse section in between lamellae of *Omphalina umbellifera* in an older developmental stage; note collapsed hyphae of the outer pileus layer (*above*) and basidia with various numbers of sterigmata. **d** Intralamellar radial section of the pileus of *Omphalina hudsoniana* (*O. luteolilacina*) with cutical hyphae, trama layers, subhymenium and hymenium with basidia of different ages, and basidiospores; note the uniform hyphal system. **e** Basidiocarps of *Omphalina hudsoniana* with basal *Coriscium* thalli. **f** Section of the thallus of *Coriscium viride*; note the leaf-like architecture of the thallus with upper and lower marginal hyphae and central algal chambers; note the short *Scytonema* trichomes, densely surrounded by hyphae

Already in 1965 Lange investigated the temperature dependence of photosynthesis, and Coxson (1987a) reported on a chilling stress which was considered to be an important factor in the limitation of distribution. Coxson (1987b,c) found an unexpected breadth of photosynthetic response under prevailing cloud conditions on La Soufrière (Gouadeloupe). The species shows no depression in net photosynthesis at saturated thallus water content, in contrast to most other lichens. Extensive ecophysiological field investigations in *D. glabratum*, carried out by Lange et al. (1994) in a premontane tropical rainforest in Panama, confirmed the observation of Coxson (1987c). The authors found that net photosynthesis was adapted to high temperatures, and when the thalli were fully hydrated the CO_2 uptake was not reduced. Desiccation of the thalli caused a decrease in net photosynthesis and influenced apparent photon yield of CO_2 fixation. Though there were days with a negative carbon balance, the total carbon gain was extremely high and estimated at 228% per year.

The C-metabolism of *D. glabratum* was investigated by Feige (1969), who found that, in contrast to all other lichens with cyanobacterial phytobionts, pentitol was synthesized and accumulated as pentitolgalactosid. The main excretory product was glycollic acid after bicarbonate incubation in liquid phase.

Larcher and Vareschi (1988) studied the variation in morphology and functional traits of *D. glabratum* from various habitats in the Venezuelan Andes. They found optimally developed thalli in the lower Paramos around 3500 m which exhibit the highest photosynthetic capacity. Specimens growing in the very high altitudes up to 4300 m have strongly reduced thalli, but the same respiration intensity as specimens from the lower Paramo, reflecting a homeostatic adjustment to the prevailing temperature regime. Thalli which grow in lower altitudes of the montane forest have approximately half the size and barely half the net photosynthetic rate in comparison to those growing in optimal Paramo regions. Thus, the increasingly unfavourable carbon budget may delimit the altitudinal distribution. Lange et al. (1994) explained the reduction of population densities of macrolichens in tropical lowland rainforests by the negative carbon balance caused by increased nocturnal temperatures at lower elevations.

The wide temperature range in photosynthetic activity is a general feature in lichens (Lange 1965). A distinct adaptation to low temperature was found in *Botrydina* and *Coriscium* by Heikkilä and Kallio (1966).

VII. Phytochemistry

Higher fungi, including the basidiomycetes, are very rich in specific chemical compounds (Gill and Steglich 1987) that often are of taxonomic importance. Likewise, secondary metabolites of ascolichens (lichen substances) are widely distributed and have been studied intensively (e.g., Huneck and Yoshimura 1996). The taxonomic use of metabolic data in lichen-forming fungi has recently been summarized by Lumbsch (1998); however, basidiolichens could not be included due to lack of data.

In a study on antibiotics in *Cora pavonia* (Mitidieri et al. 1964) no chemical substances were mentioned. Iacomini et al. (1987) isolated and characterized β-D-glucan, heteropolysaccharides, and trehalose components also from *Cora pavonia (Dictyonema glabratum)*. Again, from the same species, atranorin and tenuiorin, as well as ergosterol and ergosterol peroxide, were reported by Piovano et al. (1995). *Dictyonema* was also included in a chemical study on secondary metabolites in lichens from Bario highlands, Borneo, by Din et al. (1998). Jørgensen (1998) found no lichen acids in the new basidiolichen genus *Acantholichen*.

VIII. Ecology

In his field studies, Johow (1884) observed that *Dictyonema glabratum* appears to be confined to higher altitudes in the tropical islands of Dominica and Trinidad. His transplantation experiments to lowlands failed within a few days. Thus, the species appeared to be adapted to cool, humid and bright conditions of tropical mountain regions. In standings with disturbed or scarce higher plant vegetation, *D. glabratum* can occur in remarkably rich, ground-covering populations.

Acantholichen pannarioides is a species of moist montane regions in Central America and northern South America, where it grows on debris of vegetation, bark, bryophytes, and other lichens (Jørgensen 1998).

Multiclavula species display a remarkably diverse adaptation to ecological niches. Wet wood of huge logs is the most suitable substrate for *M. mucida*. In Central Europe preferably coniferous wood is colonized, but the species occurs also in subtropical and tropical regions where gymnosperms are lacking.

Lichenized *Omphalina* species represent the most common agarics in arctic-alpine vegetations where they are primarily lichens with a winter snow cover (Heikkilä and Kallio 1966). They also occur fully exposed in high alpine regions of tropical mountains with often extreme day and night temperature differences.

IX. Dispersal

Similarly to ascolichens, only the fungal partners of basidiolichens are able to propagate sexually. In all species basidiocarps are developed sporadically, so that production of basidiospores is limited in time. However, enormous quantities of spores are produced during the sporulation phase, also from tiny basidiocarps like those of a small *Athelia*. The high amount of basidiospores is an essential prerequisite for effective dispersal in basidiolichens, as in basidiomycetes in general. Dispersal by mitotic progagules is limited to certain taxonomic groups, e.g., the Uredinales, and is not a general feature of basidiomycetes.

Poelt and Obermayer (1990) considered thallus bulbils in *Multiclavula vernalis* as distinctive vegetative diaspores. They consist of inflated hyphae and a thin cortical layer, and contain one or a few groups of algal cells.

It is tempting to hypothesize that *Botrydina* globules can also serve as vegetative propagules in *Omphalina*. However, normally, *Botrydina* is quite solidly attached to the substrate by hyphae.

No special vegetative propagules, like isidia or soredia of ascolichens, are known in basidiolichens.

X. Distribution

Two main geographical distribution patterns can be found in basidiolichens: (1) cosmopolitan to subcosmopolitan ranges with restrictions to specific ecological niches (certain species of *Dictyonema*, *Multiclavula*, and *Omphalina*), and (2) regional to continental distributions *(Acantholichen, Cora, Dictyonema)*.

According to Petersen and Kantvilas (1986), *Multiclavula* seems to include several species which are truly cosmopolitan, e.g., *M. mucida* and *M. vernalis*.

Dictyonema sericeum is reported from all continents with tropical and subtropical regions (Parmasto 1978). *Dictyonema glabratum* is known to occur from South Florida to Cape Horn (Parmasto 1978). Thus, the species seems to be restricted to subtropical and tropical America with an extension into temperate and cool regions of South America. In contrast, *D. ligulatum* is known only from southeastern Asia and Oceania (Parmasto 1978). Besides all previous records of *D. moorei* from Japan (Parmasto 1978), this species was also recorded from Chile (Henssen 1963). *Dictyonema interruptum* was originally described from England and was additionally reported from the Azores (Coppins and James 1979) and from the Pyrenees (Etayo et al. 1995).

Acantholichen pannarioides is known from Costa Rica to the northern Andes in Venezuela and Ecuador, as well as from the Galapagos Islands (Jørgensen 1998).

Extensive fieldwork of Heikkilä and Kallio (1966, 1969) was a substantial basis to document that *Botrydina* and *Coriscium* have circumpolar distributions in artic and subarctic regions (Red-head 1989). Some lichenized *Omphalina* species appear to be cosmopolitan, e.g., *O. umbellifera (O. ericetorum)* has a circumpolar distribution and has been recorded also from New Zealand (Galloway 1985) and Australia (Lumbsch and Ewers 1992).

XI. Culture Experiments

Heikkilä and Kallio (1996) failed to obtain axenic cultures of the fungal partners in lichenized *Omphalina* species. However, in living cultures of *Botrydina* and *Coriscium*, *Omphalina* basidiocarps were developed several times. In contrast to the fungus, the phycobiont of *Botrydina* was easily cultivated on agar and in liquid media. Shape and size of the algal cells varied widely in culture. The phycobiont isolated from *Coriscium* appeared very similar.

Langenstein (1994) was successful in isolating phyco- and mycobionts of *Botrydina*, to grow them in axenic culture, and to synthesize *Botrydina* thalli in the laboratory. This was a remarkable progress in experimental lichenology. When *Coccomyxa* is present, hyphae originating from basidiospores contact the algal cells and surround them. Original compartments with single algae develop into multicellular ones by autospore production and ingrowth of fine hyphal branches into the phycobiont colony.

XII. Systematics and Evolution

Based on comparative morphology, Oberwinkler (1970) distinguished three main groups within the basidiolichens, (1) the *Athelia-Dictyonema* complex incl. of *Cora*, (2) the *Multiclavula-Lepidostroma* group with clavarioid basidiocarps, and (3) the agaricoid, lichenized *Omphalina* species. These homobasidiomycetous taxa were considered of distant relationship because of their different micromorphological characters, a view largely in agreement with a classification based on basidiocarp morphology. The multiple origin of basidiolichens is also apparent in a dendrogram based on small subunit rDNA data (Gargas et al. 1995).

In a cladistic study of corticioid fungi, Parmasto (1995) obtained a cladogram of 142 taxa of corticioid fungi based on 95% majority-rule consensus for 12000 equally parsimonious trees. *Athelia* and *Athelopsis* fall within a group "arbitrarily named at this stage Atheliaceae" (Parmasto l.c.), together with *Parvobasidium*, *Lobulicium*, *Luellia*, *Athelicium*, and *Merulicium*. Boidin et al. (1998) used the ribosomal internal transcribed spacer sequences (ITS1-5.8S-ITS2) to evaluate the phylogenetic relationships within the Aphyllophorales. They recognized an order Atheliales with *Athelia*, *Amyloathelia*, *Fibulomyces* and *Leptosporomyces*.

Parmasto (1978) united *Cora* with *Dictyonema* and accepted five *Dictyonema* species. Oberwinkler (1970) and Parmasto (1978) considered *Dictyonema* as closely related to *Byssomerulius (Meruliopsis)*, a genus of corticiaceous basidiomycetes. In Parmasto's cladistic analysis (Parmasto 1995), *Dictyonema* is the sister genus of *Gloeoporus* within the Schizophyllaceae. *Meruliopsis* is one of many genera in an unresolved assemblage. In a partial dendrogram of Boidin et al. (1998, Fig. 13) *Meruliopsis corium* groups with *Oxyporus latemarginatus*, *Irpicodon flavus*, *Erythricium* and *Athelia*. *Dictyonema* was not included in this analysis. None of these taxa is closely related with *Thelephora*, or other genera in the Thelephoraceae or Thelephorales, as explained by Oberwinkler (1970) and Parmasto (1978). The application of a wrong taxonomy, even in recent publications, may be partly due to the unfortunate title of Parmasto's monograph on the genus *Dictyonema* ("*Thelephorolichenes*").

Clavarioid basidiolichens have been assigned to different genera, e.g., *Clavaria*, *Clavulinopsis*, *Lentaria*, *Lepidostroma* and *Multiclavula*. These have been analyzed and discussed comparatively by Oberwinkler (1970) with the result that lichenized *Lentaria* cannot be separated generically from *Multiclavula*. Because of the unique thallus and the peculiar fungus-alga interaction, *Lepidostroma* was recognized as a separate lichen genus.

Lichenized and non-lichenized *Omphalina* species are structurally and micromorphologically very similar (Oberwinkler 1970). Lutzoni and Vilgalys (1995a) included molecular and morphological datasets in phylogenetic studies and compared three tests for seven lichenized and non-lichenized *Omphalina* species. Only Rodrigo's test (Rodrigo et al. 1993) suggested that the two datasets could be combined. The resulting tree groups the lichenized *Omphalina* species *O. ericetorum*, *O. velutina*, *O. luteovitellina* and *O. hudsoniana* in one cluster. In a comprehensive molecular study on the phylogeny of lichen- and non-lichen-forming omphalinoid mushrooms, Lutzoni (1997) found that four datasets representing different regions of the nuclear ribosomal unit (ITS1, 5.8S, ITS2, and 25S) obtained from 30 species of *Omphalina*, *Arrhenia*, *Chrysomphalina*, *Clitocybe*, *Gerronema*, *Hygrocybe*, *Phaeotellus* and *Rickenella*, were not identical when analyzed separately. Tests for combinability showed that pruning of the datasets was necessary to combine them in meaningful way. Five lichenized *Omphalina* species appear to represent a monophyletic group. However, only four of them, i.e. *O. hudsoniana*, *O. luteovitellina*, *O. grisella* and *O. velutina* are supported by a high bootstrap value of 98%. *Omphalina ericetorum* is outside the core group. A dendrogram resulting from maximum likelihood analyses of combined omphalinoid mushroom datasets groups *Omphalina pyxidata* together with *O. rivulicola* and *Clitocybe lateritia*.

A molecular investigation based on combined datasets of 25S, 5.8S, ITS1, and ITS2 from the nuclear ribosomal DNA of lichenized fungi and fungi associated with *Marchantia* and *Blasia* and non-mutualist fungi revealed four independent transitions to mutualism (Lutzoni and Pagel 1997). Two of these transitions refer to *Gerronema marchantiae* and *Rickenella pseudogrisella* associated with liverworts. The others are two lichenized *Multiclavula* species and a monophyletic group of five lichen-forming *Omphalina* species. The authors "demonstrate a highly significant association between mutualism and increased rates of nucleotide substitions in nuclear ribosomal DNA", and that "a transition to mutualism preceded the rate acceleration of nuclear ribosomal DNA in the lineages. The generalized genomic acceleration in the lichen-forming *Omphalina* lineage is mostly due to selection of mutations that disrupt the potential formation of thymine dimers" (Lutzoni and Pagel 1997).

XIII. Comparison with Ascolichens

The number of lichenized Ascomycetes may be as high as that of the non-lichenized ones. In Basidiomycetes the majority of symbiontic species is associated with certain groups of seed plants as ectomycorrhizal fungi. The number of basidiolichens, however, is very small. Only few species of eight genera of Homobasidiomycetes are lichenized. One of the most striking and unique features of ascolichens is the independent and convergent evolution of the thalli to morphological structures which have not been developed in fungi and algae. Such thallus evolution did not occur in basidiolichens. The macromorphology is comparatively simple. However, the fungus-alga interaction is remarkably diverse, comprising three main types. One of them, the *Dictyonema* type, is unique for all lichens.

The transition to lichen symbiosis is demonstrated as an acceleration of nucleotide substitutions (Lutzoni and Pagel 1997). This acceleration has as yet had no influence on species radiation.

Because of the limited number of species and their restriction to special niches, basidiolichens have been overlooked for a long time. However, *Dictyonema glabratum* can occur in dense populations in neotropical mountain areas, and in arctic alpine vegetations *Botrydina* and *Coriscium* develop to considerable quantities on suitable acidic substrates.

XIV. Conclusion

Though basidiolichens represent only a very small group of all obligately lichenized fungi, they are highly diverse. This is reflected by (1) structural and functional adaptations of the lichen thalli, (2) the different structures of fruitbodies which are developed by the fungal partners, (3) the cellular interactions of myco- and phycobionts, (4) the worldwide distribution in specialized ecological niches, (5) some unique ecophysiological properties, and (6) several convergent phylogenies.

Acknowledgments. I thank J. Poelt, who drew my attention to basidiolichens in 1963. Our 1-year research stay in 1968–69 in Venezuela was kindly arranged by H. Zöttl and H. Merxmüller. It was a fascinating experience for us to see species of most basidiolichen genera in the Andes and to study them as living organisms in Merida. I shared the enthusiasm of students in the field when basidiolichens were found and I acknowledge and admire the experimental work of L. Kisimova-Horovitz and B. Langenstein in our laboratory. The skilful technical assistance of E. Deml, S. Dinkelmeyer, A. Quellmalz and S. Silberhorn with TEM work years ago is gratefully acknowledged. *Acantholichen* was kindly sent by P. Blanz.

References

Acton E (1909) *Botrydina vulgaris* Brébisson, a primitive lichen. Ann Bot 23:579–585

Ahmadjian V (1958) A guide for the identification of algae occurring as lichen symbionts. Bot Not 111(4): 632–644

Arvidsson L (1979) Svampangrepp pa lavar – en orsak till lavoken. Fungus attacks on lichens and lichen deserts in cities. Sven Bot Tidskr 72:285–292

Boidin J, Mugnier J, Canales R (1998) Taxonomie moléculaire des Aphyllophorales. Mycotaxon 66:445–491

Boissière JC (1980) Un vrai basidiolichen européen: l'*Omphalina umbellifera* (L. ex Fr.) Quél. Etude ultrastructurale. Cryptogam Bryol Lichenol 1:143–149

Bornet ME (1873) Recherches sur les gonidies des Lichens. Ann Sci Nat V Bot 17:45–110

Coppins BJ, James PW (1979) A British species of *Dictyonema*. Lichenologist 11:103–108

Corner EJH (1950) A monograph of *Clavaria* and allied genera. Ann Bot Mem No 1. Oxford University Press, London 740 pp

Corner EJH (1956) A new European *Clavaria: Clavulinopsis septentrionalis* sp. nov. Friesia 5(3–5): 218–220

Coste C, Royaud A (1994) Les Basidiolichens Européens. Bull Soc Castraise Sci Nat 1994:55–59

Coxson DS (1987a) The temperature dependence of photoinhibition in The tropical basidiomycete lichen *Cora pavonia* E. Fries. Oecologia 73(3):447–453

Coxson DS (1987b) Net photosynthetic response patterns of the basidiomycete lichen *Cora pavonia* (Web.) E. Fries from the tropical volcano La Soufrière (Guadeloupe). Oecologia 73(3):454–458

Coxson DS (1987c) Effects of desiccation on net photosynthetic activity in the basidiomycete lichen *Cora pavonia* E. Fries from the cloud, mist zone of the tropical volcano La Soufrière (Guadeloupe). The Bryologist 90(3):241–245

de Brébisson A (1844) In: Meneghini (ed) Monogr Nostoch. Ital Atti R Acad Sci Torino Ser 2, 5

Din LB, Ismail G, Elix JA (1998) The lichens in Bario highlands: their natural occurrence and secondary metabolites. In: Ismail G, Din LB (eds) A scientific journey through Borneo: Bario, the Kelabit Highlands of Sarawak. Pelanduk Publications (M) Sdn Bhd, Selangor Darul Ehsan, Malaysia, pp 155–160

Eriksson J, Ryvarden L (1973) The Corticiaceae of North Europe vol 2, Fungiflora, Oslo

Etayo J, Diederich P, Sérusiaux E (1995) *Dictyonema interruptum*, new for the Pyrenees. Graphis Scripta 7(1): 5–6

Feige B (1969) Stoffwechselphysiologische Untersuchungen an der tropischen Basidiolichene *Cora pavonia* (Sw.) Fr. Flora 160:169–180

Fries E (1820) Systema mycologicum. E. Mauritius Greifswald

Galloway DJ (1985) Flora of New Zealand Lichens. P.D. Hasselberg, Government Printer Wellington

Gams H (1962) Die Halbflechten *Botrydina* und *Coriscium* als Basidiolichenen. Österr Bot Zeitschr 109(3): 376–380

Gargas A, DePriest PT, Grube M, Tehler A (1995) Multiple origins of lichen symbioses in fungi suggested by SSU rDNA phylogeny. Science 268:1492–1495

Geitler L (1955) *Clavaria mucida*, eine extratropische Basidiolichene. Biol Zentralbl 74(3–4): 145–159

Geitler L (1956) Ergänzende Beobachtungen über die extratropische Basidiolichene *Clavaria mucida*. Österr Bot Z 103(1):164–167

Gilbert OL (1988) Studies on the destruction of *Lecanora conizaeoides* by the lichenicolous fungus *Athelia arachnoidea*. Lichenologist 20(2):183–190

Gill M, Steglich W (1987) Pigments of fungi (Macromycetes). Prog Chem Org Nat Prod 51:1–317

Heikkilä H, Kallio P (1966) On the problem of subarctic basidiolichens I. Ann Univ Turku (Rep Kevo Subarct Res Stn 3) 2A:48–74

Heikkilä H, Kallio P (1969) On the problem of subarctic basidiolichens. II. Ann Univ Turku A, II 40:90–97

Henssen A (1963) Eine Revision der Flechtenfamilien Lichinaceae und Ephebaceae. Symb Bot Ups 18(1):1–123

Henssen A, Kowallik K (1976) A note on the mycobiont of *Coriscium viride* (Ach.) Vain. Lichenologist 8:197

Huneck S, Yoshimura I (1996) Identification of lichen substances. Springer, Berlin Heidelberg New York

Iacomini M, Zanin SMW, Fontana JD (1987) Isolation and characterization of β-D-glucan, heteropolysaccharide, and trehalose components of the basidiomycetous lichen *Cora pavonia*. Carbohydr Res 168(1):55–65

Johow F (1884) Die Gruppe der Hymenolichenen. Ein Beitrag zur Kenntnis basidiosporer Flechten. Jahrb Wiss Bot 15:361–409

Jørgensen PM (1998) *Acantholichen pannarioides*, a new basidiolichen from South America. Bryologist 101: 444–447

Lamoure D (1968) Parthenogenèse chez *Omphalina ericetorum* (Pers. ex Fr.) M. Lange et deux espèces affines. C R Acad Sci Paris Ser D 266:1499–1500

Lange OL (1965) Der CO_2-Gaswechsel von Flechten bei tiefen Temperaturen. Planta 64:1–19

Lange OL, Büdel B, Zellner H, Zotz G, Meyer A (1994) Field measurements of water relations and CO_2 exchange of the tropical, cyanobacterial basidiolichen *Dictyonema glabratum* in a Panamanian rainforest. Bot Acta 107:279–290

Langenstein B (1994) Experimentell-ontogenetische Studien zur Lichenisation am Beispiel von *Omphalina* (Basidiomycota) und *Coccomyxa* (Chlorophyta). Diplomarbeit Univ Tübingen 123 pp

Larcher W, Vareschi V (1988) Variation in morphology and functional traits of *Dictyonema glabratum* from contrasting habitats in the Venezuelan Andes. Lichenologist 20(3):269–277

Lumbsch HT (1998) Taxonomic use of metabolic data in lichen-forming fungi. In: Frisvad JC, Bridge PD, Arora DK (eds) Chemical fungal taxonomy. Marcel Dekker, Basel, pp 345–387

Lumbsch HT, Ewers WH (1992) Additional lichen records from Australia 10. *Omphalina umbellifera* (L. Fr.) Quélet. Australas Lichenol Newsl 30:5–6

Lutzoni F (1997) Phylogeny of lichen- and non-lichen-forming omphalinoid mushrooms and the utility of testing for combinability among multiple data sets. Syst Biol 46:373–406

Lutzoni F, Pagel M (1997) Accelerated evolution as a consequence of transitions to mutualism. Proc Natl Acad Sci 94:11422–11427

Lutzoni F, Vilgalys R (1995a) Integration of morphological and molecular data sets in estimating fungal phylogenies. Can J Bot 73 (Suppl 1):649–659

Lutzoni F, Vilgalys R (1995b) *Omphalina* (Basidiomycota, Agaricales) as a model system for the study of coevolution in lichens. Cryptogam Bot 5(1):71–81

Mägdefrau K, Winkler S (1967) *Lepidostroma terricolens* n. g. n. sp., eine Basidiolichene der Sierra Nevada de Santa Marta (Kolumbien). Mitt Inst Colombo-Aleman Invest Cient 1:11–17

Mattirolo O (1881) Contribuzioni allo studio del genere *Cora* Fries. Nuovo G Bot Ital 8:245–267

Mitidieri J, Joly S, Ferraz EC (1964) Teste de antibiose exercidapelo extrato do liquens *Parmelia tinctorum* Desp. e *Cora pavonia* (Web.) E. Fries. Rev Agr (Piracicaba) 39:119–121

Norvel LL, Redhead SA, Ammirati JF (1994) *Omphalina* sensu lato in North America 1–2. 1: *Omphalina wynniae* and the genus *Chrysomphalina*. 2: *Omphalina* sensu Bigelow. Mycotaxon 50:379–407

Oberwinkler F (1970) Die Gattungen der Basidiolichenen. Vortr Gesamtgeb Bot N F 4:139–169

Oberwinkler F (1980) Symbiotic relationships between fungus and alga in basidiolichens. In: Schwemmler H, Schenk HEA (eds) Endocytobiology endosymbiosis and cell biology vol 1. Walter de Gruyter, Berlin, pp 305–315

Oberwinkler F (1984) Fungus-alga interactions in basidiolichens. In: Hertel H, Oberwinkler F (eds) Beitrage zur Lichenologie. Festschrift J Poelt. Beiheft zue Nova Hedwigia 79. J Cramer Vaduz pp 739–774

Parmasto E (1978) The genus *Dictyonema* ("Thelephorolichenes"). Nova Hedwigia 29:99–144

Parmasto E (1995) Corticioid fungi: a cladistic study of a paraphyletic group. Can J Bot 73 Suppl:S843–S852

Parmasto E (1998) *Athelia arachnoidea*, a lichenicolous basidiomycete in Estonia [Parasiitnahkis *(Athelia arachnoidea)*, samblikel kasvav kandseen Eestis]. Folia Cryptogam Estonica 32:63–66

Petersen RH (1967) Notes on clavarioid fungi. VII. Redefinition of the *Clavaria vernalis-C. mucida* complex. Am Midl Nat 77:205–221

Petersen R, Kantvilas G (1986) Three lichen-forming clavarioid fungi from Tasmania. Aust J Bot 34:217–222

Piovano M, Chamy MC, Garbarino JA, Quilhot W (1995) Studies on Chilean lichens XXIV. Secondary products from *Dictyonema glabratum* (Basidiomycotina). Bol Soc Chil Quím 40:163–165

Poelt J (1959) Eine Basidiolichene in den Hochalpen. Planta 52(6):600–605

Poelt J (1962) Die Basidiolichene *Lentaria mucida* in Bayern. Ber Bayer Bot Ges 35:87–88

Poelt J, Jülich W (1969a) *Omphalina grisella*, ein weiterer lichenisierter Blätterpilz in den Alpen. Herzogia 1:331–335

Poelt J, Jülich W (1969b) Über die Beziehungen zweier corticoider Basidiomyceten zu Algen. Österr Bot Z 116:400–410

Poelt J, Obermayer W (1990) Lichenisierte Bulbillen als Diasporen bei der Basidiolichene *Multiclavula vernalis* spec. coll. Herzogia 8:289–294

Poelt J, Oberwinkler F (1964) Zur Kenntnis der flechtenbildenden Blätterpilze der Gattung *Omphalina*. Österr Bot Z 111(4):398–401

Räsänen V (1943) Das System der Flechten. Acta Bot Fenn 33:1–82

Redhead SA (1989) A biogeographic overview of the Canadian mushroom flora. Can J Bot 67(10):3003–3062

Redhead SA, Kuyper TW (1987) Lichenized agarics: taxonomic and nomenclatural riddles. Arct Alp Mycol 2:319–348

Redhead SA, Kuyper TW (1988) *Phytoconis*, the correct generic name for the basidiolichen *Botrydina*. Mycotaxon 31(1):221–223

Rodrigo AG, Kelly-Borges M, Bergquist PR, Bergquist PL (1993) A randomisation test of the null hypothesis that two cladograms are sample estimates of a parametric phylogenetic tree. NZ J Bot 31:257–268

Roskin PA (1970) Ultrastructure of the host-parasite interaction in the basidiolichen *Cora pavonia* (Web.) E. Fries. Arch Mikrobiol 70:176–182

Slocum RD (1980) Light and electron microscopic investigations in the Dictyonemataceae (basidiolichens). II. *Dictyonema irpicinum*. Can J Bot 58:1005–1015

Slocum RD, Floyd GL (1977) Light and electron microscopic investigations in the Dictyonemataceae (basidiolichens). Can J Bot 55:2565–2573

Tomaselli R, Caretta G (1969) Morphological and anatomical observations on *Cora pavonia* (Basidiolichenes) and hypotheses of factors determining lichen symbiosis. Atti Ist Bot Lab Crit Univ Pavia Ser 65:11–18

Tschermak-Woess E (1982) Das Haustorialsystem von *Dictyonema* kennzeichnend für die Gattung. Plant Syst Evol 143:109–115

Vainio EA (1890) Etude sur la classification naturelle et la morphologie des Lichens du Brésil. Acta Soc Fauna Flora Fenn 7:1–256

Fungal Bacterial Interactions

13 Fungal Bacterial Interactions

J.W. Bennett[1] and T. Feibelman[2]

CONTENTS

I. Introduction

Bacteria and fungi are constantly interacting in nature. Natural ecosystems depend upon the metabolism of these microorganisms to recycle nutrients such as carbon, nitrogen, and sulfur. Curiously, despite the intimate ecological relationships of these two major life forms, mycologists tend to study fungi in splendid isolation. The indices of two of the most popular English language textbooks in introductory mycology, *The Fungi* (Carlile and Watkinson 1994) and *Introductory Mycology* (Alexopoulos et al. 1996), do not contain a single entry for "bacteria." Although there are many mycologists who specialize in fungal-plant interactions and fungal-animal interactions, the topic of fungal-bacterial interactions is less commonly studied.

This chapter is written not so much as a review as a provocation. It gives a simple overview of some of the best-understood examples of fungal-bacterial interactions, and points out the disciplinary barriers that have led to the current gulf between microbial ecologists who study fungi and those who study bacteria. We hope this brief overview will stimulate mycologists to new ways of experimenting, modeling, and thinking.

II. General Taxonomic and Physiological Categories

Microorganisms (organisms too small to be seen with the naked eye) are loosely divided into five major taxonomic categories: algae, bacteria, fungi, protists, and viruses. Cellular microbes are often placed into two superkingdoms: the Eukaryotes and Prokaryotes. Algae, fungi, and protists are Eukaryotes (organisms with true nuclei) while the bacteria, including the group once known as the blue-green algae (cyanobacteria) are prokaryotes. Prokaryotes are the smallest and most abundant form of cellular life.

Recently, using a phylogenetic system based on conserved DNA sequences, especially those for ribosomal RNA genes, systematists have accepted that prokaryotes are more diverse than had been suspected. In the new system, cellular forms of life are divided into three domains: the Archaea, the Bacteria, and the Eukarya (Woese et al. 1990). In the older usage, the Archaea (methanogens, halophiles, many "extremeophiles," plankton species, etc.) were grouped with Bacteria in the Prokaryotes, so some workers prefer the terms Archaebacteria and Eubacteria to emphasize the distinction (Iwabe et al. 1989). In this discussion, however, we will use Bacteria in the Woesian sense to mean Eubacteria.

Within the Eukarya, fungi and animals are believed to represent a monophyletic lineage (Baldauf and Palmer 1993). In addition to the possession of true nuclei, fungi are characterized by their absorptive heterotrophic nutrition, apical growth, rigid cell walls, and spore formation. Although chytrids and yeasts are single-celled, the majority of fungi are multicellular and filamen-

[1] Department of Cell and Molecular Biology, Tulane University, New Orleans, Louisiana 70118 USA
[2] Microbia, Inc., One Kendall Square, Bldg. 1400W, Suite 1418, Cambridge, MA 02139, USA

The Mycota IX
Fungal Associations
Hock (Ed.)
© Springer-Verlag Berlin Heidelberg 2001

tous. Most bacteria are unicellular, although they regularly form biofilms, colonies, consortia, mats, and other intra- and interspecific associations.

Formal taxonomic hierarchies arrange bacteria into groups such as the spirochetes, gram positives, purple bacteria, green sulfur bacteria, green nonsulfur bacteria, cyanobacteria, flavobacteria, and thermatogales; and place fungi into groups such as the chytrids, zygomycetes, ascomycetes, and basidiomycetes. However, for physiologists and ecologists, taxonomic categories are often less useful than other rubrics. One alternative system uses temperature response. Psychrophiles are organisms that grow at temperatures below freezing; mesophiles grow at mid-range (15–40 °F) temperatures; thermophiles grow at higher (40–70 °C) temperatures. Another way of classifying microorganisms is by their response to gaseous oxygen. Aerobic species require oxygen for their life processes, while anaerobic species do not. Obligate anaerobes are killed by oxygen. Facultative anaerobes (e.g., the yeast *Saccharomyces cerevisiae*) can grow with or without oxygen. Until fairly recently, it was believed that all anaerobes were bacteria or archaea, and that no anaerobic fungi existed. It is now known, however, that anaerobic fungi are common among rumen flora (Theodorou et al. 1992) and it is likely that additional species of anaerobic fungi remain to be discovered from other oxygen-deprived habitats.

Energy source is another way of dividing up microorganisms. Autotrophs fix carbon. Photoautotrophs fix light energy, while chemoautotrophs are fueled by the oxidation of nitrogen, sulfur, or other elements. Examples of both photoautotrophs and chemoautotrophs exist among the bacteria; photoautotrophs are more common than chemoautotrophs and include the cyanobacteria, green sulfur bacteria, purple sulfur bacteria, and purple nonsulfur bacteria. No fungi are autotrophic; however, the most famous group of cooperative organisms, the lichens, consist of filamentous fungi growing in intimate association with an algal partner. Although lichens are dual organisms, they display distinct and stable phenotypes, and about 15000 "species" have been described. The bulk of the biomass of the lichen thallus is fungal, and lichens are being integrated into fungal classification (Hawksworth 1988). The majority of lichens are fungal-green algal associations, but approximately 8% consist of fungi associated with cyanobacteria, usually of the genus *Nostoc*. Lichens demonstrate the general advantages of associations between organisms with different life strategies. The fungus traps water, excretes organic acids and hydrolytic enzymes, and absorbs nutrients that provides a toe-hold in harsh environments, while the phototrophic alga fixes carbon dioxide, providing food for the fungus (Galun 1988).

Heterotrophs are organisms that get their energy by breaking down complex organic matter. All fungi, and most bacteria, are heterotrophs. Heterotrophic organisms are dependent on preformed organic matter for growth and energy. Heterotrophs can be further subdivided by the status of the organic matter on which they feed. Saprophytes feed on dead and decaying organisms; parasites (biotrophs) feed on living organisms; facultative parasites sometimes feed on living systems. Saprophytes are often called decomposers, i.e., organisms that break down the carbon contained in dead organisms, and release nutrients back into the environment.

When describing the interactions of two or more organisms, a common way of division utilizes perceived categories of benefit. Neutral relationships are those in which there is no perceived benefit or deficit; commensal relations occur when one organism is thought to benefit while the other remains unaffected; mutually beneficial relationships in which both organisms obtain something from the other are termed symbiotic. Lichens, of course, are the textbook example of symbiosis, even though many commentators believe that the fungal partner receives most of the benefit. All this terminology is subject to semantic ambiguity as well as to "genuinely irreconcilable points of view" (Cooke 1977, p. 3).

When an interaction between two organisms results in disease, the disease-causing organism is called a pathogen. Many parasites and some saprophytes are pathogenic, but it is important to remember that these categories are not fixed. Sometimes neutral or commensal relationships become detrimental. For example, millions of bacteria and fungi normally colonize human skin, orifices, and intestinal tracts, thereby providing essential vitamins and living harmlessly as house guests (commensals). Some of these ordinary components of the human flora are opportunistic pathogens, capable of causing disease when immune functions of the host are impaired or when antibiosis disrupts the normal population balance of microbial species. The yeast *Candida albicans* is an example of a usually benign organism that can cause terrible infections of mucous membranes, viscera, bones, and brain in compromised or debilitated hosts (Rippon 1988).

These terms are widely used by microbiologists and are useful in making generalizations about life histories and ecology. Recognizing that both fungi and bacteria are diverse and encompass thousands of species, some of these important generalizations follow.

Bacteria are more metabolically diverse than fungi in every way: temperature range, respiratory and energy-deriving capacity, and ability to grow at extremes of osmolarity, pH, and pressure. As a result, bacteria are capable of thriving in many extreme environments where fungi cannot even survive. Moreover, even in normal aquatic habitats, bacteria also seem to have an advantage over fungi, particularly in salt water and estuaries. On the other hand, fungi are especially adapted to growth in terrestrial habitats, and frequently outgrow bacteria at acid pH. Fungi are proficient spore producers and hence are more readily dispersed through air. When a spore germinates, apical growth and hyphal morphology put filamentous fungi at an enormous advantage on solid substrates, where they are able to ramify through the substrate and to transport water, carbon and other essential substances across long distances. In addition, because fungi secrete enzymes across their cell walls into the environment, later absorbing degraded foodstuffs, fungal degradation tends to be less sensitive to substrate concentration and toxicity than bacterial "digestion" which occurs inside the cell. Finally, when it comes to pathogenic processes, bacteria are more efficient at causing animal disease while fungi are the primary plant pathogens. Throughout much of the living world, however, bacteria and fungi regularly share habitats – both living and dead – and exhibit complementary and cooperative roles in the degradation of complex carbon compounds, as well as the recycling of nutrients throughout the biosphere. Many phenomena viewed with a negative impact on the human economy (e.g., biodeterioration, corrosion, agricultural wastage, etc.) are the result of mingled and manifold fungal-bacterial interactions.

III. Secondary Metabolites

Both fungi and bacteria are prolific producers of secondary metabolites. These low molecular weight compounds constitute a staggering array of diverse natural products, biosynthesized in chemical families, often by taxonomically isolated species (Bennett and Bentley 1989). Not essential to the viability of the producing species, the best-known secondary metabolites have pharmacological activity, e.g., penicillin and cephalosporin (antibacterial fungal metabolites); amphotericin B and nystatin (antifungal bacterial metabolites); and streptomycin and tetracycline (antibacterial bacterial metabolites) (Strohl 1997). The spectacular chemical diversity, and the equally impressive range of physiological activity of secondary metabolites has generated a vast chemical and biological literature (Bu'Lock 1965; Bennett and Ciegler 1983; Luckner 1984; Haslam 1985; Cane 1997).

In most cases, the benefit of secondary metabolite production to the producing species is completely unknown. There is a large speculative literature (Janzen 1977; Vining 1990; Ciba Foundation 1992; Hunter 1992). Since fungi and bacteria tend to compete for the same habitats and nutrients, it is widely believed that many secondary metabolites are biological warfare agents. For example, most known antibiotic-producing organisms have been isolated from soils where it seems likely that antibiotics would afford protection from predation and competition. According to this hypothesis, when individual microbial species establish a niche, they may protect their selective advantage by producing inhibitory or toxic secondary metabolites. Nevertheless, for a long time it was debated whether or not these antibiotics were even produced in soils and other natural habitats and some theorists went so far as to maintain that they were simply an artifact of laboratory culture (Woodruff 1980). In recent years, with improved assay and detection systems, it has been demonstrated unequivocally that antibiotics are synthesized in situ and can contribute to microbial fitness (Thomashow et al. 1997; Weller and Thomashow 1990).

Pharmaceutical companies seeking new and better antimicrobial drugs funded much of the early research on secondary metabolism. For this reason, speculations about the selective value of secondary metabolite production have been dominated by the "warfare" theory. Less attention has been paid to more subtle mutualistic and co-evolutionary advantages such as cooperative signaling for reproduction and dispersal. Botanists, writing in the context of plant-insect coevolution have demonstrated that plant secondary metabolites can be utilized as defense compounds by insects, which in return benefit plants by acting as vectors for cross-pollination (Harborne 1978). An interesting relationship involves yeasts (*Candida*, *Cryptococcus*, and *Pichia*), cactus plants, and *Drosophila* flies. When cactus plants are injured, there follows a microbial invasion in which yeasts predominate. Odiferous secondary metabolites produced during the decay process attract *Drosophila* species that lay their eggs in the rotting plant tissue. Subsequently hatched larvae feed on the resident microbial population, pupate, emerge as adults, and disperse the yeasts to distant rotting cacti. Because distinct yeast communities are associated with distinct *Drosophila* species, a coevolutionary relationship is considered likely (Starmer et al. 1991).

Workers in the field of biocontrol have discovered that plants can benefit indirectly when rhizosphere bacteria produce metabolites that suppress the growth of plant pathogens (Lynch 1987). Sometimes the effects are indirect and do not involve antibiotics or toxins. The siderophore pseudobactin, for example, is a metabolite of *Pseudomonas putida* that sequesters iron, thereby inhibiting the growth of pathogenic species (Schroth and Hancock 1982). It is plausible that myriad species of bacteria and fungi, living in close association in the rhizosphere, may be similarly using secondary metabolites for selective advantage. Experiments that simultaneously study multiple species will be required to discern these functions.

IV. Human Foods and Animal Feeds

The best-characterized fungal-bacterial associations are all human food fermentations. We will focus our discussion on three of these: the Japanese koji process for soy sauce manufacture, sour dough bread, and the unusual health drink called kombucha. A number of other human food fermentations are listed in Table 1.

Soy sauce is made from wheat, soybeans, and salt, fermented in a complex manner with filamentous fungi, bacteria, and yeasts. In Japan, at least five major kinds of soy sauce are recognized, of which the darkest form is the most popular, and is simply called shoyu. Variations in proportions of ingredients, times of fermentation, and so forth create the different forms. The following descrip-

tion is a representative process for making shoyu. Cooked soybeans or defatted soybean grits are mixed with roasted, crushed wheat. The soybean wheat mixture is inoculated with koji mold (filamentous fungi of the species *Aspergillus oryzae* or *A. sojae*). In traditional systems, the natural environment provided the inoculum. In recent times, known strains of fungi are used to produce a seed koji to act as a starter culture. Seed koji is usually pregrown on rice or wheat bran to favor sporulation.

The inoculated wheat-soybean mixture is spread in a thin layer and incubated with aeration and constant temperature for 2 or 3 days. Fungal hyphae penetrate the wheat-soybean mixture, growing throughout, excreting hydrolyzing enzymes that break down the carbohydrates and proteins. The resultant aromatic mixture of hyphae, enzymes, and partially degraded substrate is called a koji. There is no English word that is equivalent. In addition to soy sauce, koji is an integral part of the manufacture of two other important Japanese foodstuffs: miso and sake.

In soy sauce manufacture, the koji is mixed with salt and water, and the resultant mash (*moroni*) is held for 5–9 months. During the moroni period, enzymes from the koji molds further hydrolyze the protein and starch. In modern fermentations, cultures of *Pediococcus halophilus*, an osmophilic lactic acid bacterium, are inoculated. Traditional fermentation relied on natural successions. The bacteria ferment sugars into lactic acid and alcohol, simultaneously lowering the pH. The acid environment in turn favors the growth of yeasts, and inocula of *Saccharomyces rouxii* can be used to ensure standardization and success. Other salt-tolerant yeasts often occur in the late stages of fermentation and the high salt environment effectively excludes most undesirable microorganisms. At the end of the fermentation process, the final matured moromi is filtered through cloth, clarified, adjusted to standard salt concentrations, bottled, and pasteurized (Fukushima 1989; O'Toole 1997).

Sour dough contains a mixture of yeasts and lactic acid bacteria, usually of the genus *Lactobacillus*. Several different yeasts have been identified including *Saccharomyces cerevisiae*, *S. exiguus*, *Candida krusei*, and *Pichia saitori*. The bacteria do the souring while the yeasts do most of the leavening. Sour doughs are used to a limited extent for making certain wheat breads such as the famous San Francisco Sour Dough bread (a style

Table 1. Selected human foods and beverages made from mixed microbial fermentations[a] (After Campbell-Platt 1987; Steinkraus et al. 1983)

Name (synonym)	Description
Basi (njohi)	Alcoholic beverage made from sugar cane, drunk in Asia and Africa. Pressed, concentrated sugar cane sugar is fermented with both lactic acid bacteria (*Lactobacillus* sp.) and yeasts (*Saccharomyces* and *Endomycopsis*)
Blue cheese	Semi-soft or semi-hard white cheese with blue veins made with cow, goat or sheep milk, and fermented with lactic acid bacteria (*Streptococcus lactis* and *S. cremoris*), yeast (*Candida lipolytica*) and mold (*Penicillium roquefortii*). The bacteria are main fermentative species; the yeast produces lipases; the mold produces blue veins; all contribute to flavor. Many variants on basic theme, e.g., *adlelost* (Sweden), *gorgonzola* (Italy), *Roquefort* (France), *Stilton* (England)
Camembert cheese (brie, chevru)	Soft full fat cheese usually made from cow's milk and covered with white mold. Milk is fermented with *Streptococcus lactis* and *S. cremoris*; mold surface is *Penicillium camembertii*. The yeasts *Candida* and *Kluyveromyces* and the bacterium *Brevibacterium linens* may also be present
Chicha	Sour alcoholic beverage usually made from corn, drunk in South America. *Lactobacillus* sp. give characteristic acidic taste; *Saccharomyces cerevisiae* produces alcohol. Brazilian versions made from cassava are called *kaschiri* and from yucca are called *masata*. In the traditional preparation, starch was converted to sugar utilizing the amylase in saliva; women did the chewing and spitting; nowadays, malting enzymes usually replace saliva
Chicouangue	South African cassava dumpling, made with soaked, fermented, and mashed tubers that are then squeezed, pressed and boiled. Retting primarily cased by bacteria (*Corynebacterium*); acid flavor from *Lactobacillus* sp. and *Leuconostoc* sp.; in late stages of fermentation yeasts of *Candida* species are involved
Cocoa	Fermented beans of the cocoa tree are roasted and ground, used for making chocolate and coca powder. A complex; poorly understood fermentation liquefies the outer pulp of the pod surrounding the beans. Yeasts include *Hansenula* sp., *Kloeckera* sp. and *Saccharomyces* sp.; bacteria include *Acetobacter* sp. Late in the fermentation, yeasts are replaced by lactic acid bacteria (*Lactobacillus plantarum* and *L. collinoides*)
Coffee	Beans obtained by fermentation of pulp surrounding coffee beans. Like cocoa, a complex, poorly understood fermentation. Pectolytic enzymes from *Erwinia diosolvens* and *Bacillus* sp. Yeasts in the genera *Saccharomyces* and *Endomycopsis* are involved, as are lactic acid bacteria (*Lactobacillus*, *Leuconostoc*, and *Streptococcus*)
Country hams	Semi-dry cured pork, salted and dried, usually not cooked. Nitrate reduced to nitrite in curing process by *Micrococcus* and *Staphylococcus* bacteria; *Lactobacillus* sp. also active; molds may cover the surface of dried hams, usually *Penicillium nalgiovense* or *Aspergillus* sp. All of the microorganisms important in flavor production. Many local varients, e.g., *Kasseler* and *Lachsschinken* (Germany), *Virginia ham* (USA), *Yunnan ham* (China)
Italian salami	Unsmoked, undried meat sausage usually made with beef or pork; after curing bacteria reduce nitrate to nitrite (*Micrococcus* and *Staphylococcus*); lactic acid bacteria dominate the fermentation (*Lactobacillus* sp.); yeasts and molds (*Penicillium nalgiovense* and *Aspergillus* sp.) develop on dry, acidic surface, contributing to characteristic flavor. Many local variants, e.g., *chasseur* (France), *ole* (Belgium), *zweetworst* (Netherlands)
Kvas (kwas)	Alcoholic beverage usually made from watery extract of rye, drunk widely in the former USSR, and eastern Europe. Starter either yeast or sourdough bread. *Lactobacillus* sp. gives sour flavor, *Saccharomyces cerevisiae* produces alcohol
Pepperoni	Smoked, dried, flavored meat sausage usually made with beef and pork in relatively small diameter; curing and fermentation similar to salami (see above). Many local variants, e.g., *bauernbratwurst* (Germany), *rauschsalami* (Switzerland), *szalami* (Hungary), *toppen* (Norway)

Table 1. *Continued*

Name (synonym)	Description
Pulque	Mexican alcoholic beverage made by fermentation of juices from Century plant (*Agave*) or prickly pear cactus (*Opuntia*). The bacterium *Zymomonas mobilis* as well as several species of *Lactobacillus* important in fermentation; alcohol production associated with several species of yeasts: *Kloeckera apiculata*, *Saccharomyces cerevisiae*, *Endomycopsis* sp., *Pichia* sp. and *Torulopsis* sp.
Ragi (Chinese yeast)	Dried rice, millet, cassava or other starchy base, milled, mixed with herbs and spices (often pepper and garlic), water and starter from previous batch, shaped into balls, incubated in humid environment, then dried and sold as inocula. The microbial flora of *ragi* balls varies with locale and intended food use. Molds of *Aspergillus* sp., *Fusarium* sp., *Rhizopus* sp., and *Trichosporon* sp. are frequently present. Yeasts (*Candida* and *Hansenula* sp.) and lactic acid bacteria also found in *ragi*. Used as inoculum for *sake* and other fermented foods and drinks, mostly in Asia
Sake (chau, li)	Rice wine in which a rice koji with the mold *Aspergillus oryzae* produces the amylases that break down the starch; lactic acid bacteria and yeasts follow in the fermentation process. In home preparations, a *ragi* starter is often used. Many local variants in China, Japan, Korea, Philippines and other areas in SE Asia
Sapsago (green cheese)	Very low fat, strongly flavored, cows' milk cheese, pale green in color, used for grating and condiment. Initial ripening from lactic acid bacteria (*Lactobacillus* sp. and *Streptococcus* sp.). Dried clover added to give characteristic color and color; molds of species *Penicillium roquefortii* and *Mucor racemosus* also present. Local variants include *gammelost* (Norway) and *krauterkäse* (Germany)
Tempe	Soybeans bound together in a cake by mold mycelium, found especially in Indonesia, Malaysia, Thailand and Singapore. Soybeans are soaked and in initial fermentation several bacteria (*Klebsiella pneumoniae*, various lactic acid bacteria) predominate. When pH is reduced by bacterial growth, a second stage fermentation with the mold *Rhizopus oligosporus* follows in which mold proteases and lipases break down soybeans and make them more digestible

of "French" bread). More importantly, they are essential for making rye bread. Rye flour has different properties than wheat flour and unless rye flour is properly soured, it will not rise. The process can be effected by using a previous batch of sour dough as starter, or by using a commercial culture of *Lactobacillus* (Sugihara 1985).

Kombucha, sometimes called Manchurian mushroom or tea fungus (Frank 1994) is a domesticated consortium of bacteria (*Acetobacter xylinum*) and yeasts usually grown on sugared tea. In a recent taxonomic analysis of 34 German kombucha cultures, three yeast genera were identified (*Brettanomyces*, *Zygosaccharomyces*, and *Saccharomyces*) and most isolates contained more than one species (Mayser et al. 1995). The yeasts in the consortium utilize the sugar to produce ethanol, and the bacterium metabolizes the alcohol into organic acids in a manner similar to vinegar production (Swings and De Ley 1981). The bacterium also produces a polysaccharide matrix that gives the consortium its unusual, gelatinous appearance.

Proponents drink the tea as a health aid, and pass on portions of the gelatinous consortium as gifts to their friends, rather like a "biological chain letter" (Perron et al. 1995). Kombucha teas are purported to have beneficial effects on the digestive tract, the immune system, and numerous other human functions (Frank 1994; Tietze 1994).

These three examples illustrate the way in which successions of molds, bacteria, and yeasts can alter a substrate. Numerous human foods are the results of mixed bacterial and fungal fermentations. In Western culture, mold-ripened cheeses are perhaps the best known (Kosikowski and Mistry 1997). Asian and African cultures have developed many other unusual ways of making foods more palatable and more digestible (Steinkraus et al. 1983; Campbell-Platt 1987).

In addition, the nutritional value of many animal feeds can be improved by microbial enzymes that degrade soluble fibers that have antinutritional elements or that supplement the animal's own digestive capacity. Silage is formed when crude fodder is converted into an improved livestock feed through a fermentation process, usually conducted in a silo. The manufacture of silage is an agricultural "art form" and the contributing microbial consortia have been only partially characterized. There is an increasing trend in global agriculture toward consolidation, with a concomitant decrease in the use of feeds produced from locally grown raw materials. Many of these new processes utilize crude enzymes, as opposed to living microorganisms (Godfrey and West 1996). If this trend increases, there will be increased research on the consortia that operate in natural fermentations will be needed in order to characterize the appropriate enzymes. In summary, the economic incentives to standardize food and feed fermentations will ensure that these processes remain among the best-understood examples of interactions between bacteria and fungi.

V. Soils and the Rhizosphere

Soil science is a major scientific discipline with important implications for agriculture, forestry, conservation, and biodiversity (see Paul ad Clark 1990 and Kilham 1994 for useful introductions to the field). In the soil, bacteria and fungi are not only important in decomposing carbon, they are also important in recycling and retaining nitrogen, phosphorus, sulfur, and other elements in their biomass. Healthy, productive soils have high microbial populations. When microbial populations decline, nutrients are lost into ground- and surface water (Hendrix et al. 1986). In well-aerated soils, fungi constitute a large part of the total biomass, where they are particularly adept at degrading cellulose, hemicellulose, pectin, lignin, and other polymeric plant remains. The exact proportion of fungi and bacteria in soils is extremely difficult to calculate. Various techniques are used for enumeration such as the classic plate count (dilutions of soil plated on agar media); direct microscopic analysis; ergosteral analysis; PCR, etc. All techniques have advantages and disadvantages. Plating favors the growth of fungal spores and bacterial colonies while underestimating mycelial biomass. *The Ecology of Soil Fungi* (Parkinson and Waid 1960) summarizes some of the classical work on estimating size of fungal populations; a more contemporary compendium of methods for measuring growth and estimating

population size is available in the *Manual of Environmental Microbiology* (Hurst 1997). Volume 23 of *Methods in Microbiology* (Norris et al. 1991), which is devoted to techniques for studying mycorrhiza, is also an excellent resource.

Composting is one of the oldest methods for fertilizing soils and stabilizing agricultural wastes (Poincelot 1975). Compost is a mixture of decayed organic matter (e.g., bark, straw, and manure), degraded by a consortium of bacteria and fungi, used for fertilizing and conditioning soil (Biddlestone and Gray 1985). Composts vary greatly, depending on their composition and intended use. For example, cultivation of the edible mushroom *Agaricus bisporus* is often conducted with composts made from straw and horse manure, while many modern agricultural processes utilize dry bioorganic fertilizers obtained from combination of animal excreta (e.g., pigs, chickens) with agricultural plant wastes. Such biofertilizers can improve soil fertility and suppress the growth of pathogens (Williams et al. 1985).

Naturally occurring terrestrial communities consist of multiple species growing intimately in association with one another. The soil, and the vegetation it bears, are extraordinarily intricate ecosystems. The rhizosphere is the best-studied soil-microbial-plant community.

The rhizosphere can be defined most broadly as the area surrounding the roots of plants. It encompasses the soil, organisms, water and other parameters that influence, and are influenced by, plant roots and their exudates. Characteristics of rhizospheres differ with geography, climate, soil aeration, microbial flora, and plant species (Lynch 1990).

The rhizosphere is difficult to study because of its extraordinary complexity, because most of the activity takes place below the ground, and because the activities associated with gathering data may perturb normal rhizosphere activities. Fungi and bacteria live in the rhizosphere in prodigious numbers, interacting with each other and with the roots of the plants. By conservative estimate, there are between 10^5 and 10^6 fungi and between 10^{15} and 10^{17} bacteria in each gram of fertile soil (Metting 1993). Most of the fungi are aerobic and live in the top 30 cm of the soil, with the greatest number found within 10 cm of the surface. Fungi and bacteria of the rhizosphere cooperate in the decomposition of organic matter into inorganic compounds or elements, a process called mineralization. The inorganic substances are absorbed by the plant and, in turn, are reconverted into organic matter through photosynthesis. This cycle is essential for life on earth.

The rhizosphere is dynamic, and relationships among organisms vary with changing availability of nutrients, size of plant and microbial populations, and conditions of the soil. Microbes live in close proximity to the plant roots, performing similar and/or supportive functions, both competing and cooperating for resources. Physical interactions of bacteria and fungi in the rhizosphere include adherence by bacteria to the fungal hyphae and active bacterial colonization of the hyphae (Lockwood 1968). Both dead and living fungi and bacteria can provide nutrients for living fungi and bacteria. Saprophytic fungi and bacteria may work together or in succession to decompose organic compounds in the rhizosphere. For example, a limited number of species of fungi and bacteria are the only organisms capable of breaking down lignin. The most efficient lignin degraders are the white rot fungi, which produce a group of highly reactive, nonspecific lignin peroxidases. However, in nature, a succession of fungi and bacteria usually decompose lignin in a cooperative process (Kirk 1984; Deobald and Crawford 1997). Generally, white rot fungi are the primary degraders, followed by other fungi and bacteria that further break down lignin decomposition products as well as other plant polymers including cellulose and hemicellulose (Buswell 1991).

When the interactions between fungi and bacteria in the rhizosphere are described, it is important to recognize that the perspective usually (but not always) concerns the ability of microbes to help or harm plant growth. Thus, the soil microbiota may coexist in "neutral associations," have deleterious effects such as "antagonism" and "suppression of growth," or have beneficial relationships such as "cooperation for nutrients" and "stimulation of growth and reproduction." Similarly, the relationship of bacteria and fungi to one another can be described in positive or negative terms. Bacteria can maintain and improve soil attributes hospitable to fungi, and fungal mycelial growth can increase the flow of air and water through the rhizosphere. Certain bacteria "suppress" the growth of fungi by growing as parasites, taking nourishment directly from the host mycelium. If the fungus is a pathogen to an economically useful plant, the parasitic bacterium will be called a "biocontrol agent." If the fungus is an essential mycorrhizal species, the bacterial behav-

ior will be described in negative terms such as "plant growth suppression."

Mycorrhizae, a symbiotic association between fungi and plant roots, are far better studied than the interactions between fungi and bacteria in the rhizosphere. It is estimated that mycorrhizae occur in about 90% of vascular plant species (Fortin 1983). The fungus forms a network of hyphae in the soil which greatly extends the surface area of the root, increases the uptake of water and nutrients, and may serve as a pathway for nutrient exchange between plants. The fungus produces phosphatases and other enzymes which increase the efficiency of the absorption of inorganic minerals such as phosphorous, copper, zinc, potassium, and nitrogen. The fungal symbiont also produces plant growth factors, and provides protection against deleterious bacteria and other pathogens (Marx 1982). The plant provides root exudates that contain essential organic photosynthates to the fungus (Harley and Smith 1983). The hyphae of ectomycorrhizal fungi grow, with the help of pectolytic proteins, between the cortical cells of the root, and produce a characteristic internal structure called a Hartig net (Hacskaylo 1972). Mycelium also covers the epidermis near the root tip, forming a mantle or sheath. The mycorrhizal root is physically protected by the mantle and by the Hartig net. Further protection from pathogens is the result of antibiotics and other chemical deterrents produced either by the fungus itself or by the plant as a result of the mycorrhizal infection (Cooke 1977).

Mycelia of some species of fungi can be grown with plants to form mycorrhizae in the laboratory. Melin (1936) pioneered synthesis of mycorrhizae in pure culture. There are several procedures for inoculation, including placing the fungal inoculum in the soil below the seed, or dipping the roots of the seedlings into a slurry which contains the inoculum (Riffle and Maronek 1982). Nutrition, light, temperature, or competition from bacteria or other fungi can alter the effects of these inoculations.

Unfortunately, it has been difficult or impossible to cultivate many mycorrhizal mushrooms. Edible species of the Cantharellaceae, the chanterelles, have been among the most resistant. Dependable cultivation is highly desirable both for commercial production and for systematic research. There is a very low germination rate of spores, and attempts to grow pure cultures of the mycelia from basidiocarps are complicated by the coisolation of bacteria and other microorganisms (Danell et al. 1993; Straatsma et al. 1985).

Because bacteria are regularly isolated from basidiocarps, it has been hypothesized that they might be providing an essential fruiting factor. For example, when bacteria from basidiocarps of *C. cibarius* were compared to bacteria from the soil, the populations were found to be the same size ($2.4 \times 10^6 g^{-1}$), but not the same composition (Danell et al. 1993). *Pseudomonas fluorescens* biovar I predominated in isolations from *C. cibarius* mycelia, making up 78% of total bacteria in the basidiocarp, although it represented only 12% of the total bacteria in the soil (Danell et al. 1993). The bacteria in the *C. cibarius* basidiocarp were located in mucus in the spaces between the hyphae, but not within the hyphae. When *P. fluorescens* was added to artificially grown *Cantharellus* and *Picea* mycorrhizae, the bacteria did not penetrate the mantle. It is thought that the vegetative hyphae offer a favorable environment for *P. fluorescens* and that these bacteria are incorporated into the fruit body as it is formed. Effects of *P. fluorescens* on mycelial growth of *C. cibarius* were studied, but no growth stimulation was observed (Danell et al. 1993). Chanterelle basidiocarps have been successfully cultivated for the first time in a greenhouse by inoculating *Pinus* seedlings with pure cultures of *C. cibarius* (Danell and Camacho 1997). The breakthrough in cultivation of this mycorrhizal mushroom can lead to further study and a better understanding of the role of *P. fluorescens* and other bacteria in the life cycle of these fungi.

There is evidence that some bacteria in the rhizosphere can induce germination and growth of spores and hyphae of mycorrhizal fungi. Bacteria collected from basidiocarps, mycorrhizae, and soil of the rhizosphere were inoculated with spores of ectomycorrhizal fungi from the same environment. Species of the bacteria *Pseudomonas*, *Corynebacterium*, *Arthrobacter*, *Micrococcus*, *Tritirachium*, and some unidentified soil bacteria were found to stimulate germination of the spores of species of *Paxillus*, *Laccaria*, and *Hebeloma* (Ali and Jackson 1989). Germination of spores may also be enhanced by exudates of yeasts (Fries 1984). Other studies showed that citric and malic acid produced by bacteria increased the growth of *Paxillus involutus* and *Hebeloma crutuliniforme*. These bacteria were also able to metabolize and detoxify exudates of *P. involutus* that were toxic to the fungus itself (Duponnois and Garbaye 1990).

In other greenhouse studies, beech seedlings (*Fagus sylvatica*) were inoculated with *Laccaria laccata* and with the bacterium *Agrobacterium radiobacter*, both separately and together. Tree growth was increased with either inoculum, but there was no effect when both were present, perhaps due to competition between the fungus and bacterium (Leyval and Berthelin 1989).

Bacteria which have a positive effect on the establishment of mycorrhiza have been called mycorrhiza helper bacteria or mycorrhization helper bacteria (MHB) (Duponnois and Garbaye 1991a). In 1994, Garbaye defined MHB as "telluric bacteria associating with some mycorrhizal fungi and promoting the establishment of mycorrhizal symbiosis." *Pseudomonas fluorescens* BBc6, originally isolated from a sporocarp of *Laccaria bicolor*, is an example of an MHB which stimulated the mycorrhizal establishment of *Laccaria* with Douglas fir (Duponnois and Garbaye 1991b; Garbay and Duponnois 1992). Using a rifampin-resistant mutant of *P. fluorescens* called BBc6R8, Frey-Klett et al. (1997) were able to demonstrate that the mycorrhiza helper effect was observed even when the population of inoculated BBc6R8 had dropped to low concentrations of detectable colony-forming units in the soil.

VI. Mushroom Cultivation

Mushrooms have been eaten and prized by humans since antiquity. These fleshy fungi have been collected in the wild and, more recently, cultivated commercially. Substrates rich in microorganisms, such as dung, are often especially favorable environments for fungal reproduction and mushroom formation. Studies of the role of bacteria in mushroom cultivation are primarily empirical (Stamets and Chilton 1983), but the results suggest that many bacteria are helpful and some necessary for carpophore production. Although the mechanisms are not fully understood, certain bacteria provide a signal to initiate the formation of mushroom primordia, and may also stimulate an increase in the number of primordia (Hayes et al. 1969; Park and Agnihotri 1969; Eger 1972; Stanek 1974; Rainey et al. 1990). The commercially ubiquitous basidiomycete *Agaricus bisporus*, the common button mushroom, does not produce basidiomata if the cultivation material is sterile. The substratum used commercially is composted and pasteurized rather than sterilized, in order to minimize, but not eliminate, the microorganisms. A casing layer (usually consisting of peat moss mixed with lime to achieve optimal pH levels) is placed over the compost and fungal spawn to maintain a humid environment favorable to the growth of both the mushrooms and the beneficial bacteria. The casing is an additional source of helpful bacteria and is also pasteurized and not sterilized.

Pseudomonas putida was the first bacterium identified as an important factor in the initiation and development of carpophores (Eger 1961). *Bacillus megaterium*, *Arthrobacter terregens*, and *Rhizobium meliloti* have also been shown to produce metabolites that initiate primordia formation and increase yields in mushroom cultivation (Park and Agnihotri 1969). Maximum beneficial effects are often specific to the strains of bacteria and fungi used (Stamets and Chilton 1983).

Bacteria that initiate primordia formation also affect the linear growth and colony morphology of mushroom cultures (Barron 1988; Rainey 1991). In vitro studies have shown that *A. bisporus*, grown with *P. putida* has increased linear growth, less branching, and a higher ratio of aerial to submerged hyphae (Rainey 1991). Pathogens of *A. bisporus*, such as *Pseudomonas tolaasii*, that causes bacterial blotch, inhibited hyphal growth (Rainey et al. 1990).

Agaricus bisporus uses bacteria naturally found in straw/manure compost as a source of nutrients. In a study using ^{14}C-labeled *Bacillus subtilis*, *A bisporus* degraded and used carbon and nitrogen from both live and killed bacteria, even in the presence of other readily available sources of these nutrients (Fermor et al. 1991). Presence of *Scytalidium thermophilum* in compost also increased mycelial growth and mushroom yield of *A. bisporus* (Straatsma et al. 1989). Other basidiomycetes, which have been shown to utilize bacteria for nutrients, include *Coprinus quadrifidus*, *Lepista nuda*, and *Pleurotus ostreatus* (Barron 1988).

VII. Biofilms

Biofilms are a loose association of microorganisms that grow in microcolonies embedded in a polysaccharide matrix often called slime and interspersed with water channels. Confocal scanning laser microscopy has revealed a high degree of biological integration, and some researches now model biofilms as multicellular "organisms" with a primitive circulatory system. Biofilms have been studied in the context of tooth decay, implant surgery, effluent treatment, corrosion, bioreactor design, and a host of natural processes in which inert surfaces are colonized by bacteria (Costerton et al. 1987; Lappin-Scott and Costerton 1995). The bulk of the research on these physiologically integrated communities has focused on medically important issues such as the way in which biofilms exhibit resistance against antimicrobial agents (Gilbert and Brown 1995); or engineering issues such as the clogging of water purification systems (Mittelman 1995). Biofilms are usually dominated by bacterial species and bacterial polymers, and often no attempt is made to see if fungi are participants. For instance, at least one recent review on biofilms of the ruminant digestive tract entirely ignores the role of anaerobic chytrids (Cheng et al. 1995). However, in situations where filamentous fungi grow in abundance, their important contribution has been recognized. For example, the corrosion of aluminum has been associated with outgrowth of *Hormoconis* (=*Cladosporium*) *resinae* (Salvarezza et al. 1983; Videla and Charaklis 1992). Without a doubt, as more scientists become alert to the possibility of prokaryotic-eukaryotic interactions in biofilms, more examples will be forthcoming.

The economic importance of biofilms ensures that there will be continued research funding. Perhaps just as importantly, the development of biofilm research as a separate discipline has fostered a philosophical shift among the microbiologists who study them.

"If microbial ecology is to move forward, it must go beyond the reduction of the complexities of bacteria into merely isolated cell lines, enzymes, and genetic sequences. A century of pure culture studies has provided extremely detailed information on the biochemistry, physiology and genetics of bacteria. What remains to be determined is how they function as successful members of interacting communities in biofilms and how microbial communities function as components of the environment" (Caldwell 1995, p. 64).

(Note: Even this eminently admirable conception represents an example of bacteriocentric thinking!)

VIII. Experimental Techniques and Traditions

Microbiology required unique tools to develop into a science: van Leeuwenhoek's microscope, and the Pasteur-Koch perfection of pure culture technique. During the late 19th century, microbiologists using these tools discovered the organisms responsible for many catastrophic human diseases. The importance of clinical microbiology coupled with the power of Koch's postulates made the study of pure cultures a sine qua non of bacteriological technique. These bacteriological techniques were easily adapted to the study of microscopic fungi, and many of the early breakthroughs in plant pathology and medical mycology grew out of the experimental approaches developed in bacteriology laboratories.

Early environmental microbiologists also learned to apply bacteriological approaches to natural systems. Enrichment culture, a technique perfected and championed by Beijerinck and his disciples, is regularly used for finding a specific microbe that grows on a certain substrate. Enrichment cultures favor the growth of a particular species based on its nutritional requirements. In the most common application of this method, aliquots of soil, water, or other mixed inocula are placed into a medium containing the target substrate as the sole carbon source. Only organisms with appropriate degradative ability will grow. When the enrichment is conducted in a liquid medium, competition for the substrate will lead to enrichment for the microbial strain that is able to grow the fastest. When it is conducted on Petri plates, usually several species are isolated which then can be subcultured and tested further. With few exceptions, these approaches lead to the isolation of bacteria, since fungi tend to grow slower and produce fewer propagules. Moreover, the end result is the isolation of a pure culture from a heterogeneous population.

In recent years, as scientists have developed indirect methods for sampling microbial biodiversity, it has become apparent that the culture method greatly underestimates and distorts the number and types of organisms growing in nature. For example, from many habitats, only 1% of the organisms observed by microscopy are brought into culture (Amann et al. 1995). The direct sampling of nucleic acids from natural environments has provided new insights into the range of uncultivable microbes, mostly bacteria and archaea (Pace 1997); however the forces that sustain and shape microbial biodiversity are as yet poorly understood (O'Donnell et al. 1995).

All of this leads scientists to miscalculate the capacity and sometimes the necessity of microorganisms to live in close associations. It tells only part of the story to study a bacterium or a fungus in pure culture. Methodological separatism is reinforced by mental compartmentalization. There is a lack of interaction not only between bacteriologists and mycologists, but also between agricultural, clinical and industrial workers.

IX. The Future

When Nobel Laureate Selman Waksman was President of the Society of American Bacteriologists (now American Society for Microbiology) in 1942, he chastised microbiologists for failing to recognize that the great majority of microbes "carry out their normal activities not in pure cultures but in mixed populations" (quoted in Clark 1953). This failure is greater now, at the start of the new Millenium.

The extreme specialization of modern biology encourages, even requires, reductionism. In contrast, a subject as broad as fungal bacterial interactions must be based in intellectual and methodological eclecticism. It is our thesis that at this point in time, the mental obstacles may be more difficult to overcome than the methodological ones. Each subdiscipline has built order and meaning by a process of categorization and recategorization. The traditions of mycology and bacteriology have grown up independently of one another, further cementing the rift. It is remarkable how few references to bacteria are made in typical publications on fungal ecology, and vice versa. If anything, bacteriologists ignore fungi even more than mycologists ignore bacteria.

In the laboratory, all microbiologists place enormous value on pure cultures, calling any unwanted strain a contaminant. Why contaminant? Why not ally, associate, partner, or coworker? In the real world, microbial loners are extraordinarily rare; in the laboratory they are so much the rule that laboratory practice distorts the way we think about the real world.

Work on rhizosphere biology has taught us that plants cannot exist alone. Their health is dependent on the fungi, bacteria, and other organisms surrounding their roots. This same insight surely is true of many (most?) microbial populations if we choose to study them properly. We know that bacteria and fungi have evolved many antagonistic systems; it is just as probable that they have evolved mutually beneficial mechanisms for reproduction, dispersal, and survival.

In summary, this overview of selected fungal-bacterial associations is intended to stimulate both bacteriologists and mycologists to overcome their organismal isolationism, and to study more basic and applied microbial processes in terms of communities. New microscopic methods are available for immunocytochemistry and histology; new molecular methods are available for transposon mutagenesis, development of molecular probes, sampling of environmental nucleic acids; and new analytical tools are available for physical and chemical analysis (Gerhardt 1994). In order to exploit these powerful experimental techniques, microbiologists must be receptive to discoveries in population and evolutionary biology that will allow us to illuminate the complexities of microbial life across kingdoms and superkingdoms.

Acknowledgments. JWB thanks Ben Lugtenberg for the opportunity to study rhizosphere bacteria in the Dept. of Plant Molecular Biology at Leiden University, NL, and Douglas Eveleigh for help in obtaining some of the references.

References

Alexopoulos CJ, Mims CW, Blackwell M (1996) Introductory mycology. John Wiley, New York

Ali NA, Jackson RM (1989) Stimulation of germination of spores of some ectomycorrhizal fungi by other microorganisms. Mycol Res 93:182–186

Amann RI, Ludwig W, Schleifer KH (1995) Phylogenetic identification and in situ detection of individual microbial cells without cultivation. Microbiol Rev 59:143–169

Baldauf SL, Palmer JD (1993) Animals and fungi are each other's closest relatives: congruent evidence from multiple partners. Proc Natl Acad Sci USA 90:11558–11562

Barron GL (1988) Microcolonies of bacteria as a nutrient source for lignicolous and other fungi. Can J Bot 66:2505–2510

Bennett JW, Bentley R (1989) What's in a name? – microbial secondary metabolism. Adv Appl Microbiol 34:1–28

Bennett JW, Ciegler A (eds) (1983) Secondary metabolism and differentiation in fungi. Marcel Dekker, New York

Biddlestone AJ, Gray KR (1985) Composting. In: Robinson CW, Howell JA (eds) Comprehensive biotechnology, vol 4. The practice of biotechnology, specialty products and service activities. Pergamon Press, Oxford, pp 1059–1070

Bu'Lock JD (1965) The biosynthesis of natural products: an introduction to secondary metabolism. McGraw-Hill, London

Buswell JA (1991) Fungal degradation of lignin. In: Arora DK, Rai B, Mukerji KG, Knudsen GR (eds) Handbook of applied mycology, vol 1, Soil and plants. Marcel Dekker, New York, pp 425–480

Caldwell DE (1995) Cultivation and study of biofilm communities. In: Lappin-Scott HM, Costerton JW (eds) Microbial biofilms. Cambridge University Press, Cambridge, pp 64–79

Campbell-Platt G (1987) Fermented foods of the world. A dictionary and guide. Butterworths, London

Cane DE (ed) (1997) Polyketide and nonribosomal polypeptide biosynthesis. Chem Rev 97:2463–2464

Carlile MJ, Watkinson SC (1994) The Fungi. Academic Press, London

Cheng K-J, McAllister TA, Costerton JW (1995) Biofilms of the ruminant digestive tract. In: Lappin-Scott HM, Costerton JW (eds) Microbial biofilms. Cambridge University Press, Cambridge, pp 221–232

Ciba Foundation Symposium (1992) Secondary metabolites: their function and evolution (Ciba Foundation Symposium 171). Wiley, Chichester

Clark PF (1953) A half century of presidential addresses of the Society of American Bacteriologists. Bacteriol Rev 17:213–247

Cooke R (1977) The biology of symbiotic fungi. John Wiley, London

Costerton JW, Cheng K-J, Geesey GG, Ladd TI, Nickel JC, Dasgupta M, Marrie TJ (1987) Bacterial biofilms in nature and disease. Annu Rev Microbiol 41:435–464

Danell E, Camacho F (1997) Successful cultivation of the golden chanterelle. Nature 385:303

Danell E, Alström S, Ternström A (1993) *Pseudomonas fluorescens* in association with fruit bodies of the ectomycorrhizal mushroom *Cantharellus cibarius*. Mycol Res 97:1148–1152

Dawson JO (1992) Nitrogen fixation in forests and agroforestry. In: Metting FB Jr (ed) Soil microbial ecology. Marcel Dekker, New York, pp 227–253

Deobald LA, Crawford DL (1997) Lignocellulose biodegradation. In: Hurst CJ (ed) Manual of environmental microbiology. ASM Press, Washington, DC, pp 730–737

Duponnois R, Garbaye J (1990) Some mechanisms involved in growth stimulation of ectomycorrhizal fungi by bacteria. Can J Bot 68:2148–2152

Duponnois R, Garbaye J (1991a) Mycorrhization helper bacteria associated with the Douglas fir-*Laccaria laccata* symboisis: effects in asceptic and in glasshouse conditions. Ann Soc For 48:239–251

Duponnois R, Garbaye J (1991b) Effect of dual inoculation of Douglas fir with the ectomycorrhizal fungus *Laccaria laccta* and mycorrhizal helper bacteria (MHB) in two bare-root forest nurseries. Plant Soil 138:169–176

Eger G (1961) Untersuchungen über die Funktion der Deckschicht bei der Fruchtkörperbildung des Kulterchampignons, *Psalliota bispora* Lange. Arch Mikrobiol 39:313–334

Eger G (1972) Experiments and comments on the action of bacteria on sporophore initiation in *Agaricus bisporus*. Mushroom Sci VIII:719–726

Fermor TR, Wood DA, Lincoln SP, Fenlon JS (1991) Bacteriolysis by *Agaricus bisporus*. J Gen Microbiol 137: 15–22

Fortin JA (1983) Introduction, Fifth North American Conference on Mycorrhizae. Can J Bot 61:907–908

Frank GW (1994) Kombucha, healthy beverage and natural remedy from the Far East, 4th edn. Wilhelm Ennsthaler, Steyr, Austria

Frey-Klett P, Pierrat JC, Garbaye J (1997) Location and survival of mycorrhiza helper *Pseudomonas fluorescens* during establishment of ectomycorrhizal symbiosis between *Laccaria bicolor* and Douglas fir. Appl Environ Microbiol 63:139–144

Fries N (1984) Spore germination in the higher basidiomycetes. Proc Indian Acad Sci Plant Sci 93:205–222

Fukushima D (1989) Industrialization of fermented soy sauce production centering around Japan. In: Steikraus HS (ed) Industrialization of indigenous fermented foods. Marcel Dekker, New York, pp 1–88

Galun M (1988) The fungal-alga relation. In: Galun M (ed) Handbook of lichenology, vol 1. CRC Press, Boca Raton, pp 147–158

Garbaye J (1994) Helper bacteria: a new dimension to the mycorrhizal symbiosis. New Phytol 112:197–210

Garbay J, Duponnois R (1992) Specificity and function of mycorrhization helper bacteria (MHB) associated with the *Pseudostuga menziesee-Laccaria laccata* symbiosis. Symboisis 14:335–344

Gerhardt P (ed) (1994) Methods for general and molecular bacteriology. ASM Press, Washington, DC

Gilbert P, Brown MRW (1995) Mechanisms of the protection of bacterial biofilms from antimicrobial agents. In: Lappin-Scott HM, Costerton JW (eds) Microbial biofilms. Cambridge University Press, Cambridge, pp 118–130

Godfrey G, West S (eds) (1996) Industrial enzymology. Macmillan, London

Hacskaylo E (1972) Mycorrhiza: the ultimate in reciprocal parasitism? BioScience 22:577–583

Harborne JD (ed) (1978) Biochemical aspects of plant and animal coevolution. Academic Press, London

Harley JL, Smith SE (1983) Mycorrhizal symbiosis. Academic Press, New York

Haslam E (1985) Metabolites and metabolism: a commentary on secondary metabolism. Clarendon, London

Hawksworth DL (1988) The variety of fungal-algal symbioses, their evolutionary significance and the nature of lichens. Bot J Linn Soc 96:3–20

Hayes WA, Randle PE, Last FT (1969) The nature of the microbial stimulus affecting sporophore formation in *Agaricus bisporus* (Lange) Sing. Ann Appl Biol 64:177–187

Hendrix PF, Parmelee RW, Crossley DA Jr, Coleman DC, Odum EP, Groffman PM (1986) Detritus foodwebs in conventional and no-tillage agroecosystems. Bioscience 36:374–380

Hunter IS (1992) Function and evolution of secondary metabolites – no easy answers. Tibtech 10:144–146

Hurst CJ (ed) (1997) Manual of environmental microbiology. ASM Press, Washington, DC

Iwabe N, Kuma K-I, Hagesawa M, Osawa S, Miyata T (1989) Evolutionary relationship of archaebacteria, eubacteria, and eukaryotes inferred from phylogenetic tress of duplicated genes. Proc Natl Acad Sci USA 86:9355–9359

Janzen DH (1977) Why fruit rots, seeds mold and meat spoils. Am Nat 111:691–713

Kilham K (1994) Soil ecology. Cambridge University Press, Cambridge

Kirk TK (1984) Degradation of lignin. In: Gibson DT (ed) Microbial degradation of organic compounds. Marcel Dekker, New York, pp 399–437

Kosikowksi FV, Mistry VV 3rd (ed) (1997) Cheese and fermented milk foods. Kosikowski LLC, Westport, Connecticut

Lappin-Scott HM, Costerton JW (eds) (1995) Microbial biofilms. Cambridge University Press, Cambridge

Leyval C, Berthelin J (1989) Interactions between *Laccaria laccata*, *Agrobacterium radiobacter* and beech roots: influence of P, K, Mg, and Fe mobilization from minerals and plant growth. Plant Soil 117:103–110

Lockwood JL (1968) The fungal environment of soil bacteria. In: Gray TRG, Parkinson D (eds) The ecology of soil bacteria. Liverpool University Press, Liverpool, pp 44–65

Luckner M (1984) Secondary metabolism in microorganisms, plants and animals. VEB Gustav Fischer, Jena

Lynch JM (1987) Biological control within microbial communities of the rhizosphere. Symp Soc Gen Microbiol 41:55–82

Lynch JM (ed) (1990) The rhizosphere. John Wiley, New York

Marx DH (1982) Mycorrhizae in interaction with other microorganisms, B. Ectomycorrhizae. In: Schenck NC (ed) Methods and principles of mycorrhizal research. The American Phytopathological Society, St Paul, pp 225–228

Mayser P, Fromme S, Leitzmann C, Grunder K (1995) The yeast spectrum of the "tea fungus Kombucha." Mycoses 38:289–295

Melin E (1936) Methoden der experimentellen Untersuchung mykotropher Pflanzen. Handb Biol Arbeitsmeth 11:1015–1108

Metting FB Jr (1993) Structure and physiological ecology of soil microbial communities. In: Metting FB Jr (ed) Soil microbial ecology. Marcel Dekker, New York, pp 3–25

Mittelman MW (1995) Biofilm development in purified water systems. In: Lappin-Scott HM, Costerton JW (eds) (1995) Microbial biofilms. Cambridge University Press, pp 133–147

Norris JR, Read DJ, Varma AK (eds) (1991) Methods in microbiology, vol 23. Techniques for the study of mycorrhiza. Academic Press, London

O'Donnell AG, Goodfellow M, Hawksworth DL (1995) Theoretical and practical aspects of the quantification of biodiversity among microorganisms. In: Hawksworth DL (ed) Biodiversity: measurement and estimation. Chapman and Hall, London, pp 65–73

O'Toole DK (1997) The role of microorganisms in soy sauce production. Adv Appl Microbiol 45:87–152

Pace NR (1997) A molecular view of microbial diversity and the biosphere. Science 276:734–740

Park JY, Agnihotri VP (1969) Bacterial metabolites trigger sporophore formation in *Agaricus bisporus*. Nature 222:984

Parkinson D, Waid JS (1960) The ecology of soil fungi. Liverpool University Press, Liverpool

Paul EA, Clark FE (1990) Soil microbiology and biochemistry. Academic Press, San Diego

Perron AD, Patterson JA, Yanofsky NN (1995) Kombucha "mushroom" hepatotoxicity. Ann Emerg Med 26: 660–661

Poincelot RP (1975) The biochemistry and methodology of composting. Conn Agric Exp Stn Bull 754:1–14

Rainey PB (1991) Effect of *Pseudomonas putida* on hyphal growth of *Agaricus bisporus*. Mycol Res 95:699–704

Rainey PB, Cole ALJ, Fermor TR, Wood DA (1990) A model system for examining the involvement of bacteria in basidiome initiation of *Agaricus bisporus*. Mycol Res 94:191–195

Riffle JW, Maronek DM (1982) Ectomycorrhizal inoculation procedures for greenhouse and nusery studies. In: Schenck NC (ed) Methods and principles of mycorrhizal research. American Phytopathological Society, St Paul, Minnesota, pp 147–155

Rippon JW (1988) Medical mycology; the pathogenic fungi and the pathogenic actinomycetes, 3rd edn. WB Saunders, Philadelphia

Salvarezza R, de Mele MFL, Videla H (1983) Mechanisms of the microbial corrosion of aluminum alloys. Corrosion 39:26–32

Schroth MN, Hancock JG (1982) Disease suppressive soil and root-colonizing bacteria. Science 216:1337–1381

Stamets P, Chilton JS (1983) The mushroom cultivator. Agarikon Press, Olympia, Washington

Stanek M (1974) Bacteria associated with mushroom mycelium (*Agaricus bisporus* (Lg.) Sing. in hyphosphere. Proc 9th Int Cong on Cultivation of edible Fungi, Tokyo. Mushroom Sci IX, Part 1:197–207

Starmer WT, Fogleman JC, Lachance M-A (1991) The yeast community of cacti. In: Andrews JH, Hirano SS (eds) Microbial ecology of leaves. Springer, Berlin Heidelberg New York, pp 158–178

Steinkraus KH, Cullen RE, Pederson CS, Nellis LF, Gavitt BK (eds) (1983) Handbook of indigenous fermented foods. Marcel Dekker, New York

Straatsma G, Konings RNH, van Griensven LJLD (1985) A strain collection of the mycorrhizal mushroom *Cantharellus cibarius* obtained by germination of spores and of fruit body tissue. Trans Br Mycol Soc 85:689–697

Straatsma G, Gerrits JPG, Augustijn MPAM, Op Den Camp HJM, Vogels GD, van Griensven LJLD (1989) Population dynamics of *Scytalidium thermophilum* in mushroom compost and stimulatory effects on growth rate and yield of *Agaricus bisporus*. J Gen Microbiol 135:751–759

Strohl WR (ed) (1997) Biotechnology of antibiotics, 2nd edn. Marcel Dekker, New York

Sugihara TF (1985) Microbiology of bread making. In: Wood BJB (ed) Microbiology of fermented foods, vol I. Elsevier, London, pp 249–261

Swings J, De Ley J (1981) The genera *Gluconobacter* and *Acetobacter*. In: Starr MP, Stolp H, Truper HG, Balows A, Schlegel HG (eds) The prokaryotes, vol I. Springer, Berlin Heidelberg New York, pp 771–778

Theodorou MK, Lowe SE, Trinci APJ (1992) Anaerobic fungi and the rumen ecosystem. In: Carroll GC, Wicklow DT (eds) The fungal community: Its organization and role in the ecosystem, 2nd edn. Marcel Dekker, New York, pp 43–72

Thomashow LS, Bonsall R, Weller DM (1997) Antibiotic production by soil and rhizosphere microbes in situ. In: Hurst CJ (ed) Manual of environmental microbiology. ASM Press, Washington DC, pp 493–499

Tietze H (1994) Kombucha-miracle fungus: the essential handbook. Gateway Books, Bath, UK

Videla H, Charaklis WG (1992) Biofouling and microbially influenced corrosion. Biodeterior Biodegrad 29:195–212

Vining LC (1990) Functions of secondary metabolites. Annu Rev Microbiol 44:395–427

Weller DM, Thomashow LS (1990) Antibiotics: evidence for their production at sites where they are produced. In: Baker RR, Dunn PE (eds) New directions in biological control: alternatives for suppression of agricultural pests and diseases. Alan R Liss, New York, pp 703–711

Williams GH, Guido G, Hermite PL (eds) (1985) Longterm effects of sewage sludge and farm slurries applications. Elsevier, London

Woese CR, Kandler O, Wheelis ML (1990) Towards a natural system of organisms: proposal for the domains Archaea, Bacteria, and Eucarya. Proc Natl Acad Sci USA 87:4576–4579

Woodruff HG (1980) Natural products from microorgansism. Science 208:1225–1229

Biosystematic Index

Subject Index